MW00445610

Temperature-Dependent Sex Determination in Vertebrates

Temperature-Dependent Sex Determination in Vertebrates

EDITED BY

Nicole Valenzuela and Valentine Lance

Smithsonian Books

Washington

Copy editor: Bonnie J. Harmon
Production editor: Robert A. Poarch
Designer: Brian Barth

Library of Congress Cataloging-in-Publication Data
Temperature-dependent sex determination in vertebrates /
edited by Nicole Valenzuela and Valentine A. Lance.
p. cm.
Includes bibliographical references and index.
ISBN 1-58834-203-4
1. Sex (Biology). 2. Vertebrates. I. Valenzuela, Nicole.
II. Lance, Valentine A.
QP278.5.T45 2004
571.8'16—dc22 2004042882

British Library Cataloguing-in-Publication Data available

Manufactured in the United States of America
10 09 08 07 06 05 04 1 2 3 4 5

♾ The paper used in this publication meets the minimum
requirements of the American National Standard for
Information Sciences—Permanence of Paper for Printed
Library Materials ANSI Z39.48-1984.

Contents

Part 2 Thermal Effects, Ecology, and Interactions

Part 3 Evolutionary Considerations

NICOLE VALENZUELA

Introduction

This Book

Almost four decades after the discovery of temperature-dependent sex determination (TSD) in vertebrates credited to Charnier (1966), we find ourselves faced with a vast amount of information related to this mechanism that has been gathered by researchers from a variety of disciplines. The most comprehensive review on TSD was included in the well-cited book by J. J. Bull (1983), *Evolution of Sex Determining Mechanisms.* The field of TSD has come a long way since then, and twenty years after Bull's publication we felt there was a need for a new volume dedicated exclusively to this subject.

This book represents such an attempt and is intended as a comprehensive work that compiles, analyzes, and integrates existing information about this sex-determining mechanism. This volume encompasses a series of reviews of what is known about the ecological, physiological, molecular, and evolutionary aspects of TSD. We tried to bring together, for the first time, the diversity of issues related to this sex-determining mechanism and synthesize the vast literature from multiple disciplines. The book is organized in four thematic sections: Prevalence of Temperature-dependent Sex Determination in Vertebrates; Thermal Effects, Ecology, and Interactions; Evolutionary Considerations; and Conclusions: Missing Links and Future Directions. These sections correspond to basic questions about TSD: (1) Who possesses TSD? (2) How does it work? (3) How did it appear, how is it maintained, and will it prevail? (4) What do we still need to learn about it?

In Chapter 1, J. J. Bull offers an enlightening perspective in the field of temperature-dependent sex determination through an historical account of his own experience in the area. Chapters 2–7 in Part 1 encompass a series of reviews with additional new data on the incidence of TSD in fish, turtles, crocodilians, lizards, and tuataras, and on the incidence of thermal sex reversals in amphibians, including some discussion about the ecological or evolutionary significance of particular cases. Chapters 8–12 in Part 2 provide the reader with some basics about the general effects of temperature on animal biology, and then describe what is known about how temperature affects development and sex ratios under lab and field conditions for TSD species, about thermal effects on phenotypic traits other than sex, about the molecular network underlying genotypic and temperature-dependent sex determination, and about the influence of yolk steroids in sex-determination/differentiation. Chapters 13–15 in Part 3 present a new phylogenetic analysis exploring the ancestry of the sex-determining mechanisms in vertebrates, and review the postulated hypotheses explaining TSD's past and present evolution, as well as the interactions among TSD, sex-ratio production, and population dynamics, including an analysis relevant to some conservation issues.

In the second part of this introduction, I provide a brief background section about vertebrate sex-determining mechanisms and define the most commonly used terminology related to temperature-dependent sex determination. Several chapters provide more detailed descriptions of some of these terms and phenomena, but this introduction should suffice as a general guideline.

We hope that researchers, students, and other interested readers find this volume a useful reference resource, and that it fosters new investigations in this exciting field. As editors, we are indebted to each and every author who so enthusiastically contributed to this book and to the many reviewers who evaluated manuscripts, often on very short notice.

General Background

Vertebrate Sex-Determining Systems

Not all animal species determine sex in the same fashion. In many vertebrates, sex is determined by major sex factors contained in sex chromosomes (Bull 1983), which can differ morphologically from each other (heteromorphic) or not (homomorphic) (Bull 1983; Solari 1994). In some species (XX/XY), males are heterogametic (XY), whereas in other species (ZZ/ZW), females are heterogametic (ZW) (Bull 1983). Heteromorphic sex chromosomes are present in amphibians, reptiles, fishes, birds, and mammals (Bull 1983; Solari 1994 and references therein). Many amphibian and fish species possess homomorphic sex chromosomes (e.g., Winge and Ditlevsen 1948; Yamamoto 1969; Kallman 1984; reviewed in Solari 1994). Genotypic sex determination (GSD) systems different from the classic XX/XY and ZZ/ZW include vertebrates in which more than one X or Y chromosome is normally present per individual, others where three or more sex chromosome types segregate in the population, and a few species where a group of minor sex-determining genes with additive effects define the sex of the organism (Bull 1983; Basolo 1994; Solari 1994).

Mendelian segregation of sex chromosomes in heterogametic systems guarantees the production of balanced (1:1) primary sex ratios. Sex ratios also tend to unity in most cases of multiple factor systems at equilibrium, so long as fitness does not vary between the sexes (Bull 1983). However, vertebrate sex ratios can be distorted by various means, as when environmental factors override sex chromosomes. For instance, temperature-induced differential mortality (Burger and Zappalorti 1988), embryo abortion (e.g., Krackow 1992; Blackburn et al. 1998), and differential fertilization (e.g., Komdeur et al. 1997; Stockley 1999) can bias primary or secondary sex ratios (Valenzuela et al. 2003). Temperature can also alter sex differentiation and induce sex reversal in species possessing sex chromosomes (Dournon et al. 1990; Baroiller et al. 1995; Solari 1994; see also Chardard et al., Chapter 7), resulting in biased secondary sex ratios. Lastly, some vertebrate species do not display sex chromosomes; no consistent genetic differences exist between the sexes, so that sex cannot be predicted by zygotic genotype; and sex determination occurs after fertilization, by incubation temperature (Bull 1983; Solari 1994; Valenzuela et al. 2003). In this last system (i.e., temperature-dependent sex determination or TSD), primary sex ratios are not defined at conception: they are realized after fertilization, according to the incubation temperature, and can often be biased (Valenzuela et al. 2003).

Temperature-Dependent Sex Determination

Under TSD, the sex of individuals is indifferent until determined permanently after conception, by incubation temperature (Bull 1983). Among vertebrates, TSD is unknown in snakes, birds, mammals, and amphibians, is infrequent in fishes (although new reports are increasing), but is very common in reptiles—it occurs in all studied crocodilians and tuataras, is prevalent in turtles, and more infrequent in lizards (reviewed in Chapters 2–7). Although there is a wide agreement on the definitions of TSD and GSD as described, only recently has an explicit set of criteria been proposed to identify the presence of TSD unambiguously (Valenzuela et al. 2003).

Three modes of temperature-dependent sex determination are recognized, categorized according to the sex ratios produced as a function of incubation temperature (Figure I.1A). In *TSD II* (also called female-male-female or FMF pattern), low and high temperatures produce females, while intermediate temperatures produce males. In *TSD Ia* (also termed male-female or MF pattern), low temperatures produce males and high temperatures produce females, whereas the opposite is true for *TSD Ib* (also called female-male or FM pattern). These patterns are typically described from sex ratio data obtained under constant incubation conditions in the laboratory. The constant incubation temperature that produces a populationwide 1:1 sex ratio is called the *pivotal* (or threshold) temperature. Thus, TSD II is characterized by two pivotal temperatures (a lower and a higher), whereas TSD Ia and TSD Ib each have a single pivotal temperature (Figure I.1A). The range of temperatures that produces mixed sex ratios is termed the *transitional range* (TR), and again, a single transitional range characterizes TSD Ia and TSD Ib, while TSD II exhibits two

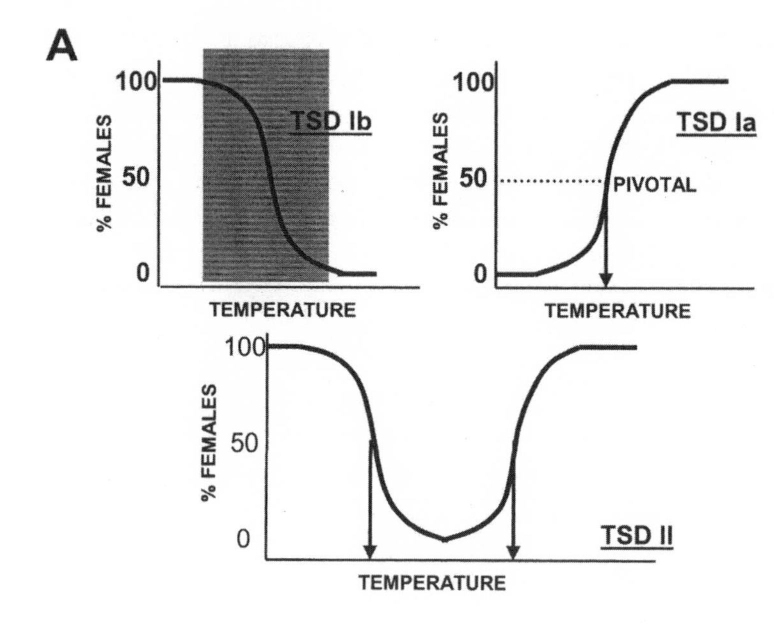

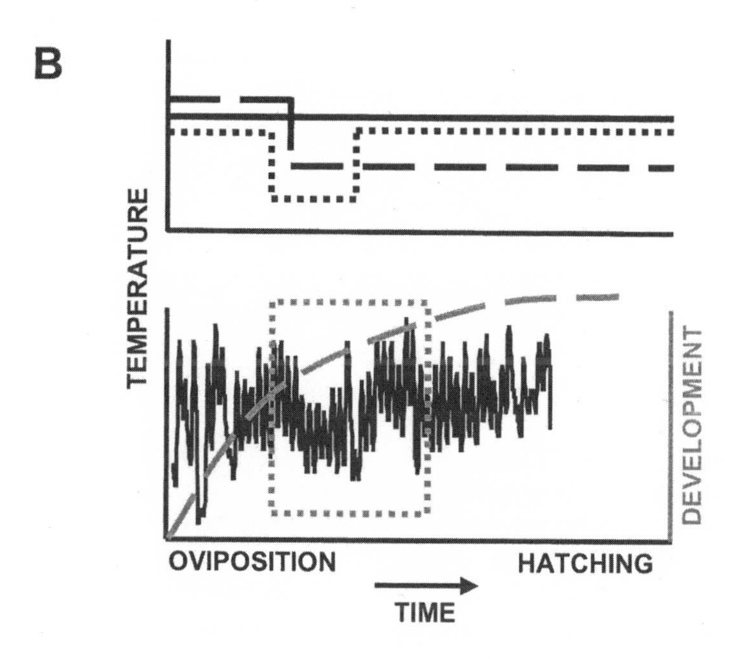

Figure I.1 (**A**) Patterns of TSD (TSD Ia, TSD Ib, TSD II) defined by the sex ratios produced as a function of constant incubation temperature. Shaded area in TSD Ib panel illustrates the transitional range. Arrows in TSD Ia and TSD II panels indicate the pivotal temperatures that correspond to the constant temperature that produces a populationwide 1:1 sex ratio. (**B**) In the upper panel the solid line exemplifies a constant temperature experiment, the dashed line exemplifies a shift-once experiment from high to low incubation temperature, and the dotted line exemplifies a shift-twice experiment from high to low to high incubation temperature. In the lower panel the solid line depicts incubation temperatures recorded from a field nest (data from Valenzuela 2001b), the dashed gray line depicts a hypothetical developmental curve, and the dotted gray line depicts the tightly associated window of time corresponding to the thermosensitive period. According to how temperature fluctuations affect developmental rate, changes in development will cause the TSP window to slide along the time axis, such that it may not coincide with the absolute TSP timing as defined by transfer experiments under constant temperatures (see also Georges et al., Chapter 9).

(Figure I.1A). TSD species and populations may vary in the range of temperatures that sustain development and produce the entire array of sex ratios, in the pivotal temperature, and in the width, slope, and symmetry of the transitional range. The effect of temperature on sex determination is exerted during a limited window of time during development called the *thermosensitive period* (or TSP). This period can be determined using transfer experiments, in which embryos start incubation at a temperature that produces a single sex, and groups of embryos are shifted at sequential times during development to the other extreme temperature that produces exclusively the opposite sex (Figure I.1B). From these experiments, a sex ratio equal to that expected at the second temperature indicates that the TSP had not started by the transfer time, whereas a sex ratio equal to that expected at the first temperature indicates that the TSP had ended by the transfer time. Sex ratios intermediate to those expected at either temperature indicate that the TSP was active at the transfer time. To further fine tune the delimitation of the TSP, shift-twice experiments can be used (Figure I.1B), in which embryos start incubation at a single-sex temperature and are then transferred to the second temperature that produces the opposite sex; groups of embryos are then transferred back to the initial temperature after periods of time of increasing length.

Results from constant-temperature incubation experiments may not mimic what occurs under field conditions, where incubation temperature can vary daily, sometimes drastically (see lower panel of Figure I.1B), in terms of both development (e.g., timing of the TSP) and sex ratio pro-duction (Figure I.1B; Valenzuela 2001b; Georges et al., Chapter 9).

1

JAMES J. BULL

Perspective on Sex Determination

Past and Future

The editors asked me to contribute to this volume, knowing full well that nearly two decades separated me from any serious work on sex determination. I am too removed from the field to write an up-to-date scholarly review, so my compromise is to offer a biased perspective on the field, with a personal and (apologetically) egocentric history. My main regret in doing this is that Claude Pieau has not shown a similar lack of self-restraint and written his own history.

Ohno's Legacy

Being one of the older contributors to this volume, I am in the unusual position of remembering a time when the thought of temperature-dependent sex determination (TSD) in reptiles was heresy. While I was an undergraduate in 1967–71, my mentor, Robert Baker at Texas Tech, pursued karyotypes of mammals with a vengeance. His enthusiasm for karyotypes was infectious, and Baker, another undergraduate (Greg Mengden), and I did a large karyotypic survey of snakes. At the same time, Baker introduced us to Susumu Ohno's 1967 *Sex Chromosomes and Sex-Linked Genes*. In a matter of a few pages, Ohno laid out a breathtakingly elegant theory for an evolutionary progression of vertebrate sex chromosomes. The "primitive" condition (we would now say "ancestral") was a lack of strict genetic control of sex determination. From that, genetic control evolved, due perhaps to as little as a single allelic difference. From there, heteromorphic sex chromosomes evolved in a series of stages that involved increasing levels of X-Y differences. Although much of this theory had been laid out for *Drosophila* half a century earlier by Hermann Muller, Ohno could uniquely demonstrate this sex chromosomal progression within the vertebrates. The earliest stages were evident in fish and amphibians, where sex-linked markers and artificial sex reversal had demonstrated genetic control of sex. Sex chromosome heteromorphisms were mostly unknown in these groups at the time, and indeed, the YY genotype was often fertile, demonstrating conclusively that sex chromosome differentiation had not occurred. Then, in reptiles—especially snakes—one saw the emergence of different magnitudes of cytological sex chromosome heteromorphism, from an absence of X-Y differences in the more (morphologically) primitive snakes to major X-Y differences in the most advanced snakes. Today, there is no reason to think that those snakes that have retained the most ancestral morphologies should also retain the most ancestral sex chromosome condition, so the mapping of this sex chromosome progression onto morphological evolution does not make sense. Nonetheless, Ohno's book, combined with my own chromosome work on snakes, inspired me to ponder the evolution of sex determination and remained my chief passion for 15 years. Even though many of Ohno's ideas about sex chromosomes and sex determination have not withstood the test of time, Ohno identified the basic questions and galvanized the field for others to think about them. His identification of the conservation of the mammalian X has held up and is sometimes referred to as Ohno's law.

Skepticism for TSD

There was no place in Ohno's evolutionary progression for a prominence of TSD in amniotes. Hermaphroditism was considered a degenerate state, and environmental sensitivity was an imperfection of strict genotypic sex determination. The best opportunity for temperature-sensitive sex determination that one could envision was as an occasional accident contributing to sloppy sex determination in fish and amphibians. Emil Witschi's work on frogs and Öjvind Winge's work on fish certainly gave the impression that the sex phenotype was developmentally unstable in some species, so a temperature effect could be interpreted as imperfection. No one seemed ready to accept TSD in reptiles. Madeleine Charnier's work was unknown to us, but a fellow graduate student (Robert Winokur) had already given me copies of Claude Pieau's work as early as 1972. We were both in the lab of a prominent turtle biologist (John Legler), and there was some discussion of testing Pieau's results with North American turtles, but nothing was done about it. I think we were open-minded but slightly dubious. Accompanying Legler on his sabbatical to Australia in 1973–74, I recovered some full-term nests of long-necked turtles and had another student look at hatchling gonads, but nothing suggested to him that they were all one sex. (It now seems that these and other chelid turtles do not have temperature-dependent sex determination.) Furthermore, another Legler student (Randy Moon) had found the first sex chromosomes in a turtle, adding justification for extension of Ohno's progression to turtles, leaving little room for TSD.

By 1976, Eric Charnov and I had developed a theory for the evolution of environmental sex determination (published in 1977). That paper discussed several known examples of environmental sex determination, but reptiles were notably absent from any mention in our paper. Before publishing, I had told Eric that there were studies purporting to show TSD in turtles, but I assured him that the findings were bogus—that the phenomenon was due either to differential mortality or to a lab artifact. From my knowledge of reptiles, there was no way that reptiles could seem to fit our theory, and my skepticism about TSD had gone from mild to extreme. In August 1977 (the day Elvis died), my first field trip with Dick Vogt was at the tail end of the turtle hatching season. Vogt knew Wisconsin habitats so well that he took us to several sites, and even that late in the season, we were able to locate several straggler map turtle hatchlings as well as one full clutch of softshell turtle eggs. Gonads of the map turtles all looked the same and seemed indistinct, but there were clear males and females in the one clutch of softshell turtles. Thus, all the evidence still pointed toward doubting TSD in reptiles. It wasn't until January 1978, when Vogt gave me two map turtle hatchlings of captive-laid eggs he had incubated at cool temperatures, that I saw obvious testes and realized that the August map turtle hatchlings had all been females.

From the perspective of today, it must be difficult to appreciate the widespread skepticism toward TSD in reptiles 25 years ago. My initial field work involved burying eggs in the coldest possible nesting sites on the study site island (deep shade, so cool in fact, that nothing hatched), and I had purchased black aquarium gravel to replace the sand over sun-exposed nest sites, to achieve the highest possible "natural" temperature. I felt that these manipulations would be necessary to show conclusively that TSD did not operate under the most extreme possible field conditions. Other anecdotes support widespread skepticism as well. My first meeting with Ohno (around 1979) was soon after he had reviewed the first Bull-Vogt paper on TSD, and I think his first comment to me was his surprise that TSD in reptiles was indeed apparently real. Ross Kiester told me that a National Science Foundation (NSF) application of his in the 1970s to study TSD was rejected, with the reviewer commenting that TSD was merely differential mortality. Vogt and I had sent Don Tinkle an early draft of our first paper; in that draft, the field manipulation awaited some data, and we had not been able to reject differential mortality under field conditions. Don's comment was that it was a shame that the critical experiment had not achieved statistical significance to reject differential mortality (the final set of results did accomplish this). Of course, there was no evolutionary reason to support differential mortality, but we—the community—needed some way to accommodate TSD as something besides natural sex determination. And although the first Bull-Vogt paper followed Pieau's and Chester Yntema's work, that paper seemed to be the one that convinced the skeptics that TSD was real, probably because we designed the study to disprove the phenomenon. It is interesting that, even today, amphibians present the same dilemma that we felt about TSD 25 years ago: Does the temperature effect occur naturally? The chapter on amphibians (Chapter 7) gives a wonderful perspective on this taxon.

A Framework

By the end of one summer's field work in 1978, Dick Vogt and I had enough evidence to show that TSD operated in nature in at least four species of emydid turtles, whereas it did not occur in softshell turtles. Differential mortality was not the explanation. Pieau was right. In my earlier work with Charnov, I had begun developing a framework for the evolution of sex-determining mechanisms. It was immediately obvious that any species with "strict sex chromosomes"—XX is always one sex, XY always the other—could not have any meaningful level of TSD, because environmental effects on sex would produce XX and/or XY of the wrong sex; chromosomal studies of reptiles with sex chromosomes had not revealed any heterotypic combinations of genotype and gender. Thus, there were two fundamentally different and extreme types of sex determination side-by-side in reptiles. In contrast to the impression one obtained from Ohno, TSD was an absolute and sufficient sex determining mechanism, with virtually no inherited basis of sex. Conversely, sex chromosome systems in reptiles were absolute genotypic mechanisms.

Sex ratio theory was blossoming at the time, and it was an obvious step to develop models explaining the evolutionary consequences of TSD. Having such a framework is one thing; applying it and testing it is another. To date, the best integration of these models with observations is from fish, and even there, only the Atlantic silverside has been studied sufficiently to provide reasonable tests of the model. Disappointingly, the framework has not been terribly useful for reptiles. Reported sex ratios are commonly excessively female (and from Girondot et al., Chapter 15 in this volume, it is not even clear if the reported sex ratios are accurate); the theory can explain non-Fisherian sex ratios, but consistent, extreme female excesses require conditions that have not been demonstrated in reptiles. Even the most basic question of why TSD is found in reptiles is still problematic. The so-called Charnov-Bull model may eventually be found to explain the "why" of TSD, but the model certainly did not lead one to predict TSD in reptiles, and even now it seems intrinsically hard to accept that the temperature during a month of development need affect the fitness of survivors 10 years later. (Perhaps once TSD starts to evolve, the physiological system evolves to reinforce and maintain it.) This is the dilemma of TSD in reptiles, however: long-term fitness effects of incubation temperature are not easy to imagine, but because many reptiles are so long-lived, only slight fitness effects are needed to stabilize TSD against genotypic sex determination (GSD), and slight fitness effects are not easily detected. Of course, the Charnov-Bull model has been used to suggest features of reptilian development that favor TSD over GSD and has inspired work toward that end. There are dozens of studies reporting lasting effects of incubation temperature in reptiles, although the effects discovered have not generalized to the level that one would hope, for purposes of explaining the widespread distribution of TSD.

Future

This volume is a welcome update on and consolidation of information on environmental sex determination in vertebrates. Over the last two decades, we have achieved a more thorough understanding of TSD's prevalence, its taxonomic distribution, and its variety. Virtually all of the original findings have been upheld by subsequent work. At the same time, many of the unsolved problems of the past are still unsolved.

Molecular Mechanisms and Physiology

One of the areas of biggest advance and greatest opportunity is the physiology and molecular biology of sex determination. Over two decades ago, Pieau and Raynaud showed that administration of steroids to turtle embryos affected sex determination. Crews soon took up this approach, using the technical improvement of applying substances to the egg surface rather than injecting them (in our hands, embryo mortality was 50–90% with injection, but was typically near 0% with topical application). Many substances, especially steroids, steroid mimics, and steroid inhibitors were shown to affect sex determination, although some compounds had only a weak effect. These manipulations support a taxonomically broad sex-determining role for estrogen (and by extension, for aromatase), as well as for other steroids and steroidogenic enzymes, but by itself, the evidence is somewhat indirect. Perhaps the most convincing evidence that estrogen is part of natural sex determination is the recent work showing that estrogen is present in eggs, and is present at different levels in different clutches. Such findings also provide an explanation for between-clutch variance in sex ratio that was formerly attributed to heritability.

The ultimate goal in this work must be to resolve the genetic hierarchy of reptilian sex determination. Here, progress has been difficult, chiefly because reptiles are not amenable to classical genetic studies, nor do they offer good model systems for gene manipulation. The work on genetics of reptilian sex determination has thus been predicated on discoveries of mammalian sex-determining genes. Extrapolations from those discoveries to look at gene expression in reptiles have not generalized well. Furthermore, if the set of genes involved in reptilian sex determination is not entirely a subset of genes involved in mammalian sex determination, then an approach that simply borrows discoveries from mammals will ultimately fail to discover the full reptilian mechanism. To rely entirely on correlations rather than experimental genetic alterations for understanding sex determination invites failure—the history of dead ends with H-Y antigen and ZFY are vivid reminders of what can go wrong in the absence of transgenic manipulations. We therefore face the prospect that a valid understanding of reptilian molecular sex determination will (1) await the development of molecular tools that achieve gene manipulation in the reptilian embryo or in cultured gonads (morpholino antisense RNAs and RNAi may be such tools) and (2) need to be based on genes derived from a reptile model system rather than a mammalian one. In the meantime, comparison of sex determination between fishes and mammals should provide useful insights to the core of conserved sex-determining genes that may minimally be expected to operate in reptiles, and zebra fish may be the best model system for this work.

Evolutionary Significance

As alluded to above, a framework exists to study the evolutionary significance and consequences of TSD. So far, fishes have proved to be the most useful organisms for application of this theory. Fishes have the advantage of relatively short generation times, which not only facilitates genetic studies, but also weakens the advantage of TSD over GSD (due to environmentally caused variation in the sex ratio). As shown so convincingly by David Conover's work, the evolutionary outcome in short-lived species is a mix of genetic and environmental effects on sex determination that responds not only to artificial selection but also to environmental gradients. In a wide-ranging fish with TSD, changes in the relative genetic contribution to sex determination that coincide with ecological habitat variation have enabled one to infer the selective basis of environmental sex determination in ways that have not been possible with reptiles. Furthermore, with short-lived species, the recent thermal history of the population is evident from the sex ratio, because most of the adults were born in the previous year. Many of the difficulties faced in studying reptilian TSD are thus absent in work on fishes.

The extreme form of TSD seen in so many reptiles does, however, offer its own advantages. Studies of the physiological consequences of sex determination and of sex-determining cues are far easier when one achieves unisexual progeny over a wide range of temperatures. Furthermore, greater population sex ratio deviations may be expected, and more novel sex ratio phenomena may occur in these extreme forms of TSD than with the modest forms found in fishes. Steve Freedberg has developed some highly original models of sex ratio evolution in reptiles involving cultural inheritance of nest sites. Although such processes may seem implausible a priori (e.g., they require strong cultural inheritance of nest site and can lead to huge female excesses), if even one species supported such a mechanism, it would constitute an extremely novel evolutionary phenomenon—evolution of a major phenotypic change due to cultural inheritance.

Amphibians are both interesting and different. Although they appear to have been key model systems in early studies of vertebrate sex determination, they are not central to sex determination work now. Yet they may well hold the key to understanding the origin of TSD in reptiles. A common pattern in amphibians seems to be strict heterogamety (XX/XY) over a wide range of conditions, but environmental extremes generated in the laboratory can override the sex chromosomes. What seems unknown at present is whether and how often the sex chromosomes are overridden in nature. This basic question about the natural history of heterogamety should be easy to resolve, because deviations from heterogamety lead to discordance between sex genotype and phenotype. The existence of discordant genotypes in natural amphibian populations can be ascertained easily by breeding wild animals to a standard lab strain and looking for violations of 50% males in the progeny. Thus, an XX male would breed all daughters, an XY female would breed ¾ sons (discordant genotypes might also be detected cytologically or molecularly). Amphibians with environmentally sensitive heterogamety would also be suited for experiments generating transitions from sex chromosomes to environmental sex determination. Exposing a lab population to temperature extremes through several generations could eventually achieve the complete loss of the X or Y. One can imagine a natural transition from aquatic to terrestrial existence causing an immediate shift in sex determination by such a route, if incubation temperatures of terrestrial nests greatly exceeded those of aquatic nests.

Temperature-dependent sex determination in reptiles remains an enigma in many ways. Yet it is an enigma that attracts widespread interest. Many of the unsolved problems will be challenging, but the rewards for solving them should continue to justify the attempt. In the next 25 years, we can look forward to possibly understanding the molecular basis of TSD and having at least a few ecological systems yielding critical insights to the adaptive basis of this unusual form of sex determination.

Part 1

Prevalence of Temperature-Dependent Sex Determination in Vertebrates

2

DAVID O. CONOVER

Temperature-Dependent Sex Determination in Fishes

The first conclusive evidence that temperature during development might influence sex determination in fishes was the work of Harrington (1967), who showed that individuals of the normally hermaphroditic *Rivulus marmoratus* would differentiate as males if reared at low temperature. Although temperature-dependent sex determination (TSD) had long been suspected also in gonochoristic fishes (e.g., Aida 1936), the first proof of its existence was found in the Atlantic silverside, *Menidia menidia* (Conover and Kynard 1981). Coming after earlier discoveries of TSD in reptiles (reviewed in Bull 1980, 1983, and Chapter 1), in a taxon where other forms of labile sex determination (e.g., sex change) were well known, and where TSD has many practical applications for aquaculture, it is surprising that the discovery of TSD did not initially stimulate much research beyond *Menidia*. From 1981 to 1995, only two species were added to the list of those displaying TSD. However, recently there has been a marked increase in the number of studies purporting to demonstrate TSD in fishes (Baroiller and D'Cotta 2001). Nearly all of these studies involve farmed fishes, where the motivation for study is manipulation of the sex ratio in hatchery environments to produce the sex with higher somatic productivity. These studies have revealed much about the interaction of temperature and genetic factors in sex determination, the critical period of temperature sensitivity, and the molecular basis of temperature-modulated gonad differentiation. Yet in many of these new cases, it is not clear if thermal influences on sex ratio truly represent TSD, instabilities in an otherwise predominately genotypic sex determination (GSD) system that are expressed only at thermal extremes, or effects of gender-dependent differences in gamete or offspring survival. Because our knowledge of the ecology of TSD in natural populations is still based entirely upon one genera *(Menidia),* the overall adaptive significance of TSD in fishes remains difficult to assess.

Here, our emerging knowledge of the diversity of TSD mechanisms in fishes and its ecological and evolutionary significance in natural populations is summarized. Many excellent reviews of sex determination and the endocrine and molecular basis of labile sex differentiation in fishes have appeared recently (Blázquez et al. 1998a; Nakamura et al. 1998; Baroiller et al. 1999; Baroiller and D'Cotta 2001; Piferrer 2001; and especially Devlin and Nagahama 2002). Rather than produce yet another review, the author's primary purpose is to point out the glaring gap in our understanding of the adaptive significance of TSD in the life history of natural fish populations and to provide motivation and advice for investigation of these topics.

Defining and Identifying TSD in Fishes

Because the array of sex-determining systems in fishes is highly diverse both within and among species, terms must be defined unambiguously. Environmental sex determination (ESD) is said to occur when the environment experienced by offspring irreversibly determines the primary sex

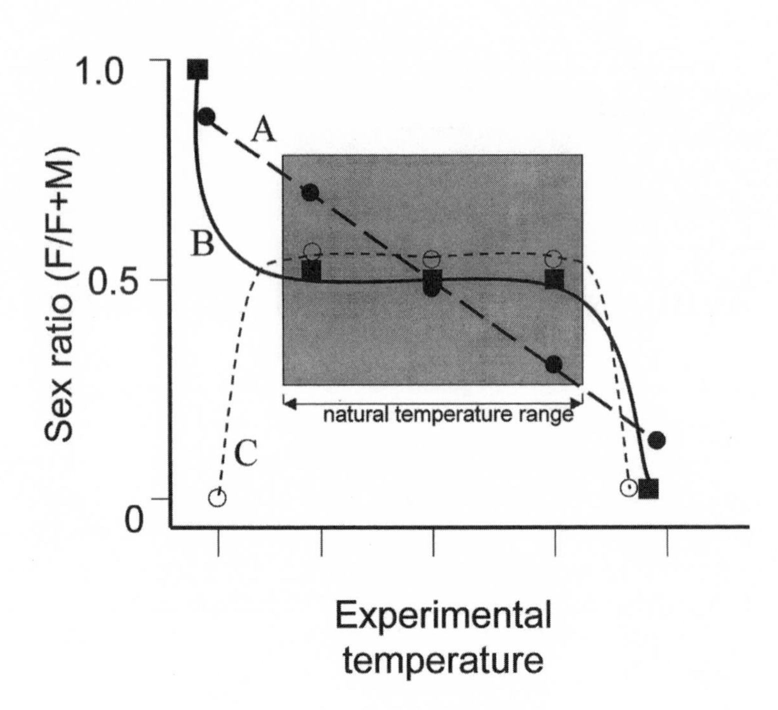

Figure 2.1 Three forms of the response of sex ratio to temperature that have been identified in fishes. Gradual changes in sex ratio such as those identified in atheriniform fishes (**A**) clearly represent TSD. Shifts in sex ratio that occur only at extreme temperatures, such as those in tilapia (**B**), and flatfishes (**C**) may or may not represent anomalous GSD. The critical factor is whether sex ratio changes occur within the range of temperatures normally experienced by developing larvae in the wild.

of the offspring (Bull 1983). GSD occurs when sex is determined by genotype at conception and is therefore independent of environmental conditions. TSD is a special case of ESD, wherein the environmental determinant is temperature. TSD appears to be the most common form of ESD in vertebrates, although it remains to be seen whether this is true of fishes. Other forms of ESD, such as pH-dependent sex determination, also occur in fishes, and it is premature to declare dominance of one form of ESD over the other, especially because both forms may have evolved for similar reasons. Hermaphroditism is yet another common form of labile sex in fishes, where individuals *function* first as one sex and then as the other.

This definition of TSD does not imply that genetic influences on gender are nonexistent. In fact, without at least some genetic influence on sex determination, TSD and the sex ratio would be incapable of evolving. In many fishes with TSD, the gender of an individual is determined by a genotype-by-environment (G×E) interaction. Many organismic traits are in fact controlled by similar forms of genetically based phenotypic plasticity, wherein phenotypic expression results from the combined influence of genetic tendencies and environmental experience at the time of trait formation. In the case of sex, however, the character is a threshold trait such that phenotypes must develop as one form or the other, thereby masking the underlying sex tendency of an individual's genotype. Hence, TSD occurs when undifferentiated offspring have *the potential* to develop as male or female, depending on environmental conditions normally encountered during development. GSD occurs when environmental influences on sex are nonexistent: that is, phenotypic plasticity for sex is nil across the range of environments that offspring normally experience.

This definition of GSD does not imply either that extreme conditions never cause disruption of the normal translation of genetic code into sex differentiation, or that sex ratios of progeny will always be 1:1. Although GSD systems often involve heterogametic (XX-XY or ZW-WW) mechanisms that produce a near 1:1 ratio under normal conditions, strict determination of gender by genotype may fail under environmental stress due, for example, to extreme temperature or to the presence of chemical factors that disrupt the normal course of sex differentiation. Instability in GSD is not ESD. Changes in sex ratio observed only at thermal extremes near the limits of species tolerance more likely reflect anomalies of an imperfect system of GSD. In practice, however, the distinction between ESD and anomalous GSD can be difficult to ascertain (Figure 2.1) because it relies on (1) an assessment of what is "normal," the standard for which must be the range of environmental factors actually experienced by progeny in nature, and (2) indirect methods involving comparison of family sex ratios, which can be distorted by a variety of other confounding factors, as explained next.

TSD is typically diagnosed by comparing sex ratios of progeny reared at different temperatures until the gonads have differentiated and sex can be identified. To control for the possibility that differential mortality of male- versus female-determining gametes (e.g., X- vs. Y-bearing sperm) might skew the sex ratio, fertilization should occur under identical conditions for all offspring, followed by random assignment of progeny to treatments. In species with internal fertilization and/or viviparity, this is not always possible. Differential mortality of the sexes after fertilization can be ruled out by recording the total number of mortalities during incubation and then calculating whether mortality alone could account for sex ratio differences among treatments assuming the worst-case, sex-biased mortality (see Conover and Kynard 1981). Both problems can be minimized by using thermal shift experiments at different developmental stages to pinpoint the critical period of temperature sensitivity.

Because TSD in fishes frequently involves G×E interaction with strong parental effects on family sex ratio, it is imperative that the offspring from each of multiple full-sib families be subdivided and compared among temperatures. Otherwise, sex ratio variation among families may obscure the thermal response. Use of several sires per dam (half-sib families) will show whether sex ratio differences among families represent genetic rather than maternal influences. The relative magnitude of additive genetic (e.g., effects of numerous minor genes), nonadditive genetic (e.g., genes of major effect, as in dominance), and environmental effects can thus be estimated (see Bull 1983; Conover and Heins 1987a). The sex ratio patterns emerging from such experiments have four general outcomes that would be interpreted as follows: (1) Family sex ratios that approximate 1:1 within and across all temperatures are consistent with standard heterogametic GSD. (2) High sex ratio variation among families within but not among temperatures suggests polyfactorial GSD. In both cases, experiments to further define the mechanism of GSD are possible but are beyond the scope of this chapter. (3) Family sex ratios that vary from near 1:0 to 0:1 across, but not within, temperatures indicate TSD with low genetic influence. (4) High among-family variance both within and among temperatures suggests TSD with both strong genetic and strong thermal effects (G×E interaction).

Finally, if one is to understand the role of TSD in sex ratio and life history evolution, and if one is to distinguish cases of true TSD from anomalies of GSD, experimental studies need to encompass the full range of temperatures that are experienced in nature and/or permit growth and survival. The various nonlinear forms of sex ratio response, and therefore an accurate diagnosis of TSD, cannot be deduced from experiments at less than four or five temperatures (see Figure 2.1). Moreover, field studies should be employed to illuminate the ecology of TSD in natural populations.

TSD and Other Thermal Sex Ratio Distortions in Fishes

Evidence of a thermal influence on sex ratio has now been documented in about 54 species of fishes, 33 of which come from a single study of the cichlid genus *Apistogramma* (Table 2.1). The other 21 cases are found among 18 genera from seven orders, all within the division Teleostei, and they span the range from primitive to highly advanced groups. In addition, there is circumstantial field evidence that TSD may exist in a very early group of agnathan (jawless) fishes (i.e., lampreys; Beamish 1993). Hence, although the sample size is yet very small, there appears to be little support for a hypothesis that TSD represents a generally primitive form of sex determination. Based on the cases identified so far, TSD is widespread throughout the fishes, and new examples are as likely to be found in one group as is any other. This lack of phylogenetic affinity applies also to the distribution of heterogametic sex chromosomes, which are found in about 10% of fishes scattered across numerous groups, but contrasts with the distribution of sequential hermaphroditism, which occurs predominately in the Perciformes, to a lesser extent in the Anguilliformes and a few other groups (Devlin and Nagahama 2002).

In most cases, the percentage of females declines with increasing temperature (Table 2.1), although in one genus of flatfishes *(Paralichthys)* the percentage of females is maximal at intermediate temperatures, and in another species *(Dicentrarchus labrax)* different investigators have reported opposite results (Pavlidis et al. 2000; Koumoundouros et al. 2002; Blázquez et al. 1998b; Saillant et al. 2002). In *Oreochromis niloticus*, XX genotypes tend to be masculinized by high temperature, while YY genotypes tend to be feminized (Abucay et al. 1999; Kwon et al. 2002). The proportion of females increases with temperature in *Oncorhynchus nerka* (Craig et al. 1996) and *Ictalurus* (Patiño et al. 1996).

Distinct thermosensitive periods (TSPs) of sex determination have been identified in several species including various atherinids (Conover and Fleisher 1986; Strüssmann et al. 1996a, 1997), European sea bass (Koumoundouros et al. 2002), flatfish (Yamamoto 1999), and tilapia (reviewed in Baroiller and D'Cotta 2001). Frequently, the TSP precedes and/or coincides with the end of the larval period and occurs prior to the onset of histological gonad differentiation. It also coincides with the period of development when ar-

Table 2.1 Fish Species Where Strong Evidence of Temperature-Dependent Sex Determination Has Been Reported

Species	Type of Data[a]	Direction of Response[b]	Evidence of G×E Interaction[c]	References
Salmoniformes				
Oncorhynchus nerka	exp	+	n.a.	Craig et al. 1996
Cypriniformes				
Gnathopogon caerulescens	exp	−	yes	Fujioka 2001
Misgurnus anguillicaudatus	exp	−	yes	Nomura et al. 1998
Carassius carassius	exp	−	n.a.	Fujioka 2002
Siluriformes				
Ictalurus punctatus	exp	+	yes	Patiño et al. 1996
Hoplosternum littorale	exp	−	yes	Hostache et al. 1995
Atheriniformes				
Menidia menidia	exp, fld	−	yes	Conover and Kynard 1981
M. peninsulae	exp, fld	−	n.a.	Middaugh and Hemmer 1987; Yamahira and Conover 2003
Odentesthes bonariensis	exp	−	no	Strüssmann et al.1996a, 1997
O. argentinensis	exp	−	yes	Strüssmann et al.1996b
Patagonia hatcheri	exp	−	n.a.	Strüssmann et al. 1997
Cyprinodontiformes				
Poecilia melanogaster	exp	−	n.a.	Römer and Beisenherz 1996
Poeciliopsis lucida	exp	−	yes	Sullivan and Schultz 1986; Schultz 1993
Perciformes				
Apistogramma (33 species)	exp	−	no	Römer and Beisenherz 1996
Dicentrarchus labrax	exp	−	yes	Pavlidis et al. 2000; Koumoundouros et al. 2002
	exp	+	yes	Blázquez et al. 1998b; Saillant et al. 2002
Oreochromis niloticus	exp	−	yes	Baroiller et al. 1995
	exp	both	yes	Abucay et al. 1999
O. aureus	exp	−	n.a.	Baras et al. 2000
O. mossambicus	exp	−	n.a.	Wang and Tsai 2000
Sebastes schlegeli	exp	−	n.a.	Lee et al. 2000
Plueronectiformes				
Paralichthys olivaceus	exp	+ then −	yes	Yamamoto 1999
P. lethostigma	exp	+ then −	n.a.	Luckenbach et al. 2003
Verasper moseri	exp	−	n.a.	Goto et al. 1999
Limanda yokohamae	exp	−	n.a.	Goto et al. 2000

[a]Exp indicates that the diagnosis of TSD is based on lab experiments; fld indicates that confirmatory field data also exist.
[b]A "−" sign indicates that available data shows that F/F+M declines with increasing temperature and a "+" sign indicates the reverse. "+ then −" indicates a dome-shaped response. Both means that some experiments indicate a "−" response and others indicate a "+" response.
[c]"n.a." means data to assess whether genetic affects interact with temperature are not available.

tificial administration of hormones may influence gonad differentiation. However, in some cases, thermal influences on sex ratio are detectable earlier, including during the embryonic period (Wang and Tsai 2000), and appear to involve cumulative effects that exert an influence throughout most of development (Koumoundouros et al. 2002).

With the exception of *Menidia*, all studies of TSD in fishes have been based entirely on experimental analyses of lab or hatchery-reared fish, with no information on thermal sex ratio effects in the wild. This has some unfortunate consequences. In many of the cases reported in Table 2.1, sex ratios at the temperatures normally used in a hatchery are compared to those at an extreme (usually warm) temperature. How this compares to the gradients in temperature that fish experience in the wild is often speculative. In the tilapia, *O. niloticus*, for example, sex ratios often become highly male biased only when offspring are reared at a relatively high temperature (Baroiller et al. 1995). How often do fish in nature experience this temperature during the critical period? Another problem in evaluating studies of TSD is that most species display very high levels of parental, strain, and population influence on sex ratio (Table 2.1), such that experiments using different parents, strains, or populations often produce different results. Hence, the results of an experiment with one set of parents from one geographic location may or may not be replicable by the same or other investigators using other parents. Below, these themes are elaborated by highlighting the fish taxons that have received the most attention in the literature.

Atherinids

The Atlantic silverside, *Menidia menidia,* is the only fish species in which the existence and effects of TSD have been demonstrated in the wild. In *M. menidia,* the proportion of females decreases with incubation temperature across a range of levels that mimic those experienced by larvae in the wild (approximately as in curve A, Figure 2.1; Conover and Kynard 1981; Conover and Heins 1987a). Progeny produced near the beginning of spring breeding season experience low temperatures that cause most of them to develop primarily as females, while most of those produced later when temperatures are warmer become male. Shifts from female- to male-biased sex ratios are evident in nature as young fish recruit to each year class (see Figure 1 in Conover and Kynard 1981), and in populations with TSD, the mean hatch dates of wild females are significantly advanced compared with those of males (Figure 2.2). There is a well-defined TSP of development coinciding with the end of larval development and preceding gonad differentiation (Conover and Fleisher 1986). Mortality during this period is too low to account for the magnitude of sex ratio changes with temperature (Conover and Kynard 1981). Four lines of evidence demonstrate that strong genetic effects also influence sex determination in *Menidia*: (1) full-sib and half-sib family sex ratios differ at the same temperature and in their temperature sensitivity; (2) local populations differ in their temperature sensitivity (Conover and Heins 1987a); (3) extreme temperatures do not produce monosex progeny, and therefore at least some temperature-insensitive genotypes are present in all populations; and (4) the sex ratio and its response to temperature evolve in response to frequency-dependent selection (Conover and Van Voorhees 1990; Conover et al. 1992). The genetic component of sex determination clearly involves the segregation of a major sex factor: family sex ratios tend to cluster around Mendelian proportions (1:0, 0:1, 3:1, or 1:1), at least in northern populations (Lagomarsino and Conover 1993). TSD also exists in at least one population of the more southern species, *M. peninsulae* (Middaugh and Hemmer 1987; Yamahira and Conover 2003).

The existence of TSD has been well established in the three South American silversides, *Odontesthes bonariensis, O. argentinensis,* and *Patagonia hatcheri* (Strüssmann et al. 1996a,b). In *O. bonariensis* there is a clearly defined sensitive period (Strüssmann et al. 1996a) that corresponds with peaks in the number of gonadotropin-producing cells in the pituitary gland, suggesting their role in eliciting early gonad differentiation (Miranda et al. 2001). There are only very weak indications of genetic influence on sex ratio (Strüssmann et al. 1996a,b), this being the only species in which sex ratio changes from nearly all female to nearly all male across temperature extremes.

Flatfishes

In the Hirame flounder, *Paralichthyses olivaceus,* sex appears to be determined by an interaction between temperature and a heterogametic sex-determining mechanism (Yamamoto 1999). Normal diploid offspring have a 1:1 sex ratio at intermediate temperatures and are strongly male biased at the extremes (of a form like curve C, Figure 2.1). Clones of gynogenetically produced offspring are nearly all female at intermediate rearing temperatures but produce an increased proportion of males at higher or lower thermal extremes. These patterns and other data summarized by Yamamoto (1999) suggest that an XX-XY system operates at intermediate temperatures, but that the sex of XX individuals is unstable and can be induced to produce male offspring, especially under conditions of extreme high temperature. Similar patterns have been reported in other flatfishes including *P. lethosigma* (Luckenbach et al. 2003), *Verasper moseri* (Goto et al. 1999), and *Limanda yokohamae* (Goto et al. 2000). Kitano et al. (1999) demonstrated that the creation of XX males in *P. olivaceus* at high temperature is associated with suppression of *cytochrome P450 aromatase* gene expression in the gonad, which at lower temperatures catalyzes the conversion of androgens to estrogens. Treatment of female larvae with an aromatase inhibitor also causes male development, clearly suggesting that atypical XX males occur when estrogen levels are reduced (Kitano et al. 2000). This knowledge is of great practical importance in aquaculture because females grow faster than males, and are therefore more desirable. Whether temperature influences the sex ratio of flatfishes in nature is unknown.

Cichlid Fishes

In *Apistogramma* spp. there is generally a gradual transition from sex ratios of about 60–90% female at 23°C to about 10–30% female at 29°C (Römer and Beisenherz 1996). Neither thermal extreme used by these investigators produced progeny of only one sex, suggesting a genetic influence on

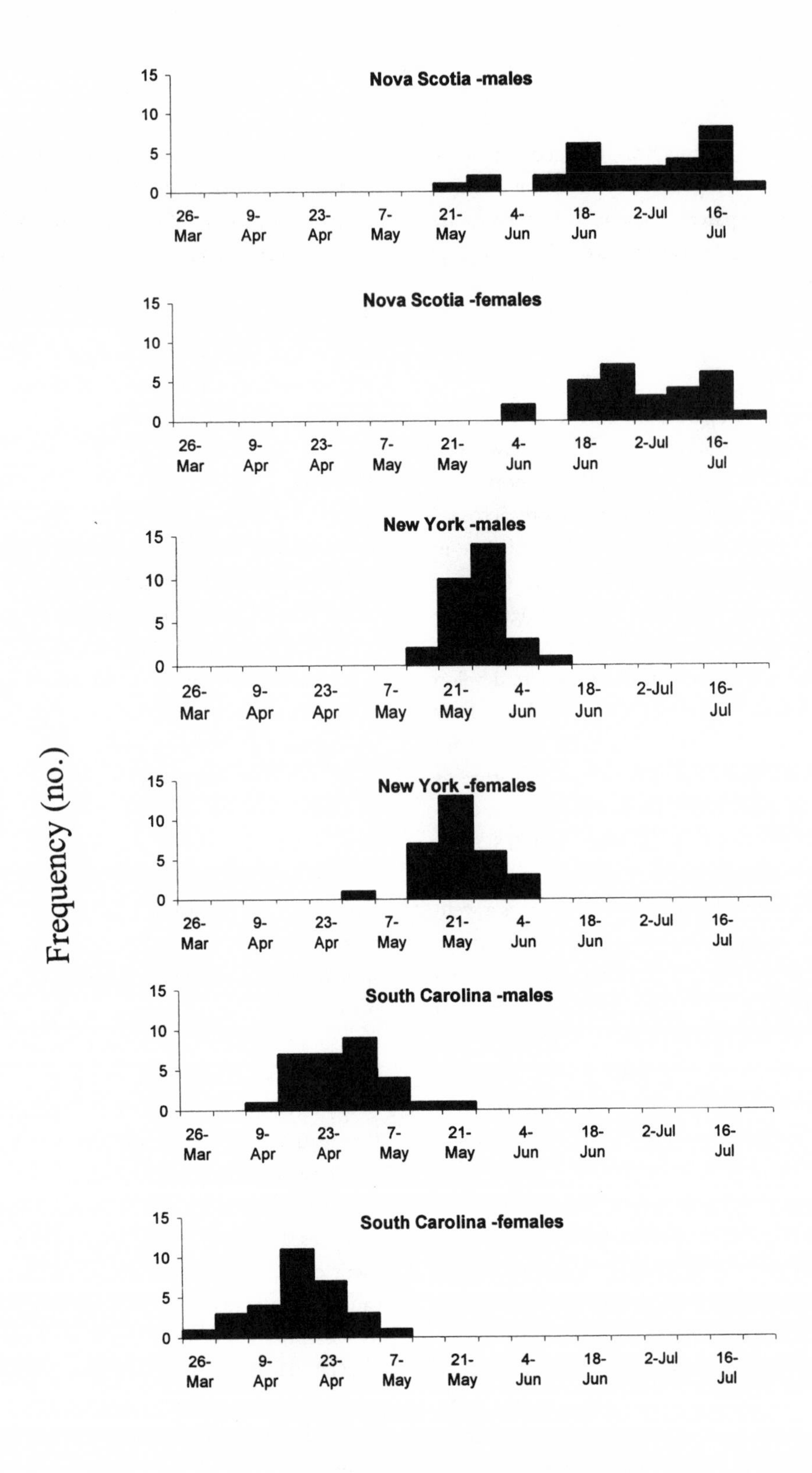

Figure 2.2 Variation in hatch dates of males and females among latitudinal populations of silversides that differ in the level of TSD. Hatch dates were estimated by counting daily rings in the otoliths of juveniles collected after the end of the breeding season (D. O. Conover, unpubl. data). In South Carolina and New York, where the change in sex ratio with temperature is high or moderate, respectively, the hatch dates of females are significantly earlier than those of males (t-test, $p < 0.05$) but not in Nova Scotia where thermal effects on sex ratio are nil.

gender, but it is not known if more extreme temperatures might have done so. There is little evidence of variation in sex ratio among broods at a given temperature, suggesting that genetic influences may be weak. Sex ratios of *Apistogramma* are also influenced strongly by pH (as also occurs in other cichlids, such as *Pelvicachromis* spp. and *Xiphophorus*, Rubin 1985). According to Römer and Beisenherz (1996), mortality was too low to account for the sex ratio changes with temperature. The TSP is relatively long, extending from just after fertilization until 30–40 days of age. Temperatures experienced by offspring of these species in the wild correspond with those used in the experiments described above, but studies of sex ratio variation with temperature in the wild have not been conducted.

Much research on sex determination has focused on tilapia species, particularly *O. niloticus*, *O. aureus*, and *O. mossambicus*, because of their importance in aquaculture. In *O. niloticus*, it was first believed that sex was determined by an XX-XY system at normal temperatures (27°C), although sex ratios were known to vary greatly among parents and strains (reviewed by Beardmore et al. 2001). Later experiments proved that extreme high temperatures (35–36°C) generally have a masculinizing effect on offspring (Baroiller et al. 1995). Subsequent work by a variety of investigators have examined thermal influences in all female XX populations (created by mating sex-reversed XX males and XX females), all male XY populations (progeny of XX females and YY males), mixed XX-XY populations, and all male YY populations (progeny of YY males and YY females) (reviewed by Baroiller and D'Cotta 2001). In general, results show that XX genotypes are masculinized by high temperature, but paradoxically many YY genotypes (and perhaps some XY genotypes) become female at high temperature (Abucay et al. 1999; Baroiller and D'Cotta 2001; Kwon et al. 2002). As in flatfishes, the *aromatase* gene appears to play a central role in sex differentiation of *O. niloticus*. Artificial application of aromatase inhibitors during the sensitive period causes masculinization (Kwon et al. 2000), and masculinized XX females and genetic males display a repression of *aromatase* gene expression when reared at 35°C (D'Cotta et al. 2001a), suggesting that temperature acts directly upon the *aromatase* gene or indirectly through transcription factors (Baroiller and D'Cotta 2001). D'Cotta et al. (2001b) showed that high temperature also stimulates expression of an unknown gene (300 bp cDNA) in masculinized genetic females. High temperatures also cause male-biased sex ratios in *O. aureus* (Desprez and Mélard 1998; Baras et al. 2000) and *O. mossambicus* (Wang and Tsai 2000).

Oreochromis would appear to be a case where a heterogametic sex-determining mechanism produces 1:1 sex ratios at normal temperatures, but is overridden in XX individuals at high temperatures, causing them to become male (i.e., as in the lower portion of curve B, Figure 2.1). Low temperatures do not appear to have a feminizing effect (Baroiller et al. 1999). However, *Oreochromis* species are mouthbrooders, and experiments have generally been conducted by removing 9- to 11-day-old fry (postfertilization) from the mouths of females. This corresponds with the beginning of the androgen-sensitive period in *O. niloticus,* as indicated in experiments where hormones were applied, and 15–27 days postfertilization also represents the developmental period when ovarian aromatase activity is downregulated in males (Kwon et al. 2001). However, Wang and Tsai (2000) have shown that high proportions of females can be induced in *O. mossambicus* by exposing progeny to low temperatures earlier in development than is required to produce masculinization of progeny at high temperatures. Correspondingly, estradiol is effective at feminizing *O. mossambicus* only if applied early in development, and methyltestosterone is most effective if applied later (Tsai et al. 2001). Hence, the timing of the sensitive periods for low-temperature feminizing and high-temperature masculinizing effects may differ. If so, thermal influences in sex ratio may extend across both low and high temperatures. The range of experimental temperatures that influence sex determination in *Oreochromis* are within those that young fish plausibly experience in nature (Baras et al. 2000), but no studies have tested whether temperature influences sex ratio in the wild.

Other Fishes

The existence of TSD has been well established in the European sea bass *Dicentrarchus labrax* (Blázquez et al. 1998b), but results vary dramatically among studies. Low temperatures produce female-biased sex ratios, and high temperatures produce mostly males in studies by Pavlidis et al. (2000) and Koumoundouros et al. (2002), but the reverse pattern has been reported by Blázquez et al. (1998b) and Saillant et al. (2002). The reasons for such wide discrepancies in results are not clear. Part of the answer is likely due to differences in the temperatures investigated and to when they were applied during development. The TSP in this species includes all ontogenetic stages up to metamorphosis, including early embryonic stages well before the period of hormone sensitivity and the histological differentiation of the gonads (Koumoundouros et al. 2002). There are also strong indications of genetic (parental) effects on sex ratio (Saillant et al.

2002). Because the brood stock in these studies originates from different geographic regions, the differences among studies may reflect population-level geographic differences in the thermal response of sex ratio.

Certain lab strains of the viviparous species *Poeciliopsis lucida* display TSD, while others do not, and breeding studies clearly show that thermal effects are genetically (not maternally) based (Sullivan and Schultz 1986; Schultz 1993). In the cyprinid honmoroko *Gnathopogon caerulescens*, the proportion of females increases with temperature, and there are also high levels of variation among parents, suggesting strong genotype-by-environment interactions (Fujioka 2001).

Summary

Thermal influences on sex determination in fishes are far more common than previously recognized. Even studies that test for but do not find TSD (Conover and Demond 1991) may reflect only the status of a particular population, not the species as a whole. But in most fishes where TSD has been reported, it is not clear whether thermal influences occur only at extreme, and therefore ecologically irrelevant, temperatures, or represent true mechanisms of TSD. If thermal influences on sex ratio occur only at extremes rarely experienced in nature, then the sex-determining mechanism is functionally GSD. For this reason, the rest of this chapter is devoted to the one genus where the ecology and adaptive significance of TSD has been studied, and the techniques for studying the ecology of TSD in other species are described.

Ecology and Adaptive Significance of TSD

Charnov and Bull (1977) proposed that ESD is adaptive when the environment experienced by offspring influences the fitness of males and females differently (see Valenzuela, Chapter 14). If some environments enhance or diminish the fitness of one sex more than the other, then a sex-determining mechanism that ensures that individuals become the sex with higher fitness in a given environment will be superior to GSD. This assumes that undifferentiated offspring cannot choose, but are capable of correctly assessing the environment they encounter with respect to expected fitness as a male or female. Studies of the Atlantic silverside have provided strong support for the Charnov-Bull model.

The Atlantic silverside is a temperate marine species ranging along the east coast of North America from northern Florida to Nova Scotia. TSD in *M. menidia* was discovered not by design or foresight, but by careful evaluation of sex ratio data from the field that was being gathered for other purposes. The key observation was that as juveniles of each year class recruited to the population, most of those produced from the early spawning period were females, while those produced later were mostly males (Figure 1 in Conover and Kynard 1981). This difference was clearly not due to differences in growth, because males and females of the same age were of similar size. Experiments then proved that TSD was the explanation for the sex ratio patterns in the wild (Conover and Kynard 1981). Females therefore tend to be larger than males, in part, because they experience a longer growing season by virtue of earlier birth. Because large size enhances the reproductive success of females more than that of males, TSD in *Menidia* appears to confirm the predictions of the Charnov-Bull model (Conover 1984). Temperature during larval development serves as a cue for the length of the growing season that offspring will experience.

Confirmation of this adaptive interpretation of TSD was achieved by comparing the ecology of TSD among populations of *M. menidia* from different latitudes. These studies showed that the response of sex ratio to temperature varied inversely with latitude: at low latitudes where the breeding and growing seasons are long, the change in sex ratio with temperature is greater than at high latitudes where the breeding and growing seasons are short (Conover and Heins 1987b; Lagomarsino and Conover 1993). In Nova Scotia, where the breeding season is only one month long, silversides almost completely lack TSD (Lagomarsino and Conover 1993). Geographic variation in TSD is adaptive because the relative benefit of being born at the beginning rather than at the end of the breeding season should decrease with the duration of the breeding season.

Further evidence that TSD in silversides evolves in response to seasonality comes from recent studies of the southern congener, *M. peninsulae*. This species is distributed along the Florida coast across a gradient from temperate to subtropical environments. In the northern populations, where there is a single protracted breeding season, TSD is evident (Yamahira and Conover 2003). But in more southern environments where the growing season is continuous and breeding occurs throughout much of the year, TSD is absent. This makes sense in terms of adaptation because when the growing and breeding seasons are continuous, there may be no advantage to one birth date over another in terms future fitness as a male or female (Yamahira and Conover 2003). Hence, in *Menidia*, both extreme northern and extreme southern populations lack TSD, while inter-

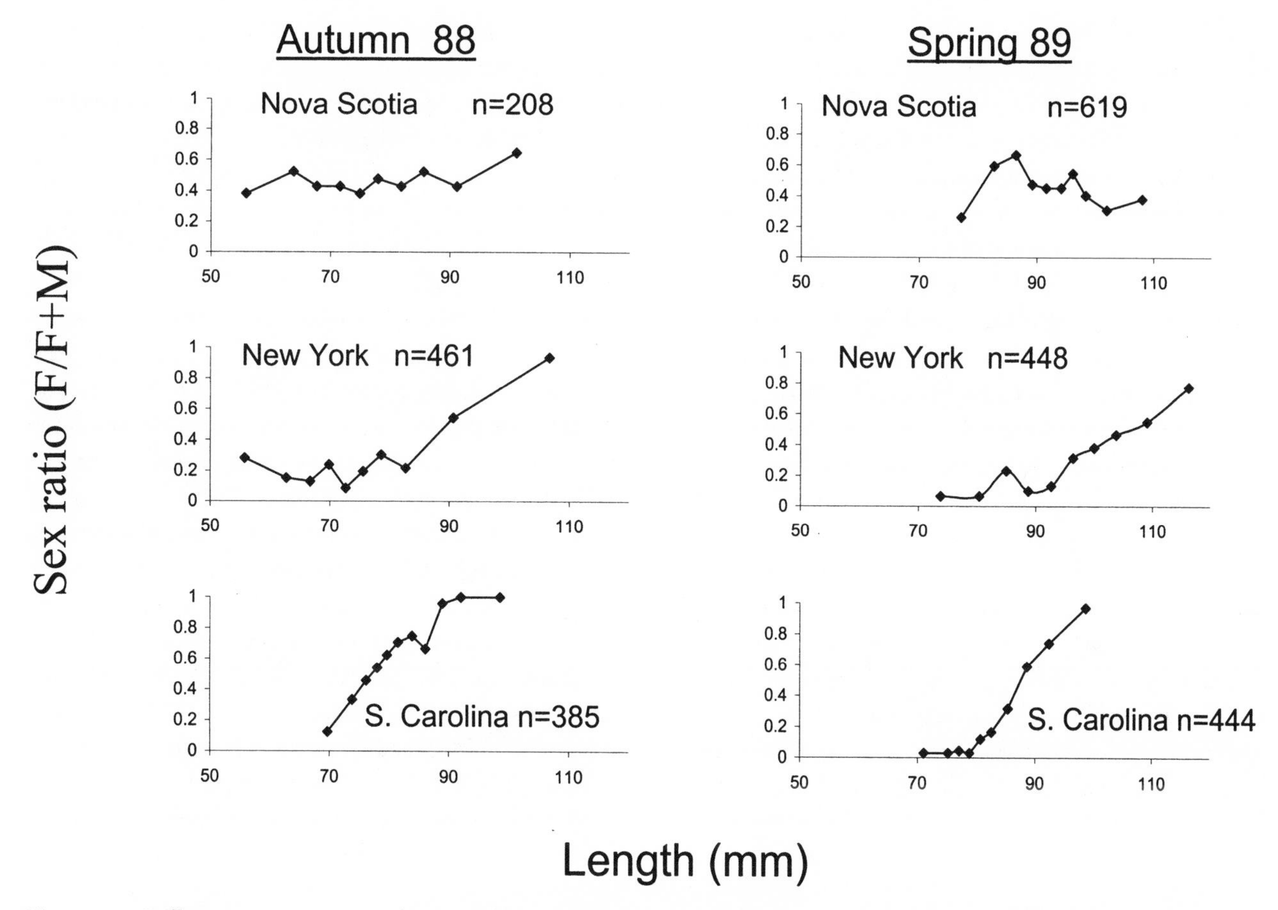

Figure 2.3 Differences in sex ratio as a function of body length among latitudinal populations of silversides that differ in the level of TSD. Samples of the 1988-year class were collected in autumn at the end of the growing season and in spring during the breeding season. The sex ratio within each quantile of the size distribution is plotted. In South Carolina (SC) and New York (NY), where the change in sex ratio with temperature is high and moderate, respectively, the smaller fish are nearly all males and the larger fish are nearly all females. In Nova Scotia (NS), however, where thermal effects on sex ratio are nil, the sex ratio does not change systematically with size (D. O. Conover, unpubl. data).

mediate populations show the maximum level of sex ratio response to temperature.

That variation in sex ratio response to temperature is capable of responding to selection was proven experimentally in *Menidia* (Conover and Van Voorhees 1990; Conover et al. 1992). During the larval period, captive populations of *Menidia* were subjected each generation to either extreme high or extreme low temperatures that skewed the sex ratio in opposite directions. Increases in the minority sex evolved over subsequent generations, as did the sensitivity of sex ratio to temperature. Coupled with the existence of a strong latitudinal cline in TSD in the wild, these experiments demonstrate that TSD in *Menidia* is not merely the plasticity of a primitive sex-determining mechanism; it is a highly evolved trait that responds rapidly to selection.

Similar explanations for TSD may exist for other species, but the absence of studies in the field for any of the other species limit generality. This gap represents an enormous opportunity for ecological research. To motivate such studies, the author next describes three ways in which the ecological signature of TSD might be visible among populations in the wild, as illustrated in *Menidia*.

First, under TSD, offspring sex ratio should vary temporally or spatially as a function of environmental conditions. In *M. menidia,* it was trends in the sex ratio of juveniles as they recruited to the population, first favoring females then males, that first motivated experimental tests for TSD (see Figure 1 in Conover and Kynard 1981). If TSD influences the sex ratio in the wild, then sex ratios of juveniles should fluctuate measurably across naturally occurring thermal gradients. Effect of pH on sex ratios could be examined in a like manner. In tropical species where seasonality is frequently driven by changes in precipitation rather than temperature, changes in pH may signify the beginning or end of the growing season and thereby provide a cue for adaptive ESD.

Second, where there are seasonal changes in temperature during the breeding season, TSD should be revealed by comparing birth dates of males and females. Analysis of daily rings in otoliths (ear bones) of fish provide a now routine method for recording birth dates and growth rates of young-of-the-year (YOY) fish in many ecological studies, but few field researchers bother to sex YOY fish. Such analyses for *M. menidia* from different latitudes reveal sex differences in birth dates that correspond with known levels of TSD: in South Carolina and New York, female birth dates occur significantly earlier than those of males, but not in Nova Scotia where thermal influences on sex ratio are minimal (see Figure 2.2). Similar studies on other species could provide direct confirmation that TSD influences sex ratio in the wild.

Finally, changes in sex ratio with body size may also be linked to the existence of TSD. Figure 2.3 shows the change in sex ratio across quantiles of the size distribution at the end of the growing season and during the breeding season for three populations of *M. menidia*. In South Carolina fish, where the level of TSD is high, fish from the smallest 10% of the size distribution are nearly all males, while those from the largest 10% are nearly all females. At the other extreme, in Nova Scotia, where TSD is minimal, the sex ratio does not change systematically with size. Hence, the ecological effect of variation in TSD is clearly detectable among populations in the wild.

Conclusions

TSD is widespread, being found now in seven orders of fishes. With the exception of *Menidia*, however, our knowledge of the ecology and evolution of TSD in nature is exceedingly poor. This contrasts markedly with the extensive literature on the ecology and evolution of sex change in fishes and with the study of TSD in reptiles. In fishes, it is often not clear whether reports of TSD represent mere anomalies of GSD that occur only at extreme, and therefore ecologically irrelevant, temperatures, or represent true mechanisms of TSD. One suspects that in many cases, TSD does have profound effects on life history and morphology in natural populations of fishes, perhaps like those of *Menidia*, or perhaps in other unique manners. Until studies of TSD move outside the hatchery and into the wild, we will remain ignorant of the ecology and evolution of one of the most profound forms of phenotypic plasticity in fishes.

3

MICHAEL A. EWERT, CORY R. ETCHBERGER, AND CRAIG E. NELSON

Turtle Sex-Determining Modes and TSD Patterns, and Some TSD Pattern Correlates

In this chapter, there are four objectives. First, current knowledge of the presence or absence of temperature-dependent sex determination (TSD) in chelonians is updated. Second, how chelonian TSD varies is examined. Special emphasis is placed on *pivotal temperature* (T_{piv}) and *transitional range of temperature* (TRT). The T_{piv} is defined as the constant temperature of sex ratio parity (1:1); TRT typically is the range of constant temperatures that yields both sexes (Mrosovsky and Pieau 1991). Application of graphic interpolation has generally sufficed to define T_{piv} and TRT. However, increased ease in applying statistical procedures now provides descriptive alternatives. The third objective is to update the relationship between taxa and plasticity in TSD patterns. Lastly, associations among pattern variation, geographic coordinates, and selected environmental variables are explored.

In exploring TSD pattern variation and correlates, special emphasis is placed on the subfamily Kinosterninae, which appears to be a close-knit group. Systematic relationships have undergone two recent and contrasting revisions (Iverson 1992b, 1998). The current arrangement recognizes a sister group *(Sternotherus),* one or two Neotropical groups of *Kinosternon*, and a principally North American group of *Kinosternon* (Iverson 1998).

Current Knowledge

Diagnosing Sex

Generally, histology provides the greatest reliability in diagnosing sex (e.g., Raynaud and Pieau 1985) and is necessary for most sea turtles (Whitmore et al. 1985; but see Hewavisenthi and Parmenter 2000). Otherwise, cool temperatures may delay morphologic differentiation through hatching (Etchberger et al. 1992b). There are nonhistological, noninvasive approaches (e.g., testosterone challenge; Lance et al. 1992). However, these techniques may need verification with histology in instances of low temperature incubation, in which delayed differentiation might suppress a hormonal response.

Which Species Have TSD?

In Table 3.1, the published data are combined with new declarations for 14 species (Table 3.2). This gives 79 species of turtles assayed through incubation at controlled temperatures. Of these 79 species, 64 have TSD and 15 have genotypic sex determination (GSD) (Table 3.1). Although sampling now represents 12 of 13 turtle families, roughly 70% of the 257–280 turtle species remain untested. The least well-studied groups with TSD expected include additional pelomedusids (especially *Pelusios*), testudinids, and Asian geoemydids. Field data on the pelomedusid *Peltocephalus dumerilianus* are ambiguous (Vogt et al. 1994). This species could lack TSD or have an atypical TSD pattern for pelomedusids. Untested Asian geoemydids include several monotypic genera, and the large genera *Cuora* and *Kachuga*. Controlled incubation for eight geoemydid species indicates TSD in each. However, one geoemydid, *Siebenrockiella crassicollis,* has heteromorphic sex chromosomes (Carr and Bickham 1981), rendering TSD unlikely (Bull 1980).

Table 3.1 Numbers of Turtle Species Tested for Sex Ratio Responses to Constant Temperature Incubation of Eggs

Family	Total Species N (N)[a]	Species Incubated[b]	TSD Species	GSD Species
Carettochelyidae	1 (1)	1	1	0
Chelidae	40 (48)	8	0	8
Cheloniidae	7 (7)	6	6	0
Chelydridae	2 (2)	2	2	0
Dermatemydidae	1 (1)	1	1	0
Dermochelyidae	1 (1)	1	1	0
Emydidae	35 (38)	25	24	1
Geoemydidae	59 (59)	8	8	0
Kinosternidae	22 (24)	16	13	3
Pelomedusidae	25 (26)	4	4	0
Platysternidae	1 (1)	none	–	–
Testudinidae	40 (45)	4	4	0
Trionychidae	25 (27)	3	0	3
Totals	257 (280)	79	64	15

Sources: For most species see Ewert and Nelson (1991), Ewert et al. (1994), Paukstis and Janzen (1990), Vogt and Flores-Villela (1992), or here (Table 3.2). Additional species include a chelid (*Elusor macrurus*, Georges and McInnis 1998), two pelomedusids (*Podocmenis expansa*, Valenzuela 2001b; *P. unifilis*, Souza and Vogt 1994), three sea turtles (*Lepidochelys kempii*, Shaver et al. 1988; *Eretmochelys imbricata*, Mrosovsky et al. 1992; *Natator depressus*, Limpus et al. 1993, Hewavisenthi and Parmenter 2000), three tortoises (*Testudo hermanni*, Eendeback 1995; *Gopherus agassizi*, Spotila et al. 1994, Lewis-Winokur and Winokur 1995; *G. polyphemus*, Burke et al. 1996, Demuth 2001), and a softshell turtle (*Pelodiscus sinensis*, Choo and Chou 1992).

[a]Tally of Iverson (1992a) (our estimate includes recently declared species less apparent synonymies and misidentified geoemydids).

[b]Includes only species with adequate documentation through incubation.

Variation in Chelonian TSD

Diversity in TSD Patterns

The two recognized TSD patterns in turtles are evident in a comparison between *Graptemys ouachitensis* (cool males→warm females = TSD Ia) and *Pelomedusa subrufa* (cool females→warmer males→warm females = TSD II; Figure 3.1a; see also, Functional Transitions). Diversity in expression of TSD occurs within the Kinosterninae. All three diagnosed *Sternotherus* species are strongly TSD II, with sex ratios declining to 0% male at cool temperatures (Figure 3.1a). All 10 *Kinosternon* species (Table 3.2, Figure 3.1c–f) depart from the *Sternotherus* by showing some males at the coolest incubation temperatures studied, which for *K. arizonense* and *K. flavescens* were the coolest temperatures that sustained development. *Kinosternon alamosae* and *K. hirtipes* appear to be strongly pattern Ia (Figure 3.1f), and *K. sonoriense* and *K. subrubrum subrubrum* (Figure 3.1c) lean in this direction. *Kinosternon arizonense*, *K. creaseri*, and *K. subrubrum hippocrepis* lean toward TSD II. *Kinosternon baurii*, *K. flavescens*, *K. leucostomum*, and *K. scorpioides* show greater ambiguity. Among these species, two partially documented populations of *K. baurii* (Ewert, unpubl. data) have moderate male biases at 22.5°C (not elaborated, but note *K. baurii* from a third population in Figure 3.1c), the coolest temperature tested. Unexpectedly, the three groups in combination suggest a GSD-like zone of temperature independence at cool temperatures (Ewert et al. 1990). In *K. f. flavescens* no temperature yields over 85% male, and *K. f. flavescens* shares with *K. arizonense* a suggestion of a dip in male bias at 25°C (Figure 3.1d). *Kinosternon leucostomum* (Figure 3.1e) uniquely has its greatest male proportion at just under 50%. Lastly, *K. scorpioides* has TSD II according to observations on one population (Ewert and Nelson 1991), but closer to TSD Ia in another (Figure 3.1e). Overall, the variation in Kinosterninae defies clean separation between TSD Ia and TSD II and, as in *K. baurii*, suggests zones of thermal insensitivity (see also Vogt and Flores-Villela 1992).

Phenotypic vs. Genetic Distance and TSD

The shapes of TSD patterns are typically similar among conspecific populations (e.g., in *Chrysemys* and *Graptemys*, Bull et al. 1982b; or *Caretta*, Limpus et al. 1985 vs. Mrosovsky 1988). In two instances, initial observations revealed variation in TSD patterns among subspecies. Both (mtDNA) now appear genetically discrete. In one instance, *Kinosternon s. subrubrum* tends toward TSD Ia, and *K. s. hippocrepis* toward TSD II (Figure 3.1c). Differences in mtDNA now support genetic disjunction (Walker et al. 1998). The other difference, originally viewed as distinguishing *K. f. flavescens* from *K. f. arizonense* (Ewert et al. 1994), now represents taxa recognized as full species (Serb et al. 2001). Cases for additional study include *K. scorpioides*, which exhibited TSD II in one population (Ewert and Nelson 1991), but closer to TSD Ia in another (Figure 3.1e). Patterns in *K. leucostomum* also suggest differences. The authors' peculiar data lacked a sex ratio greater than 0.5 male, and warm temperatures gave only females (Figure 3.1e). Another study (Vogt and Flores-Villela 1992) showed a sex ratio > 0.5 male at a cool temperature and mixed sex ratios at all three warm temperatures. Likely, the population in the latter study is a separate subspecies, or species (Vogt, pers. com.).

At a higher level, TSD patterns distinguished *Sternotherus* from *Kinosternon* (Etchberger 1991; Figure 3.1) at a time when a phylogenetic hypothesis included both genera within *Kinosternon* (Iverson 1992b). Analysis of mtDNA now separates these genera (Iverson 1998).

Table 3.2 First Time Species Diagnoses and Updated TSD Data on Several Other Species

	Constant Incubation Temperature (°C)											
Species	21.5	22.5	24	25	26	27	28	29	30	31	32	TSD Pattern
Kinosternon alamosae[a]		100		100		100	33.3	0	0			TSD Ia
		(2)		(6)		(7)	(3)	(2)	(7)			
Kinosternon arizonense			42.0	28.6	40.0	93.3	100	100	76.5	10.0	0	TSD ~II
			(25)	(21)	(20)	(15)	(17)	(13)	(17)	(5)	(11)	
Kinosternon baurii (Pen.FL)[a]	29.2	44.7	52.0	75.3	70.8	64.2	30.8	0	0		0	TSD amb[b]
	(24)	(19)	(25)	(85)	(24)	(14)	(13)	(15)	(11)		(2)	
Kinosternon baurii (W. FL)[a]		61.1	65.4		53.8	45.5	15.4		0			TSD amb[b]
		(18)	(13)		(13)	(11)	(10)		(8)			
Kinosternon creaseri[a]		7.1	42.9	27.3	64.2	36.3	0	0				TSD II
		(14)	(7)	(11)	(14)	(11)	(7)	(4)				
Kinosternon flavescens			82.8	64.0		77.8	85.0	22.7	0		0	TSD amb
			(29)	(25)		(27)	(20)	(22)	(17)		(6)	
Kinosternon hirtipes		100		100		100			16.7			TSD Ia
		(6)		(7)		(7)			(9)			
Kinosternon leucostomum		18.7	33.3	46.2	33.3	23.1	0		0			TSD amb
		(8)	(3)	(13)	(6)	(13)	(6)		(9)			
Kinosternon scorpioides		71.6	76.1	93.9	100	83.0	25.0	5.6	0			TSD amb[b]
		(58)	(44)	(66)	(41)	(53)	(20)	(18)	(14)			
Kinosternon sonoriense[a]		95.7		97.1		94.7	87.5	20.0	0		0	TSD ~Ia
		(23)		(17)		(19)	(8)	(10)	(15)		(6)	
Kinosternon subrubrum hippocrepis		28.9	47.4	57.9		84.2	50.0	5.9	0			TSD ~II
		(19)	(19)	(19)		(19)	(20)	(16)	(21)			
Kinosternon subrubrum subrubrum[a]	100	85.7		82.5		71.1			0			TSD ~Ia
	(9)	(21)		(20)		(19)			(11)			
Sternotherus carinatus		0		20.0	88.9	100	77.2	0	0			TSD II
		(10)		(5)	(9)	(21)	(22)	(14)	(6)			
Sternotherus minor	0	2.4	8.3	76.3	77.3	10.0	10.0	0	0		0	TSD II
	(18)	(41)	(12)	(38)	(22)	(40)	(20)	(6)	(36)		(3)	
Sternotherus odoratus	0	0	30.8	91.7	92.7	29.1	11.5	0	0	0		TSD II
	(14)	(65)	(26)	(60)	(110)	(55)	(26)	(18)	(51)	(6)		
Pelomedusa subrufa			0	0	0	0	0	0	60.0	72.7	63.2	TSD II[c]
			(5)	(11)	(5)	(18)	(14)	(16)	(20)	(11)	(19)	
Mauremys annamansis[a]			100	100		100	100	16.7	0			TSD Ia
			(11)	(11)		(3)	(5)	(6)	(9)			
Melanochelys trijuga			4.0	20.0	63.3	57.1	51.9	7.1	3.1			TSD II
			(25)	(40)	(15)	(28)	(27)	(14)	(32)			
Rhinoclemmys areolata			33.3	85.7		87.5	40.0		0			TSD ~II
			(6)	(7)		(8)	(5)		(6)			
Chinemys nigricans[a]				100		100			0			TSD Ia?[d]
				(4)		(3)			(1)			
Clemmys guttata		90.9	92.9	70.0	85.7	93.7	52.6		0			TSD ~Ia
		(11)	(14)	(20)	(7)	(16)	(19)		(21)			
Terrepene carolina (IN)	93.3	71.5	90.9	96.1	90.2	84.8	70.0	5.0	0			TSD amb
	(15)	(35)	(11)	(89)	(61)	(46)	(60)	(20)	(88)			
Terrepene carolina (FL)		70.6	60.0	66.7	81.8	85.7	45.5	0	0			TSD amb
		(17)	(5)	(6)	(11)	(7)	(11)	(1)	(1)			
Chrysemys picta (IN)	100	100		100		100	50.3	0	0			TSD Ia
	(84)	(10)		(83)		(33)	(313)	(28)	(78)			
Chrysemys picta (TN)	100			100	85.7	54.4	4.5	0	0			TSD Ia
	(7)			(21)	(7)	(34)	(22)	(6)	(20)			

(continues)

Table 3.2 *(Continued)*

Species	Constant Incubation Temperature (°C)											TSD Pattern
	21.5	22.5	24	25	26	27	28	29	30	31	32	
Deirochelys reticularia (FL)			100 (6)	100 (19)	100 (5)	100 (10)	90.5 (21)	55.0 (20)	12.0 (25)	0 (2)		TSD Ia
Pseudemys concinna (TN)		100 (19)	100 (11)	92.8 (69)		100 (15)	85.2 (27)	3.8 (26)	0 (65)		0 (6)	TSD Ia
Pseudemys concinna (FL)		100 (3)		100 (10)		100 (15)	75.0 (24)	0 (21)	0 (19)		0 (6)	TSD Ia
Pseudemys floridana				100 (4)		93.8 (16)	82.4 (34)	15.4 (26)	0 (4)			TSD Ia
Pseudemys nelsoni[a]						100 (7)	68.4 (19)	31.3 (16)	0 (1)			TSD Ia
Pseudemys texana[a]							100 (3)	40.0 (5)		0 (3)		TSD Ia[e]
Trachemys decorata[a]			100 (9)		100 (3)		66.7 (9)	0 (2)	0 (4)		0 (3)	TSD Ia
Emydura subglobosa[a]		50.0 (2)	55.6 (9)	44.9 (47)	50.0 (2)	50.0 (22)	53.8 (13)		50.0 (48)			GSD
Elseya novaguineae[a]		33.3 (3)	52.9 (17)	59.4 (32)	60.0 (5)	53.8 (26)	40.0 (15)	25.0 (8)	50.0 (24)			GSD
Mesoclemmys gibbus[a]				57.1 (18)		50.0 (14)			46.7 (15)		50.0 (18)	GSD
Phrynops geoffroanus[a]				52.4 (22)		40.0 (25)			52.9 (18)		68.8 (16)	GSD
Phrynops hilarii[a]				33.3 (30)		32.0 (25)			56.3 (16)		33 (9)	GSD

Note: Proportions are % male (N = number of hatchlings diagnosed). Diagnoses were conducted as for Ewert and Nelson (1991), with "intersexes" (only a few encountered) scored as ½ male (Bull et al. 1982b, Vogt and Bull 1984).

[a]First time designation for mode or pattern of sex determination in listed species.

[b]Despite the listed data, these species have ambiguous (amb) patterns as suggested from nearby populations with less complete data.

[c]Hot temperatures (not shown) yield only females (nine each from 33 and 34°C, see Figure 3.1A).

[d]The G statistic for this small sample is 6.0 ($p < 0.05$). The trend follows that in *Chinemys reevesii* (Hou 1985).

[e]The G statistic for this small sample is 8.4 ($p < 0.025$).

It was unanticipated when both TSD and GSD responses turned up in the North American genus *Clemmys* (Bull et al. 1985; Ewert and Nelson 1991; Ewert et al. 1994). Recent phylogenetic analyses have redefined the original four-species genus *Clemmys* as polyphyletic (Holman and Fritz 2001; Feldman and Parham 2002). Note that disparate patterns of sex determination again were a harbinger of this decision. In the current rearrangement of former *Clemmys*, *Glyptemys insculpta* has GSD and the small amount of data on *G. muhlenbergii* is compatible with GSD (Ewert and Nelson 1991). *Clemmys guttata* has a slightly degraded pattern of TSD (see Table 3.2). *Actinemys marmorata* has TSD Ia, which it shares with its apparent closest relatives (*Emys orbicularis, Emydoidea blandingii*; Ewert and Nelson 1991; Girondot 1999; Feldman and Parham 2002). A cross between an *E. blandingii* male (TSD) and a *G. insculpta* female (GSD) yielded two highly viable clutches of hybrids and vigorous young (Harding and Davis 1999; J. H. Harding, pers. com.). As in TSD species, *G. insculpta* lacks evidence of heteromorphic sex chromosomes (Bickham 1975).

Pivotal Temperature

Some TSD species have abrupt transitions (≤ 1.5°C) from all male to all female sex ratios, and others show mixed sex ratios across a wide range of temperatures. All TSD turtles show a cool male to warmer female (MF) transition in sex ratio, which implies a MF T_{piv} within this transition. Estimation of the MF T_{piv} has been central to much work (e.g., natural maternal effects, Janzen et al. 1998; Bowden et al. 2000; steroid effects, Crews et al. 1995; ambient gas effects, Etchberger 1991, Etchberger et al. 2002).

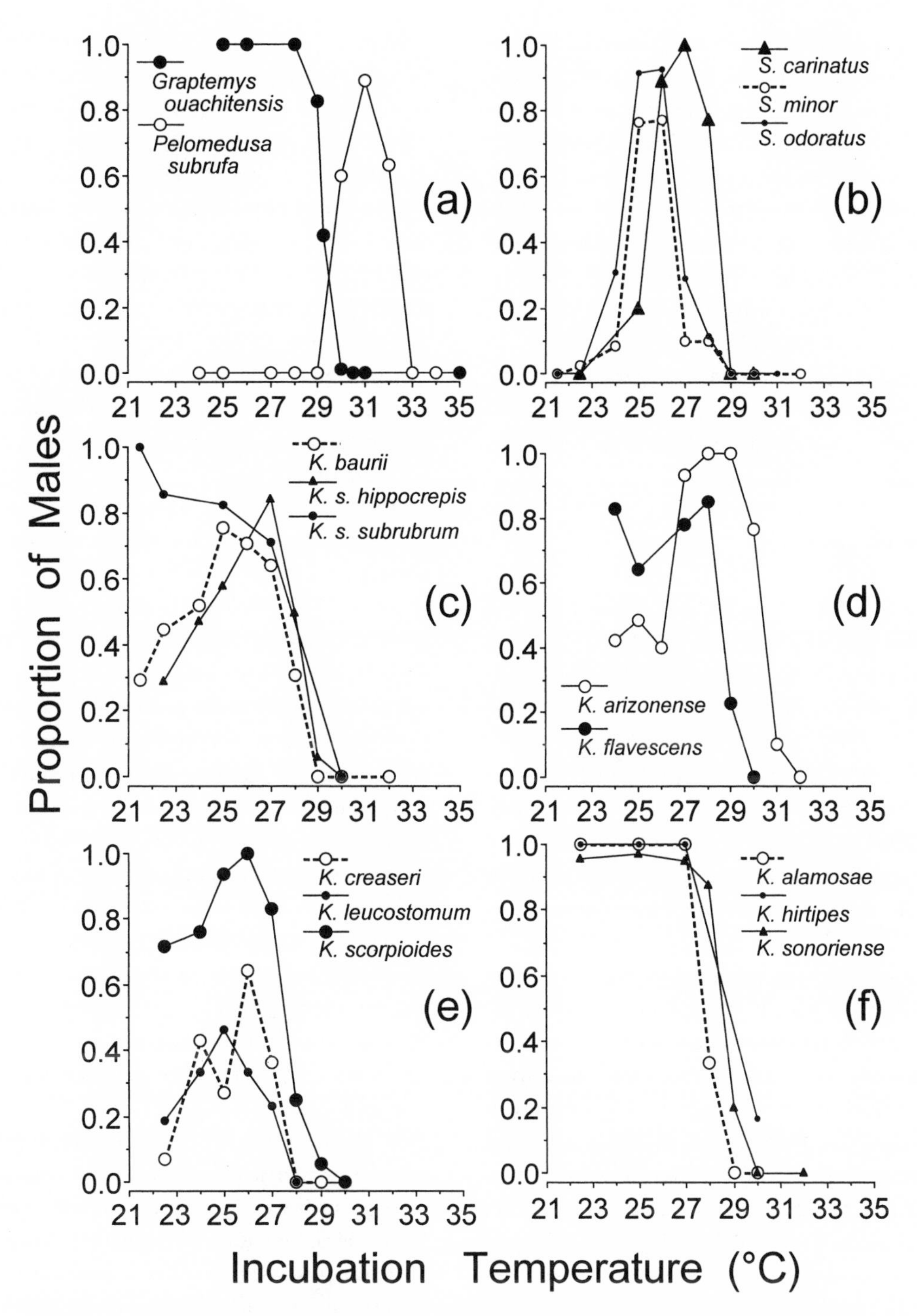

Figure 3.1 Examples of TSD patterns in turtles. (**a**) *Graptemys ouachitensis* (TSD Ia) and *Pelomedusa subrufa* (TSD II). Note: cool temperatures ≤ 28°C) yield only males in *G. ouachitensis* and only females in *P. subrufa*. (**b–f**) Kinosterninae: 13 species and 1 subspecies. (**b**) 3 *Sternotherus* sp. have TSD II. (**c**) *Kinosternon* of the SE United States: *K. s. subrubrum* has ~ TSD Ia; *K. s. hippocrepis* has ~ TSD II. (**d**) *K. flavescens* group: strongly mixed sex ratios at cool temperatures. *K. arizonense* is a hot desert species with a high T_{piv}. (**e**) Tropical *Kinosternon* of the Neotropical clade. Likely, the unevenness in *K. creaseri* represents stochastic variation. (**f**) Desert *Kinosternon* of the Neotropical clade. All have embryonic diapause and likely delay somitigenesis until the early spring following oviposition. (Data from Bull and Vogt 1979; Bull et al. 1982a, 1982b; Vogt 1980; Ewert and Nelson 1991; Ewert et al. 1994; and Table 3.2, except *P. subrufa* at 33 and 34°C: N = nine sexed/ temperature. See also Iverson 1998.)

Bull (1985) showed how sex ratios determined by variable temperatures in natural nests of *Graptemys* relate to the T_{piv} determined from constant-temperature incubation in the laboratory. Thermal acceleration of development favors the sex determined at higher temperatures. That is, more sexual differentiation is completed per unit warm versus cool exposure (Pieau 1982; Georges et al. 1994). Accordingly, apparent T_{piv} declines as nest temperature variance increases. Alternatively, particular temperatures carry sex-determining potencies. Generally, the further the departure (warmer or colder) from the constant-temperature T_{piv}, the greater becomes the potency of the temperature to influence female or male differentiation (Bull et al. 1990; Wibbels et al. 1991b).

Estimating Pivotal Temperature

Generally, plotting data on sex ratio by temperature and then interpolating temperature between the two or three points that most closely bracket a 1:1 sex ratio has sufficed to estimate T_{piv} (Mrosovsky and Pieau 1991), and has been widely used. However, Smith (1987) and Girondot (1999) argue that maximum likelihood estimating (MLE) procedures can use additional data points for a more comprehensive definition of T_{piv}. The appropriate distributions of temperature-specific sex ratios for the most effective use of MLE extend from 0 to 1 and include at least two intermediate values.

In the first of three approaches to applying MLE, Smith (1987) applied probit analysis. Then, Girondot (1999) derived a maximum likelihood equation, and Demuth (2001) applied logistic regression. In what follows, probit analysis (SAS Statistical Package) was used. In a TSD pattern with sex ratios ranging 0 to 1, MLE often yielded good best-fit (low value χ^2) sigmoid approximations of the actual data. However, calculated MLE plots approach 0 and 1 at values equidistant from the estimated T_{piv}. For some TSD patterns, particularly with TSD II, such symmetry does not adequately describe the data. Still, one can fit a limited portion of the data with probit analysis.

In typical TSD Ia, as in sea turtles, *Emys orbicularis*, and the *Chrysemys-Graptemys* group, MLE provides an estimate of the transitional range of temperatures (TRT). This is refined to TRT_{95}, with asymptotes truncated at 95% of the more abundant sex, a cool shoulder (estimated 95% male) and a warm shoulder (estimated 95% female) temperature (Smith 1987; Girondot 1999). The MF T_{piv}s and the breadth of the MF TRT_{95}s were estimated using probit MLE and simple interpolation. In some cases of TSD II, the observed maximum ("peak") male sex ratio was less than 0.95. Here, the cool shoulder temperature was estimated from values approaching the male peak temperature, using probit and extrapolation. This sometimes had the effect of extending the cool portion of a MF TRT into what otherwise might be regarded as a cool female to warmer male (FM) TRT. In some anomalous cases (e.g., *Clemmys guttata*, Table 3.2), the cool shoulder was extrapolated from values clearly within the TRT. In comparison of interpolated with probit estimates, the interpolated TRT_{95} is significantly broader (30 species, 47 populations, $z = 4.8$, $p < 0.001$, Wilcoxon signed rank test). For 12 populations that meet or reasonably approach the best probit criteria (Smith 1987), probit and interpolated estimates of TRT_{95} differ by less than 10% of the TRT value. For another 26 populations, the absolute difference exceeds 10%. For nine samples the criteria for probit are weaker.

Most often, differences in estimates of T_{piv} between different statistical techniques are <0.2°C for given data. The authors' comparisons indicated that neither interpolation nor probit systematically yields higher or lower estimates. However, in some noteworthy departures from <0.2°C, atypically high T_{piv}s estimated by interpolation became lower with probit. Four cases are *Graptemys caglei* (29.9°C by interpolation vs. 29.6°C by probit; Wibbels et al. 1991c), *G. versa* (30.4 vs. 29.7°C; Ewert et al. 1994), *Lepidochelys olivacea* (31.1 vs. 30.6°C; Wibbels et al. 1998), and *Podocnemis unifilis* (32.0 vs. 31.3°C; Souza and Vogt 1994).

Lack of control over several natural and procedural factors that affect T_{piv} and TRT may render trivial the small differences typically found between MLE and interpolation applications. A problem often ignored in applying interpolation as well as MLE (e.g., Girondot 1999; Chevalier et al. 1999) is dependence of sex-determining thresholds of individual eggs on maternal origin. Phenotypic variation in T_{piv} among clutches can be appreciable (e.g., Bull et al. 1982a; Limpus et al. 1985; Rhen and Lang 1998; Bowden et al. 2000). Thus, samples based on small numbers of clutches are at risk of distortion through sampling error. Additionally, an endogenous component, suggestively estradiol in egg yolks, leads to significant seasonal variation in sex ratio in *Chrysemys picta* (Bowden et al. 2000) and perhaps in other species.

Other potential causes of variation that may be appreciable at the scale of T_{piv} arise from imperfections in incubation equipment and thermal monitoring. Also, a reduced exchange of gases between incubation boxes and ambient air can result in buildup of biogenic carbon dioxide and feminized sex ratios (Etchberger et al. 1992a, 2002), which would give low estimates for T_{piv}.

Association of Pivotal Temperature with Transitional Range of Temperature

Moderately well documented thermal response data represent eight families, 44 species, and 73 populations. This large group shows a weak but significant inverse association in which warmer T_{piv}s associate with narrower TRTs (Pearson $r = -0.28$, $p = 0.035$; Spearman $r = -0.38$, $p = 0.0007$). The cool shoulder temperatures associate more strongly with narrower TRTs (Pearson $r = -0.66$; $p < 0.0001$; Spearman $r = -0.68$, $p < 0.0001$). The warm shoulders also increase with breadth of the TRT, but the significance of this association is ambiguous (Pearson $r = +0.31$, $p = 0.009$; Spearman $r = +0.15$, $p = 0.2$). Thus, while the slopes connecting warm and cool shoulders appear to pivot around a "universal" T_{piv}, the relationship is more complicated (Figure 3.2). The MF TRT becomes narrower at warmer T_{piv}s. In this process, the temperatures encompassing the cool shoulders become more widely spread than around the warm shoulders (see Figure 3.2a). For the 73 populations, the variance across warm shoulder temperatures is 2.06, whereas the variance across cool shoulder temperatures is 3.21. Species that lack a cool shoulder (sex ratios ≥ 95% male), in the strictest interpretation, likely affect this comparison.

Within well-represented taxa, T_{piv}s always showed inverse associations with TRT (see also Girondot 1999). According to probit values, this tendency exists at least weakly in five groups (*Caretta*, 6 populations; *Trachemys*, 7; other emydids, 20; kinosternines, 15, see Figure 3.2b,c; *Chelydra*, 7). These inverse associations are significant for *Caretta* ($r = -0.85$, $p = 0.03$), the kinosternines (Figure 3.2b), and *Chelydra* ($r = -0.94$; $p = 0.006$). The trend strengthens appreciably in comparisons of cool shoulder temperatures and TRT: all five associations are significant (respectively, $r = -0.97$, $p = 0.0015$; $r = -0.88$, $p = 0.009$; $r = -0.58$, $p = 0.0075$; Figure 3.2b; $r -0.94$, $p = 0.006$). In each of these groups, the warm shoulder either lacks a significant counterbalancing association with TRT (e.g., Figure 3.2b,c), or it has a significant inverse association (in *Chelydra*). An ecological expectation from these associations is that the array of temperatures that yields a few females among many males will be more diverse than the array leading to a few males among many females.

Taxa and Plasticity: Functional Transitions

The MF TRT and T_{piv} clearly represent the functional transition from male-biased to female-biased nests in TSD Ia turtles (e.g., *Graptemys* sp., Bull 1985; *Podocnemis unifilis*, Souza and Vogt 1994; *Chelonia mydas*, Godley et al. 2002). For some TSD II species, such as *Chelydra* and *Macrochelys*, the functional transition may be less clear. For *Chelydra*, the MF TRT seems to predominate (Wilhoft et al. 1983). However, very cool nests do occasionally yield female *Chelydra* (Hotaling 1990) and *Macrochelys* (Ewert, unpubl.).

Greater functioning of the FM transition seems likely in some other TSD II species. The FM T_{piv} in *Pelomedusa* (a TSD II species, Table 3.2, Figure 3.1a) is 30°C. Temperatures ≤ 29°C yield only females, and only warmer temperatures yield males. In contrast, the TSD Ia species *Graptemys ouachitensis* has a MF T_{piv} of 29.3°C (Bull et al. 1982a). Temperatures ≤ 28.5°C yield only males, and only warmer temperatures yield females. Male-forming temperatures in *G. ouachitensis* (< MF T_{piv}) equal female-forming temperatures (< FM T_{piv}) in *Pelomedusa*. Both species nest naturally in spring and early summer (Vogt 1980; Bull 1985; references in Boycott and Bourquin 2000). Viable thermal ranges for sustained development in both species include 24–34°C, with limited viability at 35°C, commencing with the pharyngula stages (Ewert, unpubl.). Given lack of evidence that nests are appreciably hotter for *Pelomedusa* than for *G. ouachitensis*, the FM T_{piv} of *Pelomedusa* appears to be as functional as the MF T_{piv} of *G. ouachitensis*.

Other species with at least modest FM T_{piv}s, and thus some likelihood of yielding cool females from naturally cool nests, include *Pelusios castaneus* (T_{piv} >27°C; Ewert and Nelson 1991), *Melanochelys trijuga*, (25.7°C, Table 3.2), and *Heosemys grandis* (>25°C; Ewert et al. 1994). The latter two species nest (Madras Crocodile Bank, unpubl. data), or seem to nest (Goode 1991, unpubl. data), during cool seasons. Lastly, the FM T_{piv} of *Sternotherus carinatus* is 25.4°C (from Table 3.2, Figure 3.1b). This species is associated with cool hill streams in some parts of Arkansas and Oklahoma (Iverson 2002; Ewert, unpubl. obs.). Here, eggs deposited on shaded slopes might yield cool females in cool years.

Pattern Variation and Geographic and Environmental Variables

Geographic Trends

According to the hypothesis of rare sex advantage (Fisher 1930), the local TSD pattern should evolve toward a 1:1 sex ratio if reproductive investment is similar in individuals of the two sexes. Do TSD patterns reflect adaptation to regional thermal environments or, perhaps, a resistance to change? Bull et al. (1982b) reported the first effort at identi-

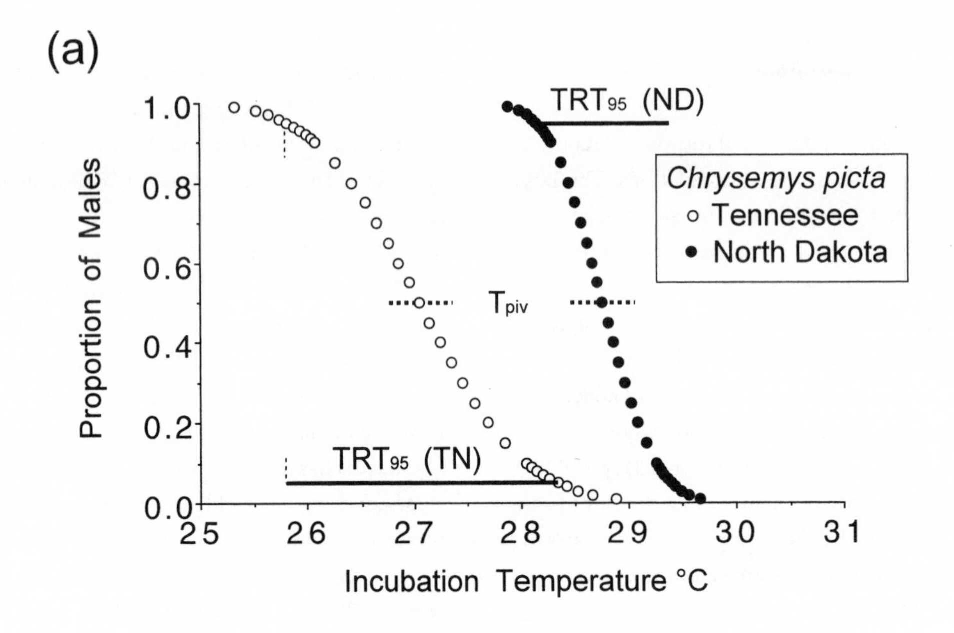

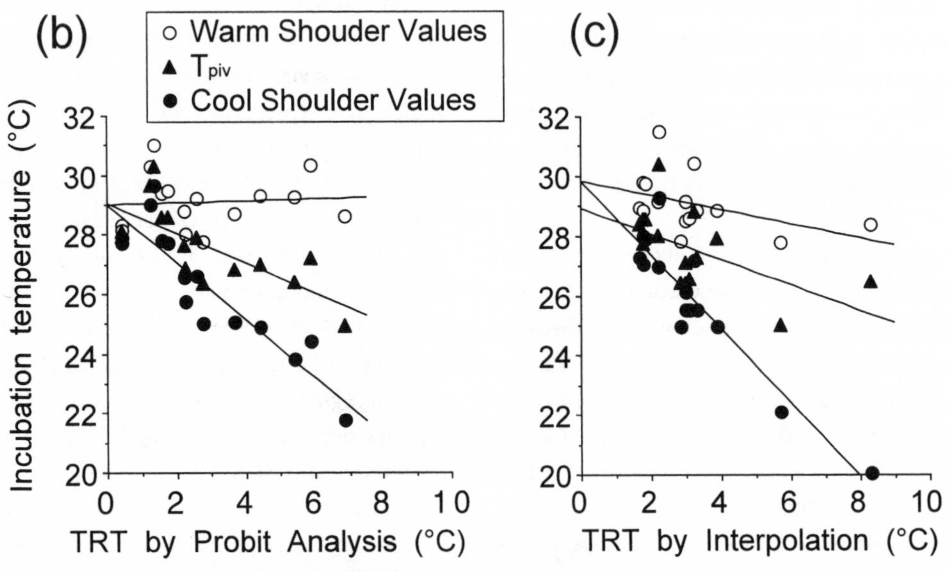

Figure 3.2 How descriptive elements of TRT (cool and warm shoulders, T_{piv}) vary along the temperature scale with breadth of TRT. The sex ratio at a cool shoulder is 95% male, a warm shoulder is 5% male, a T_{piv} is 50% male. (**a**) Schematic diagram, based on probit interpretation, shows that the broader of the two TRTs associates with cooler temperatures. (**b**) Probit values for TRT and T_{piv} in 15 kinosternine samples. Broad TRTs show declines in their cool shoulder temperatures and their T_{piv}s ($r = -0.90$, $p = 0.0001$; $r = -0.74$, $p = 0.0017$; respectively), whereas their warm shoulders remain nearly unchanged ($r = +0.07$, $p = 0.8$). (**c**) Interpolated values for TRT and T_{piv}. Relationships parallel those in (**b**) ($r = -0.93$, $p = 0.0001$; $r = -0.57$, $p = 0.027$; $r = -0.43$, $p = 0.11$; respectively). Comparisons of probit and interpolation give T_{piv}s, $r = 0.962$ ($p < 0.0001$); cool shoulders, $r = 0.874$ ($p = 0.0001$); warm shoulders, $r = 0.773$ ($p = 0.0007$). [Sources: ND *C. picta*, Rhen and Lang 1998; other data, Table 3.2; *K. baurii* (W. FL), unpubl.]

fying geographic variation. They offered that, as an alternative to altering the TSD pattern, a species might compensate for regional environmental differences by altering its choice of nest sites or its nesting phenology (e.g., nesting earlier or later in the season). TSD patterns in a given species might resist change because they best represent a balance between sex ratio and the physiological fitness of the species.

A few studies have focused on MF T_{piv}s as an indication of differences in pattern in samples from different geographic locations (Bull et al. 1982b; Mrosovsky 1988; Ewert et al. 1994). Several caveats attend data collection. Ideally, the experimental protocols should follow a common garden design. That is, eggs from designated groups should be randomized within the same incubation boxes within the same incubators. Within this array, embryos should experience their thermosensitive periods (TSPs) simultaneously. However, such precision has seldom been met. Nesting phenologies tend to vary among distant localities (Ewert 1976; Vogt 1990). Also, eggs from different localities may develop at different rates under the same temperature conditions (Ewert 1985).

Observed Geographic Trends

What appear to be the trends in geographic variation? For sea turtles, absence of any trend in T_{piv}s seems apparent. For *Caretta* nesting in Australia, Limpus et al. (1985) hypothesized variation between two local sites; however, analysis according to individual clutches denied support for this hypothesis (Smith 1987). For the southeastern United States, variation across 7° of latitude lacks consistency (Mrosovsky 1988) or association with phylogeography (see Bowen and Karl 1997). The probit values for the T_{piv} for the *Caretta* from its South Atlantic breeding grounds, Florida, and North Carolina are similar (29.15, 29.11, 29.04°C, respectively; data from Mrosovsky 1988, Marcovaldi et al. 1997). For *Chelonia mydas*, the T_{piv} for Ascension Island (28.8°C; Godley et al. 2002) shows little difference from that for Suriname (28.96°C, probit value; derived from Mrosovsky et al. 1984a and Godfrey 1997). Atlantic and Pacific populations of *Dermochelys* do not differ in T_{piv} (Binckley et al. 1998; Chevalier et al. 1999).

The first survey for continental variation (Bull et al. 1982b) concluded that T_{piv} increases with latitude in two freshwater turtles *(Chrysemys picta, Graptemys pseudogeographica)*. Currently, data on three additional species are compatible with this trend (*Chelydra serpentina*, Ewert et al. 1994; *Pseudemys concinna* and *Terrapene carolina*, Table 3.2). Although the T_{piv}s for *Chelydra* span a range of 2.5°C and those for *C. picta* span 2°C, the small variation within the other three species (0.27–0.43°C; Bull et al. 1982b; Table 3.2) could represent experimental error.

Ewert et al. (1994) suggested that high latitude populations might hasten embryogenesis during short growing seasons by seeking open, and hence, potentially warm, nest sites (see also, Vogt and Bull 1984; Janzen 1994b). Less constrained by season, lower latitude populations might seek shady situations to avoid overheating (e.g., Wilson 1998). T_{piv}s might then evolve according rare sex advantage to counterbalance sex ratios in the warm, open, northern nests versus the cooler, shady, southern nests. Data on indices of shading [scored as percent shading: (South + West + overhead)/3] support nonshaded nest sites at high latitudes. According to ranking by the same observer, nests of *Chrysemys picta* in northern Michigan (~46°N) had an average score of 5.8 ± 13.5 s.d.% shading (N = 43 nests). In more southerly regions, nests in Indiana (~39°N) and Tennessee (36.2°N) had more shading (50.5 ± 26.3% shading, N = 71 nests; 63.1 ± 33.1%, N = 15, respectively). Over this range *C. picta* does show a decline in T_{piv} (Michigan estimate of 28.5°C from 32 clutches; Indiana, 28.0°C, 91 clutches; Tennessee, 27.0°C, 17 clutches; Bowden et al. 2000; Ewert et al. 1994; Ewert, unpubl.). At this point, however, it is not known whether shading and average (versus biased) thermal trends represent the sex ratios of surviving nests.

Empirical observations also suggest a west-to-east decline in T_{piv}s in *Chrysemys picta*, and among species, there is a similar pattern for some *Graptemys* (Ewert et al. 1994). Although new data, as well as definition of T_{piv}s by probit, reduce the estimated values for *G. versa* (to 29.7°C), *G. caglei* (to 29.6°C), and *G. barbouri* (to 28.1°C), the trend persists. Confirming a prediction of Spotila and Standora (1986), *Gopherus agassizii* has a higher T_{piv} (31.0°C; see Spotila et al. 1994; Lewis-Winokur and Winokur 1995) than *G. polyphemus* (29.2°C; see Demuth 2001). In general, T_{piv}s of species of the western United States and Mexico appear somewhat higher that in related forms in the east (Ewert et al. 1994; see below).

Environmental Correlates

Interpretation of geographic coordinates assumes that they represent clinal variation in nest thermal biology. Clinal variation may be relevant for only some of several climatic parameters. For instance, air temperature records may have a direct association with nest temperature. Implying temperature less directly, rainfall keeps open ground temperatures moderately cool (e.g., Suriname beaches, Godfrey et al. 1996) or fosters growth of vegetation, which cools nests by shading them (Vogt and Bull 1984; Janzen 1994b).

The authors compared climatic data with the 14 regional samples of kinosternine turtles, each representing a recognized species or subspecies (Table 3.2, Figure 3.3). For the weather data (see references with Figure 3.3), sites close (generally, < 100 km) to the sampled populations were selected. When the sources for some of the authors' species could be approximated only to country, several sets of weather data from the apparent local geographic range of the species were averaged (Iverson 1992a). Sky cover values were estimated by averaging data from airports in two to four compass directions from the authors' sites.

The authors examined indices of annual conditions and of conditions during the estimated embryonic TSPs. Annual indices presumably reflect biome type (e.g., desert vs. tropical rain forest) and included multiyear averages of total annual rainfall and annual percent sunshine. Other indices estimated averaged conditions concurrent with estimated embryonic TSPs. These *estimated early embryonic periods* (eeeps) accounted for embryos that have diapause (Ewert 1991; Ewert and Wilson 1996; see Figure 3.3a) and commence development during the first seasonal warming following oviposition (e.g., late March–May for *Kinosternon baurii*; Wilson 1998). The estimated phenology for early development varied from February–May for the Neotropical species to July–August for *K. arizonense* (Iverson 1989, 1999). Selected indices of climate included averages of maximum, minimum, and average ambient temperatures, average range in daily temperatures, rainfall, net water balance, and hours of sunshine. More than 10 years of data were averaged to give mean values.

Most environmental variables were significantly associated in expected directions. Rainfall indices were inversely associated with indices of sunshine. Percent annual sunlight was positively associated with average daily maximum temperature. Estimated early embryonic period [eeep] maximum temperature was inversely associated with eeep rainfall. Several of the indices were also significantly associated with longitude (e.g., arcsine percent annual sunshine, $r = +0.70$, $p = 0.005$; total annual rainfall, $r = -0.69$, $p = 0.006$; eeep maximum temperature, $r = +0.68$, $p = 0.008$; eeep rainfall, $r = -0.74$, $p = 0.002$). These indices characterized aridity in the west (western Great Plains, Arizona, Mexico) and dampness in the east (Gulf region, Caribbean Central America).

Kinosternine T_{piv}s were positively associated with longitude ($r = +0.77$, $p = 0.0013$). These T_{piv}s were positively associated with an index of aridity (arcsine percent annual sunshine, Figure 3.3b) and inversely associated with total annual rainfall (Figure 3.3c). Kinosternine T_{piv}s were also significantly associated with eeep hours sunshine (Figure 3.3d), eeep rainfall (Figure 3.3e), and eeep maximum temperature, ($r = +0.72$, $p = 0.004$), but not with eeep average temperature ($r = 0.09$, $p = +0.76$).

Other aspects of the kinosternine TSD patterns, namely TRT_{95} and MF cool shoulder (by probit as well as by interpolation), were also significantly associated with measures of sunshine and rainfall (annual and eeep; all p values < 0.05 to $p < 0.001$). As expected (see Figure 3.2), TRT_{95}s were broadest, and MF cool shoulders had the lowest temperatures, in rainy climates.

These correlations suggest that TSD patterns, especially T_{piv}s, are adjusted to regional climates. T_{piv}s showed associations with the broad climatic indices that were approximately as strong as those with temporal (eeep) indices. Thus, largely nontemporal factors, such as sparse or lush vegetation, may influence nest microclimate. However, many potential sources of error in estimating the seasons of embryonic development could obscure closer associations with the temporal indices. The subject merits additional research.

Conclusions

The survey for TSD in chelonians has sampled most families but remains sparse in coverage of species. So far, all TSD turtles show either a single male-to-warmer-female (MF) transition in sex ratio biases (TSD Ia) or two transitions (FM and MF; TSD II). Some species have apparently degraded patterns that could represent evolution toward GSD, and two species of *Kinosternon (K. baurii* and *K. leucostomum)* warrant additional evaluation in this regard.

Widespread evidence indicates that temperatures across the MF transition function in determining chelonian sex ratios in natural nests, but little evidence supports influence of temperatures within the FM transition, when present. The authors identify some species, particularly the African species *Pelomedusa subrufa*, for which physiological ecology in nests seems likely to reveal "cool" females of the FM transition.

To estimate T_{piv} and TRT, the advantages of one type of maximum likelihood estimating (probit) relative to simple graphic interpolation are explored. The two approaches give similar but not identical values for T_{piv}; however, pro-

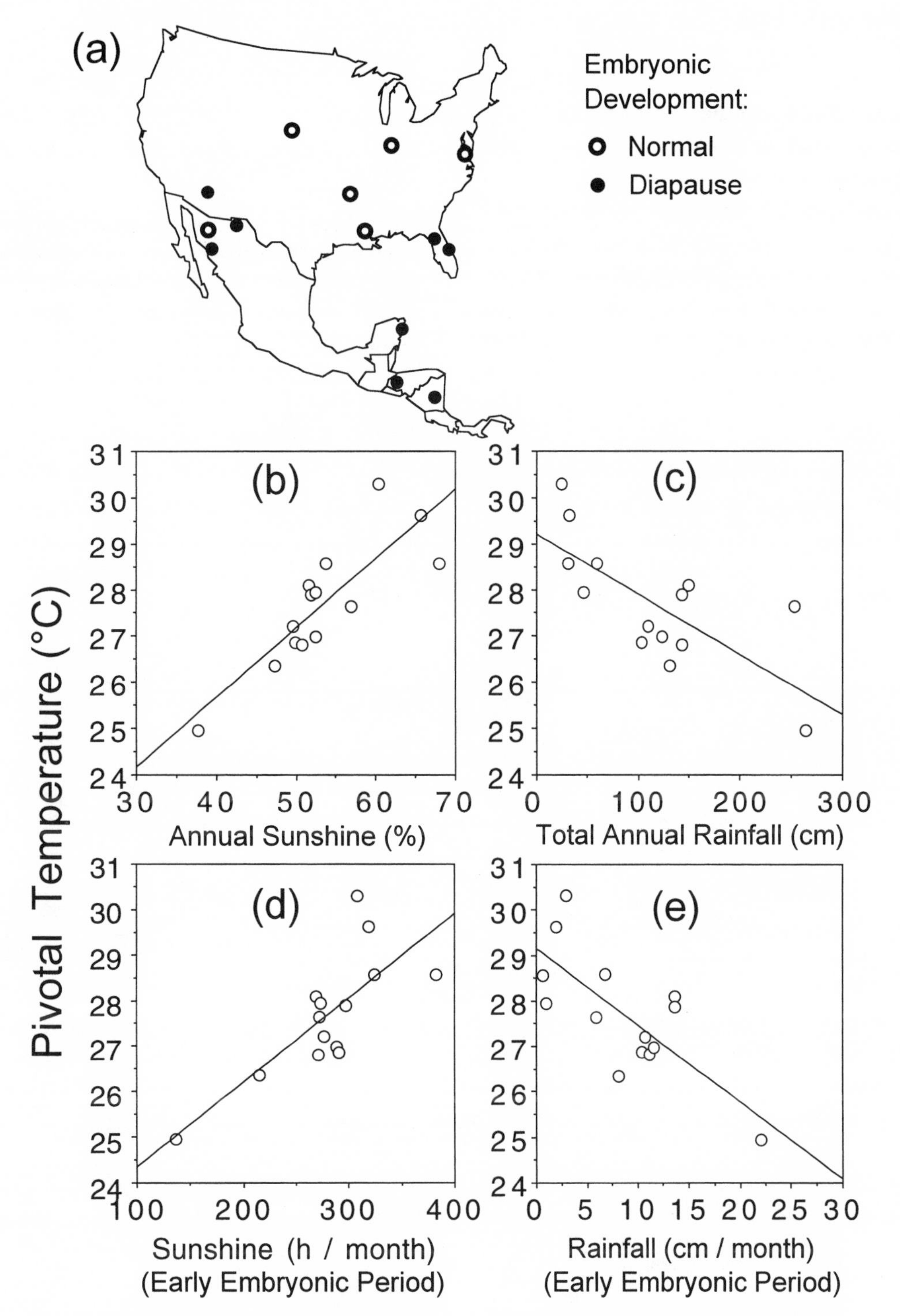

Figure 3.3 Kinosternine T_{piv}s vs. climatic variables. (**a**) Map: origins of biological samples and focal points for climatic data. *Normal*: eggs laid in the spring-summer start somitigenesis immediately. *Diapause*: eggs laid during the summer or fall-winter delay somitigenesis until the late winter-spring. (**b**) Annual percent sunshine = the fraction of total annual daylight without cloud cover; association $r = +0.84$, $p = 0.0002$. (**c**) Total annual rainfall averaged across years; $r = -0.73$, $p = 0.003$. (**d**) Sunshine during early embryonic development (eeep) = hours/month of sunshine expected during the estimated period when sex determination in embryos in nests is sensitive to temperature; $r = +0.76$, $p = 0.002$. (**e**) Eeep as in (d); average monthly values are shown; $r = -0.74$; $p = 0.002$. (T_{piv} data are derived from data in Table 3.2; Climatic data: NOAA U.S. Weather Bureau monthly summaries, especially 1980–1996; U.S. Department of Commerce 1966; Wernstedt 1959; Schwerdtfeger 1976; Rudloff 1981; Brown 1982; Coen 1983.)

bit truncates TRT_{95} values relative to interpolation. Probably, interpolation, which does not require computer processing of TSD data, suffices for much research.

The relationships found between T_{piv} and environmental variables make sense according to the principle of rare sex advantage. Still, it is not known what microclimates most of the species actually select for nesting. Overall, the authors feel that the geography of TSD offers interesting prospects for study of nest site selection relative to stasis or plasticity in TSD pattern.

Acknowlegments—The following persons and institutions contributed to this work in various ways: Columbus Zoo (J. M. Goode, R. E. Hatcher, and L. G. Burke), A. L. Braswell, P. J. Clark, J. B. Iverson, J. H. Harding, D. R. Jackson, P. V. Lindeman, P. C. Rosen, and G. R. Smith.

4

DENIS C. DEEMING

Prevalence of TSD in Crocodilians

Temperature-dependent sex determination (TSD) in crocodilians was first reported in *Alligator mississippiensis* by Ferguson and Joanen (1982). This was also one of the first studies in reptiles to demonstrate that TSD occurs in wild nests. Subsequent studies in nature and in the laboratory have demonstrated that TSD is present in 12 crocodilians examined so far and that TSD occurs in all three subfamilies of the Crocodylidae. Lang and Andrews (1994), therefore, suggested that TSD exists in all 25 species of crocodilians. Here, studies that have investigated the effects of temperature (and other environmental factors) on sex determination in this group are described. Effects of incubation temperature on crocodilian physiology and growth before and after hatching are also briefly delineated because they provide an insight into the possible evolutionary significance of TSD in crocodilians. Finally, the importance of TSD is briefly considered in relation to other archosaurs.

Species Accounts

American Alligator (Alligator mississippiensis)

Alligator mississippiensis builds mound nests of vegetation in marsh and levee sites, which are generally cool and warm, respectively (Ferguson and Joanen 1982, 1983). The effects of constant incubation temperature have been investigated in a variety of studies, and female hatchlings are produced at both low and high temperatures (Table 4.1). Discrepancies in the resulting sex ratios between studies may reflect differences in incubation conditions.

High-temperature females were found to be morphologically identical to low-temperature females and exhibited similar patterns of growth and behavior, whereas Lang and Andrews (1994) reported a high incidence of "runt" hatchlings that grew at a much slower rate. Lang and Andrews (1994) also stated that three other laboratories have produced high-temperature females in *A. mississippiensis*, and alligator eggs incubated at 34°C by Allsteadt and Lang (1995b) also produced high-temperature females (15.4% males). Smith et al. (1995) found 16.7% males in eggs incubated at 34.5°C. Phelps (1992) reported that a pulse of 37°C against a background incubation temperature of 33°C also produced females. It seems, therefore, that it was appropriate for Lang and Andrews (1994) to revise the classification of *A. mississippiensis* from the female-male (FM) pattern of TSD (Deeming and Ferguson 1989b, 1991b) to the female-male-female (FMF) pattern, that is, females are produced at both low and high temperatures within the viable range, with males at intermediate temperatures (summarized in Table 4.2).

Incubation at 32.0°C produced a range in percentage of males of 0–100%, depending on individual clutches, and at 34.0°C the range was 0–71% males (Lang and Andrews 1994). A range of 46.2–76.9% males in three clutches incubated at 32.0°C has also been reported (Allsteadt and Lang 1995b). In a further study, 15 clutches of alligator eggs were

Table 4.1 Percentage Males Recorded Following Incubation of Eggs of *Alligator mississippiensis* at Various Constant Temperatures

	Incubation Temperature (°C)															
Reference	26	28	29	29.4	30	30.5	31	31.5	32	32.5	33	33.5	34	34.5	35	36
Ferguson and Joanen 1982, 1983	0	0	–	–	0	–	–	–	13.3	–	–	–	100	–	–	100
Joanen et al. 1987	–	–	–	0	–	40.6[a]	–	74.5[b]	–	99.1[c]	–	–	–	–	–	–
Deeming and Ferguson 1988, 1989a	–	–	–	–	0	–	–	–	–	–	100	–	–	–	–	–
Joss 1989	–	–	–	–	0	–	–	–	–	–	100	–	–	–	–	–
Lang and Andrews 1994	–	0	0		0	0	0	0	66	100	100	84	35	7	0	–
Allsteadt and Lang 1995b	–	–	0	–	–	–	0	–	63.2	–	100	–	–	–	–	–
Smith et al. 1995	0	–	–	–	0	–	–	–	–	–	100	–	–	16	–	–
Dubowsky et al. 1995	–	–	–	–	–	–	–	46	–	–	–	–	–	–	–	–
Conley et al. 1997	–	–	–	–	–	–	0	–	–	–	100	–	8.3	–		
Lance and Bogart 1991	–	–	–	–	0	–	–	–	–	–	100	–	–	–	–	–

Note: Dash indicates sex ratio was not investigated at this temperature.
[a]Actual recorded value = 30.6°C.
[b]Actual recorded value = 31.7°C.
[c]Actual recorded value = 32.8°C.

Table 4.2 Average Percentage Males Recorded Following Incubation at Various Constant Temperatures

	Incubation Temperature (°C)														
Taxon	28	28.5	29	29.5	30	30.5	31	31.5	32	32.5	33	33.5	34	34.5	35
Alligatorinae															
Alligator mississippiensis[a]	0	–	0	–	0	20.3	0	40.2	47.5	99.6	100	84	47.8	11.5	0
Caiman crocodilus crocodilus[b]	–	0	–	0	0	–	0	57	97.5	95	92.5	87.5	50	–	–
Caiman crocodilus yacare[c]	–	–	0	–	0	–	0	–	13.3	–	68.8	–	93.3	–	–
Paleosuchus trigonatus[d]	0	–	–	0	0	–	8.4	–	100	–	–	–	–	–	–
Crocodylinae															
Crocodylus johnstoni[e]	0	–	0	–	3.3	–	4.3	15.1	22.4	11.6	2.3	–	–	–	–
Crocodylus moreletii[f]	–	–	–	–	–	–	0	5	33	–	55	5	–	–	–
Crocodylus niloticus[g]	0	–	–	–	–	–	0	–	–	91	–	–	81.5	–	–
Crocodylus palustris[h]	0	0	0	0	0	0	0	22	59	100	31	–	–	–	–
Crocodylus porosus[i]	–	–	–	–	–	–	30.9	–	88.1	–	51.9	–	–	–	–
Crocodylus siamensis[j]	0	–	0	–	–	–	0	–	0	100	60	–	–	–	–
Gavialinae															
Gavialis gangeticus[k]	–	–	–	–	–	–	–	0	89	20	15	–	–	–	–

Note: Dash indicates sex ratio has not been investigated at this temperature.
[a]Averaged values from Table 4.1.
[b]Lang et al. 1989, Lang and Andrews 1994.
[c]Pinheiro et al. 1997.
[d]Magnusson et al. 1990.
[e]Webb et al. 1983, 1987, 1990, 1992.
[f]Lang and Andrews 1994.
[g]Hutton 1987.
[h]Lang et al. 1989.
[i]Webb et al. 1987.
[j]Lang 1987.
[k]Lang and Andrews 1994.

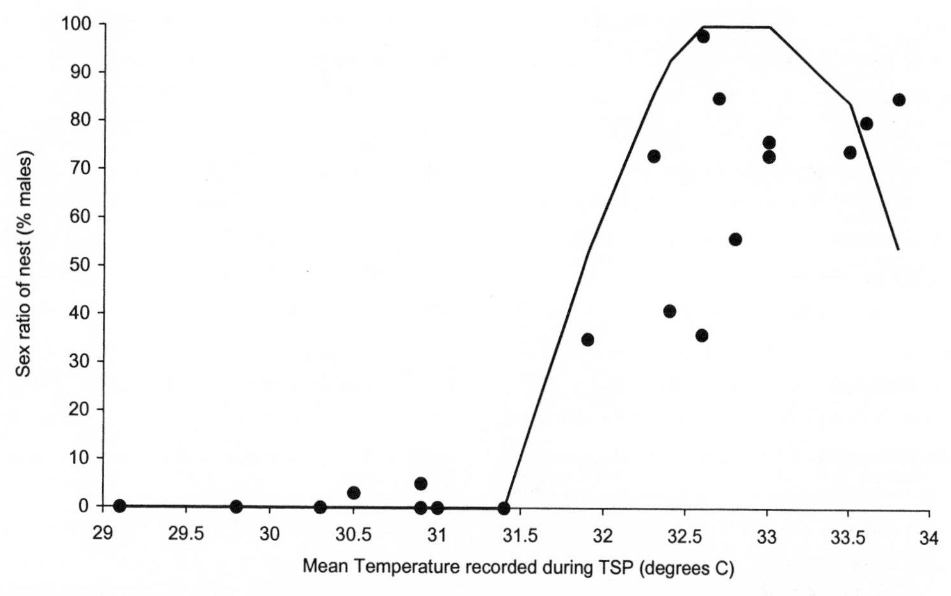

Figure 4.1 The relationship between the mean temperature recorded in 20 alligator nests during the TSP and the sex ratio of hatchlings from the nest. There is a significant correlation between the variables (Spearman's rank correlation $r = 0.856$, $p < 0.001$). The solid line is the sex ratio predicted from constant temperature experiments. (Data from Rhodes and Lang 1996.)

split equally among three temperatures within the male-female transition ranges (Rhen and Lang 1998). Clutch effects were observed with 0–82% males at 31.8°C, 60–100% males at 33.8°C, and 0–100% males at 34.3°C. Reported data of 40.6% male hatchlings at a temperature of 30.6°C (Joanen et al. 1987) were not considered by Lang and Andrews (1994). Perhaps this was also an effect of the clutches incubated? Clutch effects reflect genetic variability in the population (Rhen and Lang 1998) or may relate to variability in the amounts of steroid hormones invested in the egg during yolk formation (Conley et al. 1997).

Nest maps for *A. mississippiensis* show the gender of individual embryos in relation to temperatures recorded by probes in the nest (Ferguson and Joanen 1982, 1983). Dry levee nests had consistently high temperatures (32–34°C) and produced 93.3% males. Wet marsh nests were cooler (28–30°C) and produced only one male egg (4.2% of total), which developed close to a temperature probe registering 32°C. Two dry marsh nests exhibited more thermal heterogeneity, and 15–17% males developed.

Ferguson and Joanen (1983) also collected eggs after five weeks of incubation (after the presumptive thermosensitive period) from every nest within a 2,000-acre area and artificially hatched them. The hatchlings were grown to 1.5 years under constant conditions until they could be externally sexed. The mean sex ratio of this population was five (± 0.7) females to each male (16.7% males). Hatchlings from 23 nests in South Carolina ranged from 0–100% males, with an average of 24.2% males (Rhodes and Lang 1995). Fourteen nests produced no males, and only four nests produced more than 50% males (range of 59–100% males). Nests located on high elevations had higher average temperatures (31.7°C) and produced 29% males, whereas cooler nests on low elevations produced fewer males (4%). Rhodes and Lang (1996) recorded mean temperatures of alligator nests during the presumptive thermosensitive period (TSP) for sex determination and then examined the sex ratio of the hatchlings. Contrary to the suggestion by Rhodes and Lang (1996), the match between the actual and predicted values at high temperatures was not very good (Figure 4.1). Indeed, none of the nests produced 100% males; high temperatures that should have produced sex ratios biased towards females were more biased towards males; and at certain low temperatures some males were produced.

Temperature switch experiments have been carried out to determine TSPs, that is, times at which the gender of individuals can be changed from that of the preceding temperature. Hence, Ferguson and Joanen (1982, 1983) shifted eggs from 30°C (100% females) to 34°C (thought at the time to produce 100% males) and from 34°C to 30°C, at different times during the developmental period and observed the effects on the hatchling sex ratio. That a temperature of 34°C fails to produce 100% males (see Table 4.1) does tend to undermine the conclusion that there was a TSP between 14 and 21 days for the 30 to 34°C switch and a TSP between 28 and 35 days for the switch from 34 to 30°C. Later switch experiments using a temperature of 33°C (100% males) allowed Lang and Andrews (1994) to conclude that the TSP was between embryonic stages 21 and 24, irrespective of temperature.

Shift periods of seven days of one temperature imposed during incubation at another temperature did not cause a complete change in the gender of hatchlings (Deeming and Ferguson 1988, 1989b). Seven-day periods at a higher temperature (33°C) only produced a few males for the period between 7 and 21 days, but a drop in temperature (to 30°C) during any seven-day period from day 14 to day 49 produced a few females. It is also possible to produce 100% females, mixed sex ratios, and 100% males, depending on the timing and duration of the high-temperature pulse (Lang and Andrews 1994). By contrast, a pulse of a lower temperature invariably produced a mixed sex ratio except when the pulse was prolonged, in which case 100% females were produced. Lang and Andrews (1994) also carried out multiple, progressive shifts in temperature, and the results were largely similar to their double-shift experiments. Clutch effects were also observed in shift experiments (Lang and Andrews 1994).

Thermal heterogeneity is a feature of reptilian nests, and Ferguson and Joanen (1982, 1983) demonstrated that the position an egg occupied in the nest mound was correlated with the gender of the hatchlings. Alligator eggs incubated in isolation from each other were compared with eggs incubated within a cluster, and it was shown that metabolic heat production raised egg temperature above the incubator set point by 0.5–2.5°C during the TSP (Ewert and Nelson 2003). There was an increase in the fraction of male hatchlings: 73.1% in the egg cluster compared with 15.4% in the eggs incubated in the open arrangement. However, Ewert and Nelson (2003) thought that this effect had little relevance in wild nests.

Deeming and Ferguson (1991a) reduced the effective surface area of the shell of alligator eggs incubated at 30°C and 33°C to investigate whether restricted gas exchange could influence sex determination. Although embryonic growth was retarded, there was no effect on the sex ratio. The possibility that respiratory gas tensions in the egg could influence the sex ratio at a temperature that normally produces 50% males has not been studied.

A. mississippiensis has been a popular subject for studies on the factors affecting TSD in crocodilians (e.g., Lance and Bogart 1991; Lang and Andrews 1994; Matter et al. 1998). Lance and Bogart (1994) reviewed the pharmacological effects of various agents on sex determination in alligator embryos. It is notable that none of the treatments so far attempted have caused masculinization of embryos developing at 30°C.

Chinese Alligator (Alligator sinensis)

Temperatures within the mound nests of *Alligator sinensis* have yet to be recorded. Sex determination has only been studied once, with eggs incubated at 33–35°C producing at least 91% male hatchlings and those incubtated at below 28°C producing 100% females (Chen 1990).

Spectacled Caiman (Caiman crocodilus crocodilus)

Caiman crocodilus crocodilus is a mound-nesting species with nest temperatures in the range of 25–32°C (Magnusson et al. 1985). Originally classified as producing a FM pattern, (Lang et al. 1989) *C. c. crocodilus* has now been shown to exhibit the FMF pattern (Lang and Andrews 1994). At 31°C and below no males were produced, but at higher temperatures males developed, although as the temperature continued to increase the percentage of males decreased (see Table 4.2). Embryonic viability was greatly reduced at temperatures above 33.5°C. Different clutches produced varying proportion of males (up to 100%) at temperatures between 31.5–33.5°C.

Yacare Caiman (Caiman crocodilus yacare)

Caiman crocodilus yacare builds its mound nests in both forests and flooded areas of the Pantanal, where the nests float on the water surface. Campos (1993) showed that eggs from nests with originally high temperatures (> 31.5°C) produced 100% females if translocated to the laboratory before day 40 of development. By contrast, similar nests translocated after 40 days of development produced varying numbers of males. Nests with a temperature of 30.5°C produced 100% females, but nests at 30.5–31.5°C produced 10–15% males. In nests with temperatures above 31.5°C, 50–100% males were produced. Forest nests were more stable in temperature than the floating nests, which were more exposed to the elements. Hence, the estimated sex ratio of forest nests was similar in the two study years, but that of the floating nests was significantly different (81.57% males were produced in 1989 and 30% males in 1990). Pinheiro et al. (1997) showed that no males were produced at constant incubation temperatures below 31°C, but as temperatures rose there was a progressive increase in the percentage of males, with nearly all eggs producing males at 34°C (see Table 4.2).

Schneider's Dwarf Caiman (Paleosuchus trigonatus)

Nests constructed by *Paleosuchus trigonatus* are of loosely compacted leaf litter commonly located against or on top of large termite mounds (Magnusson et al. 1985). Heat generated by the termite mound and insulation by the leaf litter serve to maintain nest temperature at about 30°C, a level that would be impossible under the forest canopy of the typical habitat of this species. Magnusson et al. (1990) showed a FM pattern (see Table 4.2) but nests in *P. trigonatus* do not appear to reach temperatures higher than 32°C, which may explain the absence of high-temperature females. Contrary to the report in Lang and Andrews (1994), Campos (1993) did not describe sex ratios in *P. trigonatus*.

Johnston's Crocodile (Crocodylus johnstoni)

Crocodylus johnstoni is a hole-nesting species with average nest temperatures around 29°C (Webb et al. 1983). This species exhibits the FMF pattern of sex determination (Webb et al. 1983, 1987, 1990, 1992) but differs from all other species yet investigated in having a very low proportion of males produced at any single constant temperature; the average value at 32°C is only 22.4% (see Table 4.2) and the maximum recorded is 39% (Webb et al. 1990). Nevertheless, nests with 100% male hatchlings have been reported (Smith 1987). Contrary to the suggestion by Lang and Andrews (1994), the values in Table 4.2 indicate that the range of male-producing temperatures was quite narrow in this species and that the percentage of males was higher. Webb et al. (1990) showed that incubation under a fluctuating temperature regimen (31–33°C with an average of 32°C) produced males at a similar frequency to incubation at a constant 32°C (41% and 39% males, respectively). Progressively increasing temperatures produced more males (Webb et al. 1992). The proportion of males increased as the initial temperature was increased and as the final temperature increased; that is, incremental rises in the range 28–31°C produced 3% males but 30–35°C produced 80–100% males. This effect was diminished once the initial temperature was 31°C. This result suggests that metabolic heating of eggs (Ewert and Nelson 2003) may well be significant in natural nests.

Studies of the sex ratio of adult *C. johnstoni* caught in a sample of McKinlay River polls in Australia showed only 20% males, with a range of 0–39% males (Webb et al. 1983). The sex ratio of hatchlings collected in the field also varied from near parity (51% and 55% males) to values heavily skewed towards females (26% and 18% males).

Morelet's Crocodile (Crocodylus moreletii)

TSD in *Crocodylus moreletii* (which builds nest mounds) has been investigated at the Madras Crocodile Bank, India, by incubation of five clutches of eggs at different temperatures (Lang and Andrews 1994). The pattern was FMF but the highest percentage of males was only 55% at 33°C (Table 4.2). It is possible that a higher percentage of males would be produced at a temperature between 32 and 33°C.

Nile Crocodile (Crocodylus niloticus)

Crocodylus niloticus is a hole-nesting species with mean nest temperatures that vary between 28 and 30°C, although maximum nest temperatures can be well above 31°C (Hutton 1987). The sex ratio of embryos from clutches of eggs was skewed towards females (89–91%), and there were distinct clutch differences in the percentage of males produced at 34°C. The position of an egg in a nest also affected the rate of embryonic development, with embryos in eggs at the lower, cooler parts of the nest being significantly reduced in body length. This variation in size appeared to suggest that only 50% of viable eggs would be able to hatch.

Under artificial incubation, eggs from captive stock incubated at 28 and 31°C produced no males, but at 34°C there were 88% males. In one clutch collected from a wild nest and incubated artificially, 0% males were found at 31°C and 91% and 82% males at 32.5°C and 34°C, respectively.

Recent work at Lake St. Lucia, South Africa, suggests that colonization of potential nesting sites for *C. niloticus* by an alien plant, *Chromolaena odorata*, is causing a 5–6°C drop in soil temperatures (Leslie and Spotila 2001). Such temperatures could prevent normal embryonic development, or at least produce entirely female clutches, which could adversely impact the population structure of the Lake St. Lucia area.

Mugger (Crocodylus palustris)

Crocodylus palustris constructs a hole nest that has temperatures within the range of 28–33°C (Magnusson et al. 1985). Temperatures at or below 31.0°C produced 100% females, but temperatures between 31.5 and 33°C produced varying amounts of males (Lang et al. 1989; Lang and Andrews 1994).

A temperature of 32.5°C produced 100% males (Lang et al. 1989); although Lang and Andrews (1994) reported a value of 93% males, it was unclear whether this value represented additional information. Hence, *C. palustris* exhibits the FMF pattern, but at any particular male-producing temperature there were significant clutch effects on the proportion of males hatching.

The average production of males in a total of 52 field nests was only 24% (Lang et al. 1989). There was also a weak trend towards production of more males in "late" nests than in "early" nests. Although the soil temperatures were higher for late nests (33–34°C in early May compared with 27–28°C in early February), these were located in shaded sites rather than in the more open locations of early nests. Hence, the anticipated rise in the percentage of males was mitigated by changes in nesting behavior. Lang et al. (1989) also moved eggs to artificial nest sites that produced fewer males than natural nests in apparently similar positions (6% vs. 44% males, respectively). This appeared to be related to lower nest temperatures in the artificial nest sites. Periods of rain lowered nest temperatures such that more females than expected were produced from the typical nest temperatures during dry periods.

Although no one temperature produced 100% males, temperature switch experiments were used to define an approximate TSP as stages 21–24, which corresponds to 28–42 days of incubation at 32°C, similar to the TSP in *A. mississippiensis* (Lang and Andrews 1994).

Topical application of estradiol 17β to the shell of *C. palustris* eggs just before the TSP (25–30 days) caused complete feminization of the embryos despite incubation at a male-determining temperature (Lang and Andrews 1994).

Saltwater (Indopacific) Crocodile (Crocodylus porosus)

Crocodylus porosus is a mound-nesting species with average nest temperatures between 30 and 33°C (Magnusson et al. 1985; Webb and Cooper-Preston 1989). The FMF pattern is observed (Webb et al. 1987; Table 4.2). Therefore, at 30°C and below no males are produced, but at higher temperatures various proportions of males were produced (Webb et al. 1987; Table 4.2). Other reports of this work (Webb and Cooper-Preston 1989; Lang and Andrews 1994) are inaccurate because they incorrectly combined the number of males and females from the two methods employed by Webb et al. (1987) and then determined the sex ratio of the totals. Mohanty-Hejmadi et al. (1999) split a single clutch of *C. porosus* eggs in two and incubated one half at 33°C and the other at 34°C. At the lower temperature 100% females hatched, compared with 100% males at 34°C, probably reflecting the effect of temperature on this particular clutch.

Webb and Cooper-Preston (1989) investigated nests and examined embryos older than an equivalent morphological age of 52 days at 30°C. Sex ratios were correlated with spot temperature recordings measured at a point two to three eggs deep in the nest. Below 29°C and above 34°C no males were produced, but varying levels of males were observed at intermediate temperatures: 5.9% for 29.0–29.9°C, 40.7% for 30.0–30.9°C, 84.4% for 31.0–31.9°C, 25.2% for 32.0–32.9°C, and 22.2% for temperatures between 33.0 and 33.9°C.

A series of temperature switch experiments were carried out in order to elucidate the TSPs of *C. porosus* (Webb et al. 1987). In summary, early switches produced the sex associated with the final temperature, but the gender of embryos switched later in incubation was unaffected by the change in temperature. The experiments showed that the TSP was encompassed by development stages 18–23 (17–52 days of incubation at 30°C).

Siamese Crocodile (Crocodylus siamensis)

TSD has been reported in mound-nesting *Crocodylus siamensis* by Lang (1987) and Lang and Andrews (1994). The pattern is FMF, with 100% males being produced at 32.5°C and only 60% males at 33°C (Table 4.2), but these results were based on a small sample of eggs.

Gharial (Gavialis gangeticus)

Gavialis gangeticus is a hole-nesting species in which nest temperatures can range from 24 to 37°C (Magnusson et al. 1985). Lang and Andrews (1994) showed that there was a FMF pattern with no single temperature producing 100% males, although the male-producing temperatures appeared to be lower than those observed for both crocodiles and alligators (Table 4.2). Unfortunately, the study was based on small sample sizes (between 3 and 13 eggs) and was further compounded by the difficulty in sexing young gharials.

Influence of Incubation Temperature on Other Posthatching Phenotypic Characteristics

Incubation temperature affects the rate of embryonic development in crocodilians (Deeming and Ferguson 1989a, 1991b; Lang and Andrews 1994) and has long-lasting posthatching effects on pigmentation patterns, rates of growth, survival, and thermoregulation. These factors may affect the reproductive fitness of an individual and so are briefly described here (see also Rhen and Lang, Chapter 10).

Incubation temperature affected the pigmentation pattern on hatchling *A. mississippiensis* (Murray et al. 1990). On average, there were two more white stripes ($p < 0.001$) along the length of the body and tail in male hatchlings (from eggs incubated at 33.0°C) than in females (30°C), despite similar body lengths ($p > 0.05$). Other male and female hatchlings, produced under intermediate incubation conditions, had the same numbers of stripes, which were between those produced at 30 and 33°C. Murray et al. (1990) suggested a cell chemotaxis model, proposing that the length of the animal at pigment deposition determines how many stripes can be accommodated along the body. Data for *A. mississippiensis* embryos supported this model. At equivalent developmental stages immediately before pigmentation patterns were recognizable, male hatchlings (at 33°C) were longer than females (at 30°C).

According to a privately published report by Manolis et al. (2000), the number of scale rows on the belly skin of *C. porosus* hatchlings is a function of temperature. High constant temperatures produced more scale rows, and pulse experiments show that lowering the temperature decreased the scale row count, while increasing the temperature produced more scale rows, compared with a background temperature of 32°C. The period between 10 and 30 days was important in determining the number of scale rows that develop.

The incubation temperature of *C. siamensis* eggs affects the thermal preference of animals posthatching (Lang 1987). Within a temperature gradient, males from eggs incubated at high temperatures (32.5–33.5°C) consistently selected a higher body temperature than females from eggs incubated at lower temperatures (27.5–28.0°C).

Posthatching growth rates of *A. mississippiensis* are influenced by incubation temperature (Joanen et al. 1987). Hatchlings from eggs incubated at a range of temperatures were raised under identical conditions at a temperature of 30°C. Both males and females from intermediate temperatures (30.6 and 31.7°C) grew faster than the animals from the extreme temperatures (29.4 and 32.8°C). For *C. porosus,* males grew faster posthatching than females incubated at the same temperature, and the growth rates of high-temperature females were higher than those of females incubated at low temperatures (Webb and Cooper-Preston 1989). Hutton (1987) reported that in *C. niloticus* hatchling length was affected by incubation temperature, with the smallest animals hatching at the highest temperature (34°C—87.6% males). However, by three months of age the crocodiles from the highest temperature had outgrown those from lower temperatures.

Posthatching survival to two years of age has been shown to be related to incubation temperature in *C. porosus* (Webb and Cooper-Preston 1989). Survival rates for hatchlings incubated at 29°C are only around 45%, compared with 85% for hatchlings incubated at 32–33°C. By contrast, incubation temperature had little effect on posthatching survival of *A. mississippiensis* hatchlings, with at least 89% of hatchlings surviving to two years of age (Joanen et al. 1987).

A General Overview of TSD in Crocodilians

The generalized pattern of TSD in twelve species in the three subfamilies of crocodilians, albeit derived from a variety of studies, is shown in Figure 4.2. Alligators and caimans have a broad range of male-producing temperatures, and 32.5–33.0°C temperatures usually produce 90–100% males. It is unclear to what extent the low percentage of males produced at 30.5°C is an atypical observation. Crocodile species have a narrower range of male-producing temperatures, and the temperature that produces the highest percentage of males is lower (Figure 4.2). Limited studies in the gharial suggest that its main male-producing temperature is lower still and that males occur within a narrow temperature range, but more information is needed. It is likely that all crocodilian types conform to the FMF pattern for TSD, and it is probable that these values are representative of all crocodilians. This idea is supported by the fact that heteromorphic sex chromosomes are absent in crocodilians (Cohen and Gans 1970). However, it would be interesting to investigate further the effects of temperature on some of the species described here and to the FMF pattern in the other 13 crocodilian species.

Although there are variations in the sex ratios recorded in wild populations of crocodilians (e.g., Lance et al. 2000), many reports do show that there is often a significant bias towards females (e.g., Ferguson 1985; Hutton 1987; Webb et al. 1983). This has led to attempts to model crocodilian populations in relation to skewed sex ratios at hatching (Phelps 1992; Woodward and Murray 1993). Space precludes any discussion of these models save that they do seem to provide a good explanation of the observed situation (but see Valenzuela, Chapter 14; Girondot et al., Chapter 15). Interestingly Phelps's (1992) prediction that *A. mississippiensis* should have high-temperature females was subsequently shown to be valid (see Table 4.2).

Considerable effort has been made to understand the mechanism of TSD in a variety of animals. Crocodilians

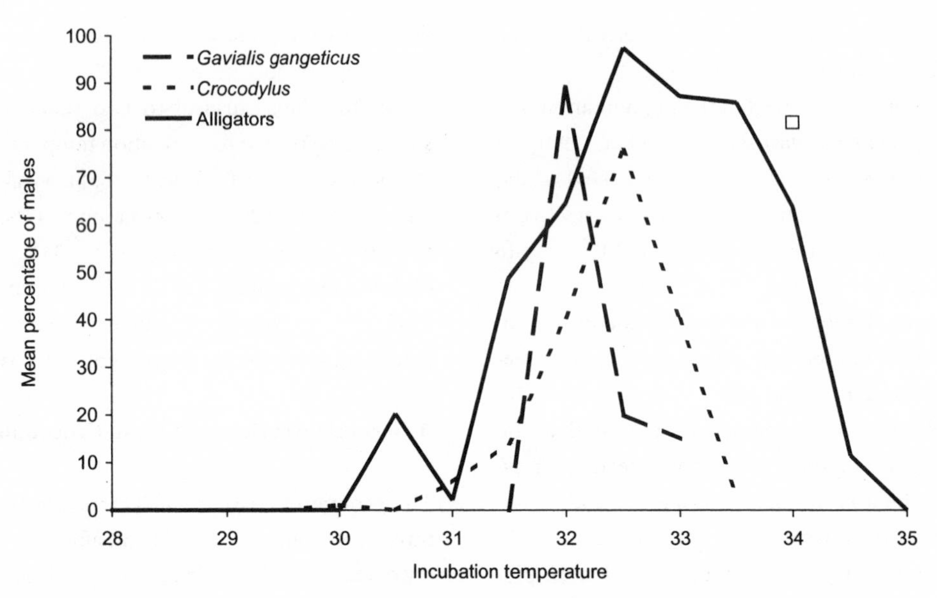

Figure 4.2 Influence of incubation temperature on production of males in different crocodilian lineages. With the exception of *G.gangeticus*, values are means calculated from data for species of *Crocodylus* and from data for alligator and caiman species, described in Table 4.2. The square indicates a value derived from *Crocodylus niloticus*.

differ from other reptiles in being archosaurs, so they are more closely related to birds and dinosaurs. The range of incubation temperatures producing viable hatchlings may be wider for crocodilians than for birds (28–34°C vs. 36.5–38.5°C), but it is usually considerably narrower than those observed in turtles and squamates. Furthermore, unlike in turtles and lizards, TSD is also considered to be prevalent in all crocodilian species. These factors set crocodilians apart from other groups, and perhaps their mechanism of TSD is group specific, so that adopting a broad view of the mechanism for TSD for all reptiles may be inappropriate.

Various ideas have been put forward to explain TSD in crocodilians, including developmental asynchrony (Joss 1989; Smith and Joss 1994). Alternatively, Deeming and Ferguson (1988, 1989b) suggested that production of a male depended on the temperature-dependent production of a "male-determining factor" (MDF) that had to be maintained above a threshold for an allotted period of time, perhaps equivalent to the TSP. Failure to maintain the appropriate dose of the MDF for the correct length of time would result in an individual developing into a female by default. This hypothesis may not have received a lot of attention, but it does explain many of the elements of TSD in crocodilians, and subsequent work has not detracted from its merits. In particular, it is predicted that any disruption of the incubation conditions for maleness will lead to development of females. Hence, pharmacological treatments of eggs are likely to feminize embryos at male-determining temperatures but do not induce development of males at other temperatures (Lance and Bogart 1994). Moreover, Deeming and Ferguson (1988, 1989b) suggested that there would be variation within a population in the threshold for the MDF and in the length and timing of the TSP. It can be argued that the reports of clutch effects on sex determination at male-producing temperatures (Lang and Andrews 1994; Rhen and Lang 1998) provide evidence of genetic variability within the TSD mechanism (but see also Valenzuela, Chapter 14).

Deeming and Ferguson (1988, 1989b, 1991b) were unable to suggest the exact nature of the MDF, and many of the candidate molecules have been shown either to not be involved in TSD or to not be the MDF. Neither zinc finger genes (Valleley et al. 1992), aromatase (Gabriel et al. 2001), nor *Sf1*, *Wt1*, and *Dax1* genes (Western et al. 2000) have been shown to have temperature-dependent expression. However, the *Dmrt1* gene has temperature-dependent differential expression in gonadogenesis in alligator embryos during the TSP (Smith et al. 1999a) and is a candidate for the MDF.

Deeming and Ferguson (1988, 1989b, 1991b) proposed that gender was linked with embryonic growth and also with various posthatching characteristics of crocodilians. Moreover, it was suggested that TSD is a critical aspect of the overall life-history strategy of crocodilians. It was proposed that the hypothalamus was perhaps involved in determining gender, establishing patterns of pigmentation, setting rates of posthatching growth and establishing behavior patterns such as thermoregulation. Hence, the incubation environment establishes baseline characteristics that may influence posthatching life in crocodilians. The gender-specific advantage of this effect is that the temperatures that produce a majority of males are those that maximize rates of embryonic growth (see also Rhen and Lang, Chapter 10). Therefore, these male hatchlings are preprogrammed to grow very quickly if optimal environmental conditions occur posthatching. Larger males have more mating opportunities (Lance 1989), so those developing under ideal conditions maximize their reproductive fitness by achieving a large size as quickly as possible. By contrast, egg laying is not restricted by female body size. Therefore, because relatively few male crocodilians get to mate and optimal conditions for male development are restricted, it is advantageous to the individual developing at a suboptimal, low or high temperature to be female (see also Valenzuela, Chapter 14).

Further advantage to the system lies in the variability in temperatures producing both genders (Deeming and Ferguson, 1988, 1989b, 1991b). Hence, there will be male and female hatchlings preadapted to a range of environmental conditions. Therefore, should the environment change in the short term, there will be males and females within the population that will find the conditions optimal and will have higher fitness than their cohorts. Any hatchlings produced during these altered environmental conditions will be preadapted to the new environment and will have improved fitness. Should the altered environmental conditions persist over the long term, populations of crocodilians will quickly adapt to the new environment. Deeming and Ferguson (1989b) suggested that the remarkable survival of the crocodilians over evolutionary time is a function of the ability to link gender, growth, and other physiological process with the prevailing environment.

TSD has been of interest in both birds and dinosaurs for a variety of reasons. The relatedness of crocodilians to birds has prompted considerable interest in whether TSD could be applicable to any avian species, particularly commercial poultry. To the author's knowledge there has been no convincing evidence of long-term, temperature-induced sex reversal in birds, and this is probably due to the genetic basis of the linkage between gender and posthatching characteristics (see also Valenzuela, Chapter 14).

In dinosaurs, it has been proposed several times (e.g., Standora and Spotila 1985; Gardom 1993) that TSD played a role in the extinction of the group at the end of the Cretaceous Period, 65 million years ago (but see Girondot et al., Chapter 15). Changing climatic conditions are supposed to have skewed the sex ratio towards a single gender and thus proved fatal to the group. However, Deeming and Ferguson (1989c) suggested that the success of large dinosaurs depended on their large size, which would need to be attained as soon as possible, and that growth rate was probably under genetic control. Hence, there would be little significant advantage in linking sex with the rate of embryonic development. Placing growth rate under genetic control probably proved too inflexible for a rapid response to changing environmental conditions at the end of the Cretaceous, and dinosaurs died out. By contrast, Cretaceous crocodilians probably had TSD, were able to rapidly alter their body size to match the prevailing environmental conditions, and were able to survive into the Tertiary and to the present day.

Acknowlegments—Many thanks go to Glenn Baggott, Nicole Valenzuela, Zilda Campos, Grahame Webb, Adam Britton, and Bill Magnusson for providing some of the articles referred to in the text. Many thanks to Glenn Baggott, Nicole Valenzuela, and an anonymous referee for their constructive comments on previous drafts of this chapter.

5

PETER S. HARLOW

Temperature-Dependent Sex Determination in Lizards

In the mid-1960s biologist Madelaine Charnier in Dakar, Senegal, was investigating the development of the reproductive system in the common agamid lizard *Agama agama*, using hatchlings from eggs incubated in the laboratory. From the eggs left to incubate in the cool sand of her lizard cages, 46 hatchlings emerged, but only one was male. From the warmer laboratory incubator, 30 eggs hatched, and all were male. Charnier suggested that the different incubation temperatures had somehow affected the gonadal sex of the hatchlings. These curious results were published in a small paper (Charnier 1966) and largely ignored, as sex determination in all vertebrates was assumed to be genetically controlled.

Then in 1971, a similar story emerged from Claude Pieau's laboratory in France. He was investigating the development of the reproductive system in hatchlings of two species of European turtles, and his different egg incubation temperatures produced very different hatchling sex ratios (Pieau 1971). Biologists began looking at other reptiles, and soon temperature-dependent sex determination (TSD) was identified in a eublepharid gecko (Wagner 1980), the American alligator (Ferguson and Joanen 1982), and eventually in the tuatara (Cree et al. 1995). Thus, in all surviving reptile lineages, and in three of the five major lineages of amniotes (Testudines, Lepidosauria, and Crocodylia), at least some species use this mechanism of sex determination.

Within the squamate reptiles, TSD has so far only been convincingly confirmed (but see Valenzuela et al. 2003) in a few species in two lizard families in the Gekkota (Eublepharidae and Gekkonidae; Viets et al. 1994), in one family within the Iguania (Agamidae; Harlow 2001), and in the family Scincidae (Robert and Thompson 2001). Within each of these families, genotypic sex determination (GSD) has also been shown to occur in other species (Viets et al. 1994). Thus TSD and GSD co-occur within all these families. There are limited data and anecdotal reports to suggest that TSD may also occur in some species in the lizard families Anguidae, Chamaeleontidae, Diplodactylidae, Iguanidae, Lacertidae, and Varanidae, but at present no single species in these families that is suspected of having TSD has been rigorously investigated (Viets et al. 1994). Future research will undoubtedly confirm TSD in some species in these and other lizard families. Genotypic sex determination, however, remains the dominant form of sex determination in lizards, and Viets et al. (1994) provide evidence that only GSD occurs in all species so far investigated in a further five lizard families (Anolidae, Basiliscidae, Crotaphytidae, Sceloporidae, and Teiidae).

All TSD species in the families Agamidae, Eublepharidae, and Gekkonidae so far investigated over the full range of viable incubation temperatures show a similar sex-determining pattern: only females are produced at high and low incubation temperatures, and varying proportions of males hatch at intermediate temperatures (Tokunaga 1985; Viets et al.

1994; Harlow and Shine 1999; Harlow 2000; Harlow and Taylor 2000; El Mouden et al. 2001). This is often referred to as pattern II (Ewert and Nelson 1991), or as pattern FMF (i.e., female-male-female; Lang and Andrews 1994).

Several lizard species in which the TSD pattern was originally described as all females from low incubation temperatures and mostly males from "high" temperatures were subsequently found to also produce all females at even higher incubation temperatures than those used in the initial incubations (*Gekko japonicus*, Tokunaga 1985; *Eublepharis macularius*, Viets et al. 1993). A similar extension to the previously reported sex-determining pattern has also been shown in crocodilians, after incubation temperatures higher than the mostly male-producing incubation temperature were later investigated and found to produce all females (e.g., Lang and Andrews 1994).

Review of Lizard Families Reported to Have TSD

Anguidae

Based on a statement by Langerwerf (1984) ("*Gerrhonotus multicarinatus* also produces more males at 27–28°C, but this is not very obvious"), the Anguidae are listed by Janzen and Paukstis (1991a) as a lizard family containing a questionable TSD species and referenced as a family with TSD in a later publication (Pough et al. 2001). As no sample sizes or hatchling sex ratios were given by Langerwerf (1984), the Anguidae family needs more investigation before it can be classified as containing species with TSD.

The Gekkota

Temperature-dependent sex determination has been reported in all three of the main familial groups within the Gekkota.

Diplodactylidae

Five species in the New Caledonian gecko genus *Rhacodactylus* are reported to have TSD (*R. auriculatus, R. chahoua, R. ciliatus, R. leachianus,* and *R. sarasinorum;* Seipp and Henkel 2000). Although no sample sizes or hatchling sex ratios are given, Seipp and Henkel's (2000) data are suggestive of a FMF pattern of sex determination, similar to that seen in the Eublepharidae and Gekkonidae (Tokunaga 1985; Viets et al. 1993). In *Rhacodactylus,* mostly females are produced at low incubation temperatures (20–26°C), mixed sexes or male-biased sex ratios at intermediate temperatures (23–30°C), and mostly females again at higher temperatures (26–30°C; Seipp and Henkel 2000).

The Diplodactylidae are sometimes called the "southern geckos" and are restricted to Australia, New Caledonia, and New Zealand. This family also includes the flap-footed or "legless" pygopod lizards (often treated as a separate family, the Pygopodidae), for which there are currently no data available on the sex-determining mechanism for any species.

Eublepharidae

Since the initial discovery that *Eublepharis macularius* has TSD (Wagner 1980), the leopard gecko has become the preferred laboratory animal for almost all studies on TSD in lizards. The African fat-tailed gecko *Hemitheconyx caudicinctus* is another eublepharid gecko in which TSD is well documented (Anderson and Oldham 1988), and it has an FMF pattern of sex determination similar to that of *E. macularius* (Viets et al. 1993, 1994). Three species of the eublepharid genus *Coleonyx* have been investigated for their sex-determining mechanism by laboratory incubation *(C. brevis, C. mitratus,* and *C. variegatus),* and all clearly show GSD (Viets et al. 1994; Bragg et al. 2000).

Gekkonidae

Tokunaga (1985) established that *Gekko japonicus* had TSD, and more recent incubation studies on species within the genera *Phelsuma* and *Tarentola* show that TSD is widely distributed in the Gekkonidae. Viets et al. (1994) present incubation data that suggest at least four species of *Phelsuma (P. guimbeaui, P. guentheri, P. lineata,* and *P. madagascariensis)* and at least two species of *Tarentola (T. boettgeri* and *T. mauritanica)* have TSD.

The Iguania

Agamidae

The family Agamidae contains well over 300 species in at least 34 genera worldwide (Greer 1989). Although the first vertebrate recorded with TSD was the African agamid *Agama agama* (Charnier 1966), until recently the sex-determining mechanism was only known for six agamid species worldwide. Four species of the African–central Asian genus *Agama (A. agama, A. caucasia, A. impalearis,* and *A. stellio)* have been reported to have TSD (Charnier 1966; Langerwerf 1983, 1988; El Mouden et al. 2001), while the Asian *Calotes versicolor* and the Australian *Pogona vitticeps* have been reported to have GSD (Ganesh and Raman 1995; Viets et al. 1994).

SEX-DETERMINING MECHANISMS IN THE AUSTRALIAN AGAMIDAE

In Australia, there are 65 species of agamids (usually called "dragons" or "dragon lizards" in Australia) recognized by Cogger (2000), currently placed in 13 genera. Australian agamids occur in all habitats from tropical rainforests to alpine woodlands, although the greatest species diversity occurs in the arid zone (Cogger and Heatwole 1981). The Australian agamids vary considerably in body size, from small species such as *Ctenophorus fordi*, with an adult body size of 45 mm and mass of 4 g (Cogger 1978), to *Physignathus lesueurii* at 275 mm SVL (snout-vent length) and a mass of almost 1 kg (Thompson 1993).

While all Australian agamids are oviparous (Greer 1989), viviparous and parthenogenetic species do exist elsewhere (Hall 1970; Tryon 1979). Reproduction appears to be highly seasonal in all Australian species (James and Shine 1985; Greer 1989). The general pattern for most Australian agamids is that vitellogenesis and mating occur in the spring (August–November), and oviposition from early to late summer (November–February). In tropical Australia, where no pronounced cold winter occurs, oviposition in most species occurs during these same months and coincides with the beginning of the annual wet season. All species so far investigated appear to be able to produce multiple clutches within a season, at least in years with abundant food resources (e.g., Greer 1989; Harlow and Harlow 1997; Dickman et al. 1999). Clutch size varies from a mean of two eggs in small species like *Ctenophorus fordi* (Cogger 1978), to at least 41 for large species like *Pogona barbata* (P. S. Harlow, pers. obs.).

The author and his coworkers have recently identified the sex-determining mechanism in 20 Australian agamid lizard species, based on the sex ratios obtained from laboratory incubation of eggs over a wide range of incubation temperatures (Harlow and Shine 1999; Harlow 2000, 2001; Harlow and Taylor 2000; Harlow et al. 2002). Eggs were mostly obtained from wild-caught gravid females and incubated on moist vermiculite in constant- or cycling-temperature incubators (for incubation details, see Harlow and Taylor 2000). During this study, 1,707 eggs obtained from 20 agamid species were incubated; 93% (1,580) of the eggs produced hatchlings that could be reliably sexed. Hatchlings were sexed by hemipene eversion, usually under a low-power microscope (Harlow 1996). Agamids are easy to sex in this way, and with a little experience this harmless technique is close to 100% accurate, and it has been verified by histology and laparoscopy (Harlow 1996).

Nine agamid species from five genera were found to have TSD, and a further 11 species from six genera have GSD (see Table 5.1). Perhaps the most surprising result from this study is to find both types of sex-determining mechanism in what appear to be closely related species. In the genera *Ctenophorus* and *Amphibolurus*, both forms of sex determination co-occur. The genus *Ctenophorus* currently consists of 23 species, and phylogenetic analyses based on mitochondrial DNA suggest that, with the addition of one other species *(Rankinia adelaidensis)*, this clade forms a monophyletic group within the Australian agamids (Melville et al. 2001). The genus *Amphibolurus*, however, is based on morphological similarities and is a "small group of closely related species" (Cogger 2000). There are currently only three species in this genus; one has TSD, and two have GSD. In particular, the GSD species *Amphibolurus norrisi* is so similar to the TSD species *A. muricatus* that the two taxa were regarded as conspecific prior to the formal description of *A. norrisi* (Witten and Coventry 1984). They are allopatric species that are taxonomically distinguished by only a few subtle differences in scalation and color, though no comparative molecular studies have yet been done.

SEX-DETERMINING PATTERN IN THE AUSTRALIAN AGAMIDAE WITH TSD

The sex-determining pattern in TSD agamids is FMF with increasing temperature, similar to the pattern shown in the two species of geckos with TSD in which the full range of viable incubation temperatures have also been investigated (Tokunaga 1985; Viets et al. 1993). Preliminary agamid egg incubations at constant temperatures showed that the upper 100% female–producing incubation temperature for TSD species consisted of a narrow temperature range, usually extending over about 1 or 2°C only. Embryo mortality rapidly increased as constant incubation temperatures increased above this 100% female–producing temperature threshold. There is considerable interspecific variation in these maximum viable incubation temperatures. For example, in the TSD species *A. muricatus* the highest constant incubation temperature utilized (32°C) produced 70% female hatchlings, although 100% females were produced at higher cycling incubation temperatures (Harlow and Taylor 2000). Hatchling sex ratios for *A. muricatus* incubated at a constant 30°C were not significantly different from a 1:1 sex ratio (Harlow and Taylor 2000). However, for the sympatric TSD species *Physignathus lesueurii*, a constant 30°C and above produced 100% females (see Figure 5.1), but at temperatures of 31°C and above hatchling mortality signifi-

Table 5.1 Hatchling Sex Ratios Obtained from Constant- and Cycling-Temperature Incubation for 20 Australian Agamid Lizard Species

Agamid Species	Sex - Determining Mechanism	Incubation Temperature	Males	Females	Unsexed	χ^2 Value	χ^2 p Value
Amphibolurus muricatus	TSD	23°C	1	11	0	27.01	0.0014
		25°C	0	18	0		
		26°C	4	14	3		
		28°C	3	7	0		
		29°C	8	12	3		
		30°C	4	8	2		
		31°C	2	1	0		
		32°C	9	19	0		
		27 ± 5°C	11	9	0		
		33 ± 5°C	0	14	2		
Amphibolurus nobbi	GSD	23°C	4	1	0	4.09	0.2524
		24°C	10	3	2		
		25°C	5	7	0		
		27.5°C	2	1	0		
Amphibolurus norrisi	GSD	24°C	9	3	1	4.51	0.2116
		25°C	6	6	2		
		27.5°C	3	7	1		
		23°C–25°C [a]	2	2	0		
Chlamydosaurus kingii	TSD	26°C	0	12	1	19.03	0.0003
		29°C	10	7	1		
		32°C	6	3	4		
		33 ± 5°C	0	8	4		
Ctenophorus decresii	TSD	25°C	0	15	0	12.96	0.0237
		26°C	1	14	4		
		27.5°C	4	9	4		
		29°C	0	2	1		
		30°C	2	5	4		
		32°C	1	0	1		
Ctenophorus fordi	GSD	25°C	3	5	0	1.54	0.6732
		26°C	7	5	0		
		29°C	6	5	1		
		32°C	6	3	3		
Ctenophorus ornatus	TSD	25°C	0	10	0	10.00	0.0016
Ctenophorus pictus	TSD	25°C	6	15	3	13.10	0.0003
		26°C	1	4	0		
		27°C	2	13	0		
		29°C	1	1	2		
		32°C	0	1	1		
Diporiphora albilabris	GSD	26°C	6	6	1	0.08	0.7815
		32°C	1	0	0		
Diporiphora bilineata	GSD	25°C	1	1	1	4.10	0.2509
		26°C	13	11	1		
		29°C	3	7	0		
		31°C	2	8	1		
Hypsilurus spinipes	GSD	20°C	12	15	0	10.12	0.3406
		23°C	22	18	2		
		25°C	6	11	0		

(continues)

Table 5.1 (*Continued*)

Agamid Species	Sex - Determining Mechanism	Incubation Temperature	Males	Females	Unsexed	χ^2 Value	χ^2 p Value
		26°C	18	23	2		
		27°C	4	7	1		
		28°C	4	4	0		
		20 ± 4°C	3	4	0		
		22 ± 5°C	0	9	0		
		23 ± 10°C	3	5	0		
		27 ± 5°C	2	2	0		
Lophognathus burnsi	TSD[b]	25°C	0	9	0	8.69	0.0692
		26°C	2	8	2		
		27°C	1	4	0		
		29°C	6	5	0		
		32°C	4	5	3		
Lophognathus gilberti	TSD	25°C	0	2	0	17.20	0.0018
		26°C	1	13	1		
		30°C	0	2	1		
		32°C	4	0	0		
		33.7 ± 5°C	2	8	3		
Lophognathus temporalis	TSD	25°C	0	5	0	44.44	0.0001
		26°C	0	24	0		
		27°C	2	11	0		
		29°C	14	11	0		
		32°C	17	11	0		
		33°C	8	5	0		
		33.7 ± 5°C	0	16	0		
Physignathus lesueurii	TSD	20°C	0	1	0	321.65	0.0001
		22°C	0	6	0		
		23°C	0	24	3		
		25°C	0	31	3		
		26°C	10	17	0		
		27°C	58	13	4		
		27.5°C	7	0	1		
		28°C	107	15	1		
		29°C	3	34	1		
		30°C	0	27	2		
		31°C	0	42	1		
		32°C	0	24	0		
		33°C	0	17	0		
		25-33°C[c]	0	22	0		
Pogona barbata	GSD	23°C	0	1	0	1.65	0.8946
		25°C	1	2	0		
		26°C	5	7	1		
		27°C	41	32	7		
		29°C	6	7	1		
		32°C	32	31	25		
Pogona minor	GSD	25°C	3	9	1	1.47	0.2248
		29°C	5	5	1		
Pogona vitticeps	GSD	26°C	5	1	0	5.44	0.0659
		29°C	3	4	0		
		32°C	1	5	0		

Table 5.1 (*Continued*)

Agamid Species	Sex - Determining Mechanism	Incubation Temperature	Males	Females	Unsexed	χ^2 Value	χ^2 p Value
Tympanocryptis diemensis	GSD	23°C	6	5	2	3.05	0.5489
		24°C	2	5	2		
		27°C	4	6	0		
		29°C	0	2	0		
		31°C	2	5	0		
Tympanocryptis tetraporophora	GSD	26°C	4	5	2	1.81	0.6119
		29°C	6	3	1		
		32°C	4	6	0		
		27 ± 5°C	2	1	0		

Sources: Harlow and Shine 1999, Harlow 2000, 2001, Harlow and Taylor 2000, Harlow et al. 2002.

Note: Data are shown as hatchling males/females/unsexed at each incubation temperature. Unsexed refers to eggs that failed to develop or embryos that died during development. The χ^2test was used in contingency table analyses to investigate if the hatchling sex ratios were significantly different between different incubation temperatures. All cycling-temperature incubators (denoted by ±) were programmed to follow a 10-step daily sinusoidal curve around the stated mean. Note that some temperatures within species, and some species overall, have very small sample sizes or few temperature treatments—conclusions derived from such data should be viewed as tentative.

[a]Incubated at 23°C for the first 30 days, then 25°C until hatching.

[b]*Lophognathus burnsi* is tentatively classed as a TSD species here even though small sample sizes at some incubation temperatures resulted in no significant effect of incubation temperature on hatchling sex ratio across all incubation temperatures. The overall sex ratio was female biased, and against the null hypothesis of a 1:1 sex ratio $\chi^2 = 7.36$, and $p = 0.007$.

[c]Incubated at constant 25 and 33°C for 11 hours each day, with a one-hour temperature ramp between these two constant temperatures.

cantly increased with increasing incubation temperature (Harlow 2001).

The minimum viable constant incubation temperature for Australian agamid lizards appears to be within the lower 100% female–producing temperature zone for all TSD species investigated. For those species from the temperate southeast of Australia and for some arid zone species, 25°C was found to be an effective incubation temperature for establishing the sex-determining mechanism. For example, for the 49 hatchlings from the two TSD species identified from the southeast of Australia that were incubated at 25°C (*A. muricatus* and *P. lesueurii*), 100% were female. At 26°C these same species produced 22 and 37% males, respectively, and the percentage of males further increased with increasing incubation temperatures (see Figure 5.1 and Harlow and Taylor 2000). By comparison, for the two species identified with GSD from the southeast of Australia (*Hypsilurus spinipes* and *Tympanocryptis diemensis*), the combined sex ratio for 102 hatchlings incubated at 25°C and below was not significantly different from a 1:1 sex ratio (48 males to 54 females; see Table 5.1).

For tropical species and for some arid zone species, which presumably experience higher mean nest temperatures than temperate zone species, 26°C was often found to be the lowest constant temperature that allowed normal development and produced all females in the TSD species. For species from these hotter regions, 25°C was often, but not always, too low to allow normal embryonic development, and eggs either failed to develop (e.g., *Ctenophorus nuchalis*, unpubl. data) or embryos died at full term (e.g., *Ctenophorus ornatus*, Harlow 2000). For a total of 50 hatchlings from eggs incubated at 26°C in the three tropical TSD species investigated (*Chlamydosaurus kingii, Lophognathus gilberti, L. temporalis*), only one was male. In the tropical frillneck lizard (*C. kingii*), even 26°C was seemingly too low to allow normal embryo development. In this species, a significant number of the all-female hatchlings from 26°C showed serious neurological abnormalities when compared with siblings (of both sexes) incubated at 29°C (Harlow and Shine 1999).

This overemphasis on laboratory incubation temperatures close to the minimum temperature at which viable embryos develop simplified the identification of TSD species from GSD species. The hatchling grand sex ratio for the 11 GSD species identified from the laboratory incubations was not significantly different from the null hypothesis of a 1:1 sex ratio (295 males to 321 females, $\chi^2 = 1.097$, $p = 0.295$, $df = 1$). The preferential use of low incubation temperatures, however, resulted in the overproduction of females in the TSD species identified here. The grand sex ratio of hatchlings from the nine TSD species was significantly different from a 1:1 sex ratio (320 males to 644 females, $\chi^2 = 108.9$, $p < 0.0001$, $df = 1$). Please note that tests of parity of grand sex ratios are not tests of TSD versus

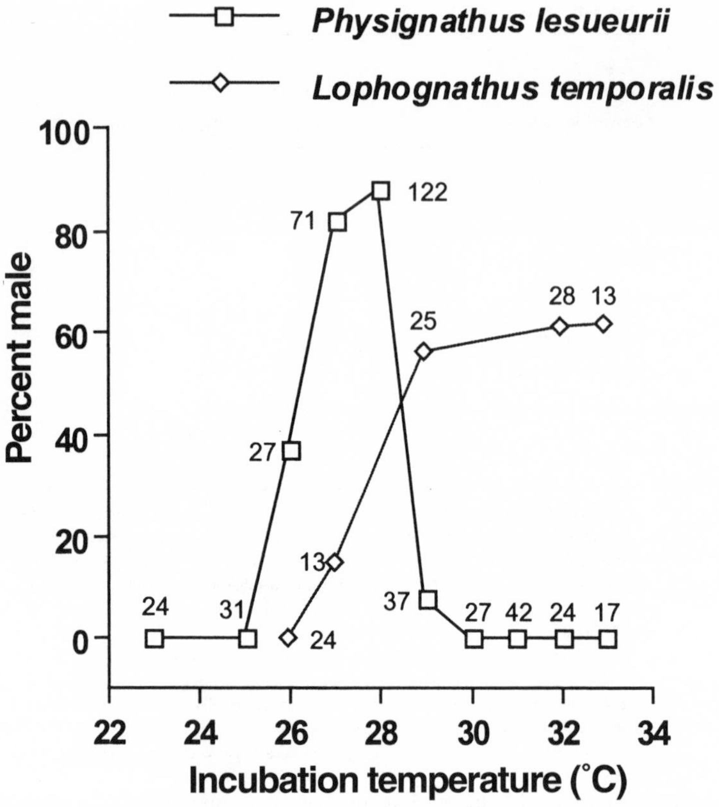

Figure 5.1 Hatchling sex ratios as a function of constant incubation temperatures in two Australian agamid species with temperature-dependent sex determination, a Temperate Zone population of *Physignathus lesueurii* and the tropical species *Lophognathus temporalis*. The sample size at each incubation temperature is shown.

GSD, which include tests of deviations from parity at each temperature treatment separately and the general trend of such deviations with increasing temperatures.

Chamaeleontidae

Although TSD is not yet confirmed in any species of chameleon, anecdotal information suggests that it may occur in this group. For example, in their book on chameleon captive care and breeding Schmidt et al. (1994a) state that the sex ratio of offspring *Chamaeleo lateralis* is dependent on the temperature during development. In their book on chameleons, Schmidt et al. (1994b) mention that hatchlings of the common European chameleon *(Chamaeleo chamaeleon)* incubated at a constant 27–29°C were sexed after three months and found to be all females. They commented, "We are yet to hear of a single case of males bred in captivity." Whether these reports are due to difficulty in correctly sexing offspring or are really indicative of TSD will need to be verified by future studies.

Iguanidae

The iguanas are listed as containing TSD species in one herpetology textbook (Pough et al. 1998), presumably based on reanalyses by Viets et al. (1994) of Muth and Bull's (1981) hatchling sex ratio data for the iguanid lizard *Dipsosaurus dorsalis*. The supposition of Viet et al. (1994) that this may be a TSD species is based on the fact that the hatchling sex ratio was significantly different from 1:1 at one incubation temperature (34°C) out of five that produced 13 or more hatchlings. There are currently no other data available to suggest that this or any other species of Iguanidae have TSD.

Lacertidae

Based on very dubious data, the Lacertidae are listed as a family containing species with TSD in several herpetology textbooks (Pough et al. 1998; Zug et al. 2001), as it seems that the original source (Eichenberger 1981) was misinterpreted in a review on TSD (Janzen and Paukstis 1991a). As pointed out by Viets et al. (1994), *Podarcis pityusensis* eggs incubated by Eichenberger (1981) at 29°C produced 1 male to "10–15" females. Again, these data are far from conclusive and are in need of future verification.

Scincidae

A somewhat different form of TSD has been reported in a few species of skink. In the small Australian montane skink *Bassiana duperreyi*, eggs incubated at cool cycling temperatures (15 ± 7.5°C) produced > 70% males, whereas eggs incubated at warmer temperatures produced sex ratios at 1:1 (Shine et al. 2002). The sex of hatchlings in this study was determined by hemipene eversion and has not yet been verified by gonad histology. If this species (which has strongly heteromorphic sex chromosomes) is ultimately shown to have TSD, then low incubation temperatures must override or deactivate developmental pathways that genetically determined sex, as is the case in some salamanders and fish (Baroiller et al. 1995; Dorazi et al. 1995; Wang and Tsai 2000).

A similar mechanism may occur in some viviparous skinks, as in the Australian alpine water skink *Eulamprus tympanum,* where gravid females forced to maintain their developing embryos at unnaturally high and constant temperatures in the laboratory produce 75% males at 30°C and 100% males at 32°C (Robert and Thompson 2001). In this species, gravid females maintained at 25°C in the laboratory and females captured in the field just prior to parturition produced offspring at 1:1 sex ratios (Robert and Thompson 2001). Recent work on *Eulamprus tympanum* suggests that maternal visual or olfactory cues may somehow be involved in orchestrating sex determination (or sex allocation) in this species. Gravid females in field enclosures isolated from any contact with males after mating produced significantly more male offspring, while those sharing enclosures with adult males produced mixed-sex offspring (Robert, et al. 2003).

Females that can produce sons or daughters facultatively, in response to populationwide shortages in males or females, will have increased fitness during times of severe population sex ratio skewness, compared with females that can only produce offspring in a 1:1 ratio. Facultative sex allocation of offspring appears to occur in several species of skink, and TSD is one of several possible mechanisms that may allow a female to "choose" the sex of her offspring in these species. Olsson and Shine (2001) found that the Australian snow skink *Niveoscincus microlepidotus* can produce male-biased litters in years when adult males are scarce, while the reverse was true in years when adult males were relatively more common than adult females.

In another study of a viviparous alpine skink, the Australian spotted snow skink *Niveoscincus ocellatus*, females appear to "select" their offspring's sex based on basking opportunities, and in the laboratory they overproduce male offspring when basking is limited and overproduce female offspring when basking opportunities are not restricted (Wapstra et al. 2004). Although a field population of this species produced an overall 1:1 offspring sex ratio, males were overproduced in litters born late in the reproductive season.

Alternative explanations for the observed sex ratio bias in offspring of *Bassiana duperreyi* (with sex chromosomes) and the viviparous species above (without sex chromosomes) are yet to be investigated. Although TSD may be the underlying process here, further studies may show that differential fertilization, differential embryo mortality, or sex reversal is responsible for the sex ratio bias in offspring seen in these species (Valenzuela et al. 2003).

Varanidae

Two very different sex ratios have been reported for hatchling *Varanus salvator* incubated at similar temperatures. Hairston and Burchfield (1992) report a sex ratio of 21 males to 4 females for hatchlings from incubation temperatures between 30 and 32.2°C (over four breeding seasons), while Viets et al. (1994) report a hatchling sex ratio of 0 males to 16 females for eggs incubated by H. Andrews at 31–32°C "or cooler." The lizards reported by Hairston and Burchfields (1992) were sexed by hemipene eversion at hatching, while those reported by Viets et al. (1994) were sexed by an unspecified technique more than one year after hatching.

Without histological verification of sex, these reports are not convincing evidence for TSD in the Varanidae. Monitor lizards are notoriously difficult to sex by hemipene eversion, as females have hemipene-like structures called hemiclitores (Böhme 1995; Ziegler and Böhme 1996). This difficulty in sexing varanids is probably the reason that some reptile keepers wrongly insist that all varanids hatch as male and that some individuals sex reverse to females as they approach maturity.

Embryo Development at Oviposition in TSD and GSD Agamid Lizards

In most squamate reptiles, eggs are laid at an advanced state of embryonic development. Agamid lizards are typical in this regard, and embryonic development at laying has been reported to be within the range of stage 26 to stage 31 by Dufaure and Hubert (1961) in all agamid species previously examined (Shine 1983). A typical stage 30 agamid embryo is between about 2 and 5 mm in length and has limb buds and large, slightly pigmented eyes.

In the TSD gecko *Eublepharis macularius,* eggs are also laid at a similar embryonic stage of development (29–31; Bull 1987). By switching eggs between one incubation regime known to produce 100% females (26°C) and one known to produce a maximum of males (32°C), Bull (1987) estimated that the earliest stage of embryonic development at which temperature affected sex determination was stage 32. The first seven days of incubation at 32°C and the first 14 days at 26°C had no effect on the future sex of the embryo. Bull (1987) showed that at the higher and mostly male-producing

Table 5.2 Stage of Embryonic Development at Oviposition in Agamid Lizards

Species	Embryonic Stage at Oviposition	Number of Clutches	Reference
Agama impalearis (TSD)	28	–	El Mouden et al. 2001
Amphibolurus muricatus (TSD)	30	1	Harlow 2001
A. muricatus (TSD)	31	2	Shine 1983
Calotes versicolor (GSD)	26	–	Thapliyal et al. 1973
C. versicolor (GSD)	27	–	Muthukkaruppan et al. 1970
Ctenophorus decresii (TSD)	27–28	1	Harlow 2000
Hypsilurus spinipes (GSD)	29–30	2	Harlow 2001
H. spinipes (GSD)	30–31	1	Shine 1983
Lophognathus burnsi (TSD)	27–28	2	Harlow 2001
L. gilberti (TSD)	28	1	Shine 1983
Pogona barbata (GSD)	27–28	1	Harlow 2001
P. barbata (GSD)	29	–	Packard et al. 1985
P. barbata (GSD)	26–30	2	Badham 1971
P. barbata (GSD)	26–30	2	Shine 1983
P. vitticeps (GSD)	29–30	2	Shine 1983
Physignathus lesueurii (TSD)	28–30	3	Harlow 2001
P. lesueurii (TSD)	26–30	3	Shine 1983
Tympanocryptis tetaporophora (GSD)			
(as *T. lineata*)	30	1	Greer 1989
T. tetaporophora (GSD)	28	1	Harlow 2001

Note: Stage numbers refer to Dufaure and Hubert (1961), where hatching occurs at stage 42. The sex-determining mechanism indicated for each species is based on the hatchling sex ratios in Table 5.1, except for *Calotes versicolor,* which is from Ganesh and Raman (1995), and *Agama impalearis,* which is from El Mouden et al. (2001).

temperature (32°C), sex was determined between stages 33 and 37, while at the all female-producing incubation temperature, sex was irreversibly determined between stages 32 and 34. Although no study has yet investigated the stages of embryo development during which sex is irreversibly determined in agamid lizards, all evidence suggests it should be similar to that described for the TSD gecko above. In crocodilians and turtles, eggs are laid at an earlier stage of embryo development than in squamates, but sex is determined during approximately the same morphological stages of development as in *Eublepharis macularius* (Bull 1987). The thermosensitve period (TSP) of embryo development in all TSD reptiles overlaps with gonad morphogenesis (Bull 1987).

Stage 32 of Dufaure and Hubert (1961) may represent some kind of developmental threshold for squamate reptile embryos. No oviparous species appears to carry eggs beyond this stage of embryo development (Shine 1983). If agamids with TSD were capable of withholding eggs in utero beyond stage 32, then by selective maternal thermoregulation, offspring sex determination would be under maternal control. To compare the stage of embryonic development at oviposition in agamids with TSD and GSD, the author opened a freshly laid egg from each of seven species and staged the embryo under a low-power microscope using the staging criteria of Dufaure and Hubert (1961).

Table 5.2 gives the stage of embryonic development at laying for 11 agamid species reported in the literature and from this study. Six species have TSD, while five have GSD. Using the mean value for each record, the grand mean for eight records for the six TSD species is stage 28.6 (standard error [SE] = 0.45) of Dufaure and Hubert (1961), while for 11 records for the five GSD species the mean stage of embryonic development at oviposition is 28.5 (SE = 0.41). An unpaired *t*-test shows that there is no significant difference between the mean stage of embryonic development at laying in the TSD and GSD agamid species examined here ($t = 0.276$, $p = 0.786$, $df = 17$). In all of these agamids the eggs are laid well before stage 32 of embryonic development, the earliest stage at which temperature is known to influence sex determination in a TSD species (Bull 1987).

Discussion

The considerable evolutionary plasticity in sex-determining mechanisms seen both within and between different families of lizards suggests that TSD is unlikely to be a plesio-

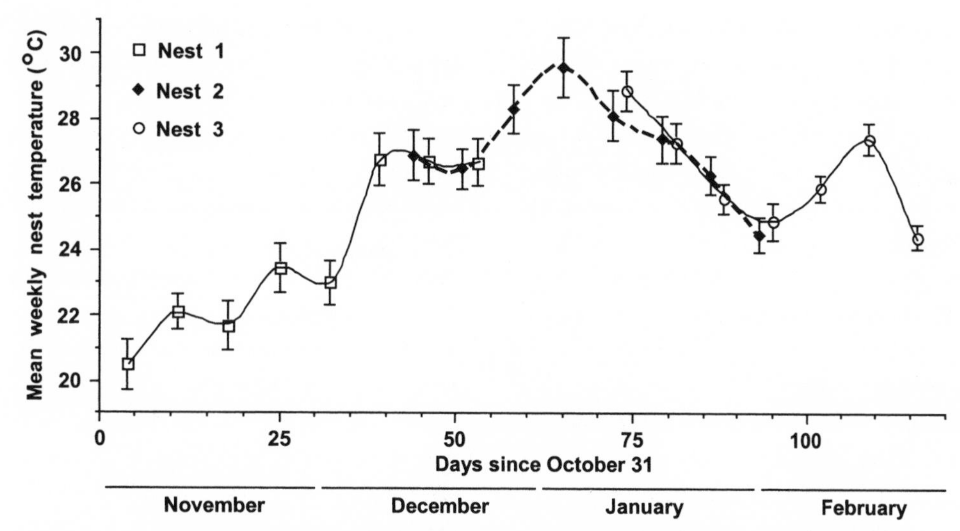

Figure 5.2 Mean weekly temperatures (± 2SE) recorded in three jacky dragon *(Amphibolurus muricatus)* nests over one season (1998–1999). [From P. S. Harlow and J. E. Taylor, "Reproductive ecology of the jacky dragon *(Amphibolurus muricatus)*: An agamid lizard with temperature-dependent sex determination," *Austral Ecology* (2000), 25:640–652. Figure 6, reproduced with permission. Journal website: http://www.blackwell-science.com/aec]

morphic condition with no current adaptive value (Janzen and Paukstis 1991a; Burke 1993; Mrosovsky 1994), as may well be the case in other lineages of reptiles, such as the Crocodylia. In the light of recent studies, the consensus is leaning more towards alternative explanations, and an increasing body of data suggests that TSD in reptiles may be correlated with sex-specific traits that affect individual fitness and therefore inhibit the evolution of GSD (see Valenzuela, Chapter 14).

Charnov and Bull (1977) were the first to attempt to explain environmental sex determination (ESD) within a theoretical framework. Although their hypothesis was formulated to explain the abundance of ESD among diverse plant and invertebrate groups, it applies equally to that specific form of ESD that occurs in reptiles: temperature-dependent sex determination. The differential fitness hypothesis of Charnov and Bull is underpinned by the assumption that the embryonic environment is spatially heterogeneous or patchy in resources, and that "patch quality" (i.e., incubation temperature in the case of TSD reptiles) will affect male and female fitness differently.

That different incubation temperatures can affect the hatchling phenotype via differences in size, color, growth rates, locomotor ability, and many other characteristics that may ultimately influence survival or reproductive fitness is well recorded in the reptile literature (e.g., Joanen et al. 1987; Bobyn and Brooks 1994; O'Steen 1998). These effects are seen in both TSD and GSD species (Rhen and Lang, Chapter 10). How these hatchling traits may ultimately affect lifetime reproductive success remains the major unanswered question in the TSD debate.

Many phenotypic determinants of reproductive success (e.g., survivorship, body size, and age at maturity) differ markedly between males and females (Andersson 1994). Thus, the variance in reproductive success between individuals of one sex may be much greater than for individuals of the opposite sex. In polygamous animals, for example, if a few males can monopolize mating with many females, the variance in reproductive success will be low in females but high in males.

One hypothesis suggests that in TSD species there are seasonal differences in hatchling sex ratios. This idea relies on the simple concept that predictable seasonal changes in egg incubation temperatures will produce predictable seasonal changes in hatchling sex ratios in TSD species. Seasonal changes in hatchling sex ratios have been reported in several reptile taxa with TSD, including freshwater and marine turtles (Vogt and Bull 1984; Mrosovsky 1994), a crocodilian (Webb and Smith 1984; Smith 1987), and two species of agamid lizard (El Mouden et al. 2001; J. E. Taylor, pers. com.).

In fast growing, early maturing, and short-lived lizard species, the time of hatching within the reproductive season will translate into significant differences in body size at the beginning of the next reproductive season (Jun-Yi and Kau-Hung 1982). In the cool temperate region of Australia, where several TSD agamid species occur, lizards that hatch from early summer nests (in December) will have a four-month growth period before hibernation, compared with perhaps only a few weeks for late summer (March–April) hatching lizards. These early hatching lizards reach maturity the following spring, while late summer hatching individuals will have to postpone reproduction until their second summer (e.g., *Amphibolurus muricatus*, Harlow and Taylor 2000; *Ctenophorus ornatus*, Bradshaw 1986, Harlow 2000).

Reproductive fitness in many lizard species may be enhanced by larger body size: in females due to a fecundity advantage and in males due to a body size advantage for larger males during male-male combat in territorial species (Stamps 1983; Olsson 1992). A scenario that produces male-biased sex ratios from early summer (medium-temperature) nests, progressing towards female-biased sex ratios for midsummer (hot-temperature) nests may well occur in some TSD lizards. Within season increases in mean nest temperatures suggest that this may be the case in the jacky dragon, *Amphibolurus muricatus* (see Figure 5.2). In this territorial agamid, males mature at 17% of their maximum body mass (Harlow and Taylor 2000). A male has little to gain by reaching sexual maturity per se, unless it is linked to large body size. By being better able to defend a territory containing several females, a large male may disproportionately increase his reproductive fitness compared to that of a large female.

If TSD is adaptive in some male-larger sexually dimorphic and territorial lizards, why is it not found in other strongly territorial lineages such as iguanids? Many iguanids are ecologically similar to agamids, yet all species so far investigated have GSD and many have evolved differentiated sex chromosomes (Olmo 1986; Viets et al. 1994). The selective forces that lead to the evolution of differentiated sex chromosomes are not well understood (Charlesworth et al. 1987), but it is plausible that chromosomal evolution in these groups may have reached a point where transition to TSD is impossible.

Only field studies investigating how both incubation temperature and the seasonal timing of hatching affect adult body size, and how body size affects lifetime reproductive success in both sexes of closely related TSD and GSD lizard species will ultimately show if there is an adaptive explanation for the occurrence of TSD in some lizards.

Acknowledgments—This research was supported in part by grants from The Peter Rankin Fund for Herpetology (Australian Museum, Sydney), and a Macquarie University postgraduate student grant. Jean Joss and Rick Shine gave invaluable assistance and logistic support throughout my study on sex determination in Australian agamids. Although many people assisted in collecting gravid agamids and are acknowledged elsewhere, Nathalie Apouchtine, Gavin Bedford, and Jennifer Taylor deserve special thanks for their exceptional help and continuing support of my project. B. Viets, M. Ewert, G. Talent, and C. Nelson's 1994 work on sex-determining mechanisms in squamate reptiles was a landmark publication which stimulated me to investigate the sex-determining mechanisms in Australian agamids.

6

NICOLA J. NELSON, ALISON CREE, MICHAEL B. THOMPSON, SUSAN N. KEALL, AND CHARLES H. DAUGHERTY

Temperature-Dependent Sex Determination in Tuatara

Tuatara (*Sphenodon* spp.) are medium-sized, sexually dimorphic reptiles with temperature-dependent sex determination (TSD) (Dawbin 1982; Cree et al. 1995). Now restricted to offshore islands of New Zealand, the two extant species *S. punctatus* and *S. guntheri* are of high conservation importance (Gaze 2001) and are biologically significant as the sole living representatives of the order Sphenodontia (Benton 2000). They are oviparous, with the eggs being laid early in embryonic development, prior to any organogenesis (Moffat 1985). Mean clutch size of *S. guntheri* is significantly smaller (6.5 eggs; Cree et al. 1991) than that of *S. punctatus* (10.6 eggs; Thompson et al. 1991), while mean egg mass for both species is approximately 4.9 g (Cree et al. 1991). Average hatchling mass of *S. punctatus* is 4.6 g (Nelson 2001). Eggs have been collected and artificially incubated to produce founders for new captive and wild populations and to augment rare existing populations (Daugherty 1998; Nelson 1998). These long-lived animals have low fecundity, do not exhibit secondary sexual characteristics until several years of age, and do not reach sexual maturity until they are at least 11 or more years old (Dawbin 1982; Castanet et al. 1988; Cree 1994; Thompson et al. 1996).

Pattern of TSD in Artificial Incubation Conditions

Since 1985, eggs of both species of tuatara have been artificially incubated routinely at constant temperatures (Thompson 1990; Cree et al. 1995). Incubation temperatures have ranged from 15 to 25°C (see Table 6.1), but hatching success is very low at 15°C and at 25°C and above (Thompson 1990; Cree et al. 1995). The incubation period of *S. punctatus* at constant artificial conditions varies with temperature; the mean incubation period at 21°C is 183 days (Nelson 2001). Hatching success is generally very high at constant temperatures between 18 and 22°C (e.g., 93% for *S. punctatus*; Nelson 2001). Eggs of *S. guntheri* have also been incubated at variable regimes in which temperature was increased, by 0.5–1°C intervals every two weeks, from 18 to 23°C, then decreased in the same stepwise manner to 18°C, mimicking mean daily ambient temperatures in midsummer on Stephens Island, the most populous of all tuatara islands (Newman 1982; Cree et al. 1995; Thompson et al. 1996). Temperatures during the middle third of the variable incubation regime (the thermosensitive period, TSP, in other reptiles) were between 22 and 23°C (Cree et al. 1995).

Both laparoscopy and histology have been used to sex juveniles (see Figure 6.1). The pattern of TSD in artificial incubation conditions appears to be female-male (FM) for both species, that is, females are produced from cooler incubation temperatures than males (Cree et al. 1995). However, the limited range of constant incubation temperatures for which sex ratios are available (18–23°C) does not yet allow exclusion of other patterns (e.g., female-male-female, FMF). Sex ratios of juveniles produced under constant incubation conditions by Nelson (2001) vary slightly from those incubated by Thompson (1990) (see Table 6.1; Cree et

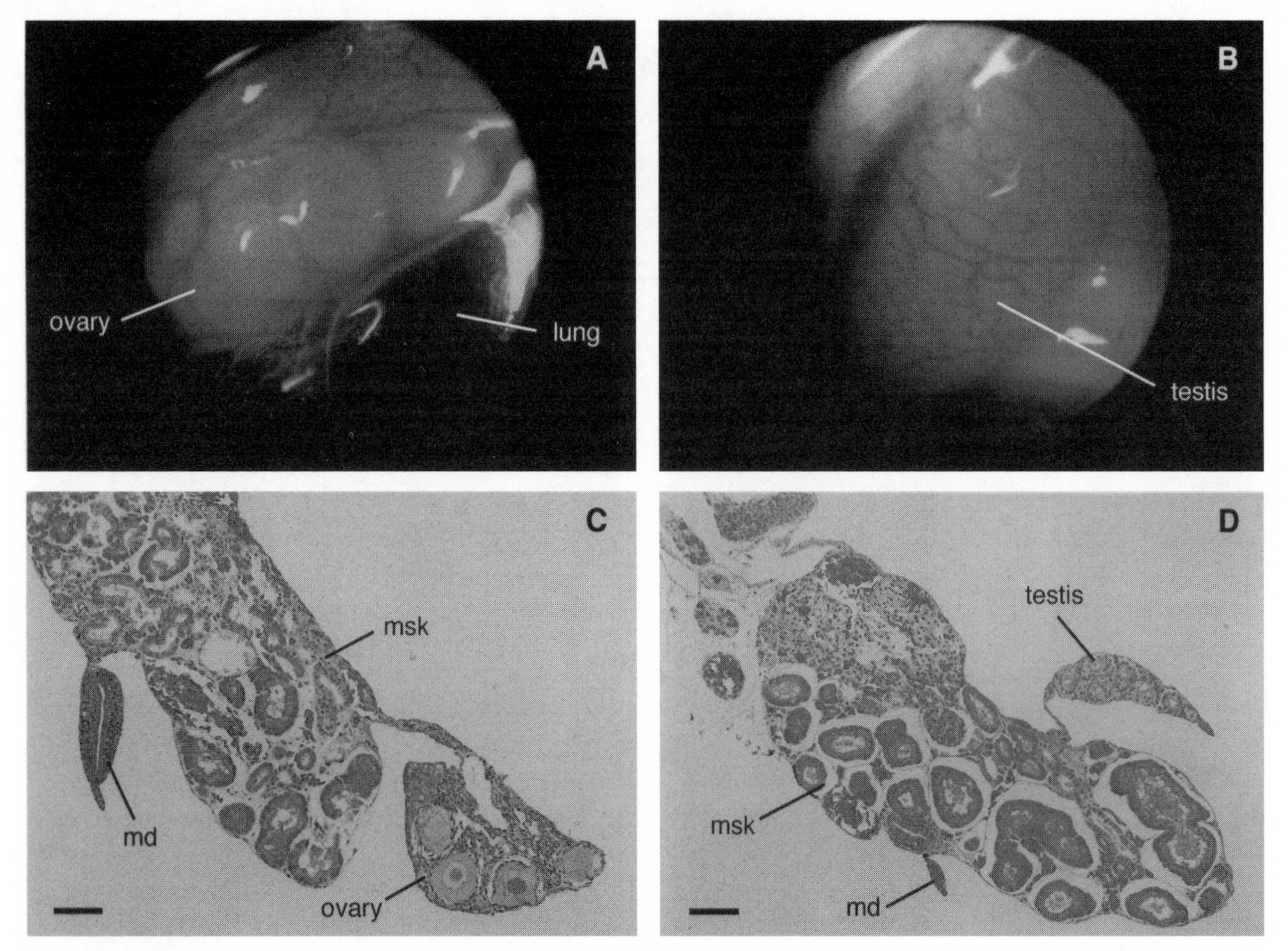

Figure 6.1 Features used to sex juvenile tuatara *(Sphenodon punctatus)*. (**A, B**) Laparoscopic views of the gonads of live juveniles. (**A**) Female, snout-vent length (SVL) 136 mm. The ovary contains numerous immature follicles (opaque, white, < 3 mm diameter). (**B**) Male, SVL 156 mm. The testis is a cream-colored, elongate and disc-like structure covered by anastomosing blood vessels. (**C, D**) Histological sections of the gonads and adjacent reproductive ducts of smaller juveniles. Both specimens were preserved after dying of heat stress. Gonads were embedded in Paraplast, sectioned at 6 μm, and stained with Mallory's trichrome stain. Scale bar in lower left corner = 100 μm. (**C**) Female, SVL 65 mm. Note ovary with follicles, a well-developed Müllerian duct (md) differentiating into an oviduct, and mesonephric kidney (msk). (**D**) Male, SVL 58 mm. Note testis with seminiferous tubules, mesonephric kidney and regressing Müllerian duct.

al. 1995), possibly due to increased reliability of incubators and sophistication of temperature recording since Thompson's incubation experiment. The pivotal temperature for sex determination of *S. punctatus* eggs lies between 21 and 22°C (Nelson 2001).

Pattern of TSD in Natural Nests

Conditions for eggs in natural nests of tuatara are noticeably different from those in artificial incubators. Female tuatara nest once every two to five years, and incubation takes on average one year to complete (Cree et al. 1991; Thompson et al. 1996), exposing eggs of a single cohort to both seasonal and annual fluctuations in temperature. The nesting season on Stephens Island occurs throughout November and December (early summer), when nesting females congregate in exposed rookeries (Thompson et al. 1996). Tuatara nests are unsuccessful under full forest cover because temperatures are too low for embryonic development (Thompson et al. 1996).

To determine whether there is evidence for TSD in nature, the authors investigated the relationship between nest temperatures and hatchling sex on Stephens Island *(S. punctatus)*. Hourly temperatures were recorded throughout incu-

Table 6.1 Percentage of Male Hatchlings from Tuatara Eggs Incubated at Different Artificial Incubation Temperatures

		Incubation Temperature (°C)					
Species	Data Source	18	20	21	22	23	Variable (18–23)
S. punctatus unnamed ssp. (Stephens Is.)	MBT[a]	0	7		70	[b]	
		(14)	(44)		(30)		
	NJN[c]	0		4	100		
		(105)		(80)	(113)		
S. p. punctatus (Little Barrier Is.)		0		0	21	100	63
		(24)		(1)	(14)	(7)	(8)
S. guntheri	[a]	0			0	[d]	100
		(19)			(5)		(10)

Note: Numbers in brackets refer to the number of eggs that hatched. MBT refers to eggs incubated by M. B. Thompson in 1985–87. NJN refers to eggs incubated by N. J. Nelson in 1998–99.

[a]Updates the data in Cree et al. (1995); no embryos hatched from 15°C.

[b]One unmatched embryo of Stephens Is. *S. punctatus* from 25°C was male (A. Cree).

[c]See Nelson (2001).

[d]Incubation of *S. guntheri* at 23°C produces males (pers. comm., Nicola Mitchell; see Conclusions, Future Directions).

bation at one location in the nest using waterproof Stowaway® TidbiT® temperature recorders (dimensions: 30 × 41 × 17 mm; Onset Computer Corporation, Massachusetts). Gonadal sex of hatchlings was determined using laparoscopy at about one year of age. Twenty-five nests were selected to encompass the diversity of nest characteristics. Nest entrances were covered with wire mesh to protect them from other nesting females. Most nests had a northeasterly aspect, and mean depth from the top egg to the soil surface was 103 mm (range = 40–200 mm). Eggs were laid in clusters of one to three layers, and mean clutch size was 9.2 eggs. Mean number of days for incubation was 365, and hatching success was 65% (Nelson 2001). Incubation temperatures throughout the year ranged from 2.9 to 34.4°C, with the magnitude of daily fluctuations ranging between 0.5 and 15°C in summer (December to March). Four representative values of temperature were selected for each nest: minimum, mean, and maximum temperature of nests during summer, and mean constant-temperature equivalent (CTE) in February.

Mean CTE accounts for the variability of a nest as well as its mean temperature. The CTE is defined as the temperature above which half of embryonic development occurs (Georges 1989; Georges et al. 1994). For tuatara, production of males is predicted if the CTE of a nest exceeds the threshold temperature for sex determination of males in artificial conditions; production of females is predicted from nests where the CTE is lower than the threshold. For each nest, mean CTE during February was calculated using Georges' model (Georges 1989; Georges et al. 1994) with the following inputs: estimated developmental zero equal to 11.1°C (based on incubation duration at constant temperatures and assuming a linear relationship below 18°C; incubation by N. J. Nelson), reference temperature equal to 18°C with an incubation period of 264 days (incubation by N. J. Nelson), pivotal temperature estimated as 21°C (see Table 6.1), and hourly temperature records for each nest during February. Temperatures during February were selected because February was assumed to include the thermosensitive period (TSP) for sex determination.

The overall proportion of male hatchlings was 0.64, and sex ratios varied significantly among nests (see Table 6.2, Figure 6.2). Males were produced from warm nests, females from cool nests, and mixed sex ratios from intermediate temperatures (see Figure 6.3). Sex ratios of nests were correlated with incubation temperature and nest depth, indicating that TSD occurs in nature.

Possible explanations for a male-biased sex ratio from a single nesting season include a bias from sampling or sex-biased mortality. Alternatively, the authors' sample may be representative of the incubation temperatures normally experienced on Stephens Island, suggesting that the majority of nests on this island exhibit high (male-producing) incubation temperatures in a majority of years.

The authors selected nests on a stratified random basis to capture the variability among and within rookeries, thus

Table 6.2 Clutch Size, Hatching Success, and Sex Ratios of Tuatara Eggs in Naturally Incubated Nests on Stephens Island 1998–99

Nest[a]	Clutch Size	No. Eggs Failed in Nest	No. Hatchlings	No. Juveniles Sexed	Sex Ratio (M:F)
1	11	4	7	7	7:0
2	6	0	4	6[b,c]	6:0
3	11	0	11	10	5:5
4	8	1	7	7	7:0[d]
5	13	0	13	12	12:0
7	10	1	8	8	1:7
8	9	0	9	9	9:0
9	12	4	8	7	7:0
10	10	2	8	8	7:1
11	3	2	1	1	1:0
12	4	1	1	1	0:1
13	9	8	1	1	1:0
14	9	0	9	9	1:8
15	9	8	1	1	1:0
16	11	11	0	–	–
17	10	2	8	8	7:1
18	7	1	6	6	0:6
19	13	6	7	7	3:4
20	8	0	8	8	8:0
21	11	11	0	–	–
22	11	4	7	7	0:7
23	11	1	10	9	0:9
24	11	1	10	8	7:1
25	4	4	0	–	–
Total	221	72[e]	144[b]	140[c]	90:50

[a]Nest 6 not relocated.

[b]Embryos from five eggs died after excavation of nests but prior to hatching (two were punctured during excavation; three died during subsequent incubation).

[c]138 live juveniles were sexed; six juveniles could not be sexed; the embryos from two punctured eggs were sexed.

[d]Temperature data logger failed in this nest.

[e]Of the 72 eggs that died during incubation in nests, 34 suffered beetle damage, 16 showed no embryonic development, 7 died at early developmental stages, 10 had mid- to late-stage embryos, and 2 were rotten and could not be assigned to any of these categories. Three eggs had possibly hatched and juveniles had escaped.

attempting to sample representative nests from the whole population. However, artificial nest protection may preclude females that nest later from digging up existing nest sites. The authors cannot speculate on the effect the temperatures in the subsequent nest may have on the sex ratio, or on whether individual female tuatara nested in warmer sites in the sampling year, but they do know that rookeries were located in similar positions to those sampled previously (but hatchlings were not sexed; Thompson et al. 1996). Tuatara eggs are laid in exposed locations in relatively shallow nests, resulting in temperature gradients within nests (Thompson et al. 1996), by which mixed sex ratios are likely to be produced. Temperature fluctuations between years could explain a biased sex ratio in any one year. During the summer of 1998/99, when sex was likely to be determined in embryos in the marked nests, air temperatures (minimum, mean, and maximum daily averages) were higher than the averages for years 1990–2001. For example, mean air temperature in January 1999 was 17.2°C versus between 15.2 and 16.5°C for January in 1990–1998 and 2000–2001. Even minor fluctuations in temperatures in natural conditions may alter sex ratios if they occur close to the pivotal temperature for sex determination (see Figure 6.3). It seems likely that the 1998/99 summer was more favorable for production of males, and preceding and following years would have produced more females.

Although we do not know whether female tuatara are philopatric, or even if individuals nest on the same rookery in subsequent nesting seasons, bias through sampling rookeries with high nesting densities cannot account for the male-biased sex ratio that we observed. We chose nests in high density nesting areas (rookeries) because very few nests were located elsewhere. Hatching success was high in these areas, but this was partly a result of the nest protection provided in this study. Nonetheless, no female-biased sex ratio was achieved on any rookery, and only on two rookeries was the sex ratio as low as 0.5.

Differential mortality as a function of incubation temperature has been reported in pine snakes, but death of embryos occurred late in development (Burger and Zappalorti 1988). We discount egg mortality as a cause for the bias in hatchling sex ratio of tuatara. Insect larvae destroyed half of the unsuccessful eggs in this study. In nests where some eggs were destroyed by beetle larvae, other eggs hatched successfully. We cannot be sure whether insect larvae were the cause of death or ate eggs containing embryos that had already died. If insect larvae do kill embryos, then we assume this is independent of sex. If insect larvae eat eggs that have already died, then this could occur any time between laying and excavation of nests, when most eggs had still not hatched. We have evidence of only 10 embryos that died at a stage late enough for sex determination to have occurred. Even if these were all females, the sex ratio would still be significantly greater than 0.5.

Sex ratios in natural nests in this study provide support for the FM pattern observed at constant artificial conditions (see Figure 6.3). As temperatures in nests reached 34.4°C and nests with CTE during February greater than 22°C produced predominantly male hatchlings, it seems likely that tuatara do not have a FMF pattern of TSD.

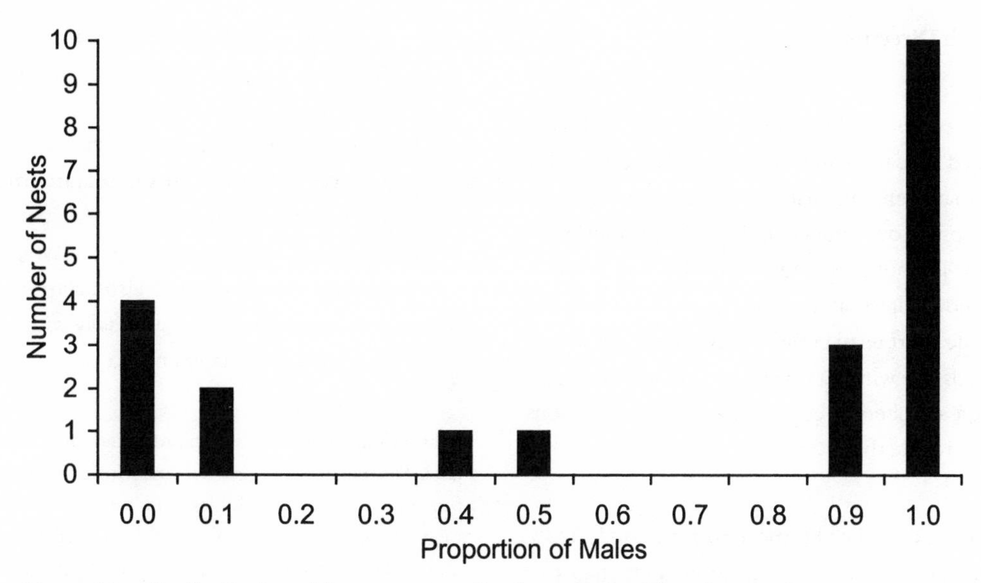

Figure 6.2 The distribution of the proportion of male tuatara hatchlings in natural nests on Stephens Island. These data indicate the existence of TSD in natural nests.

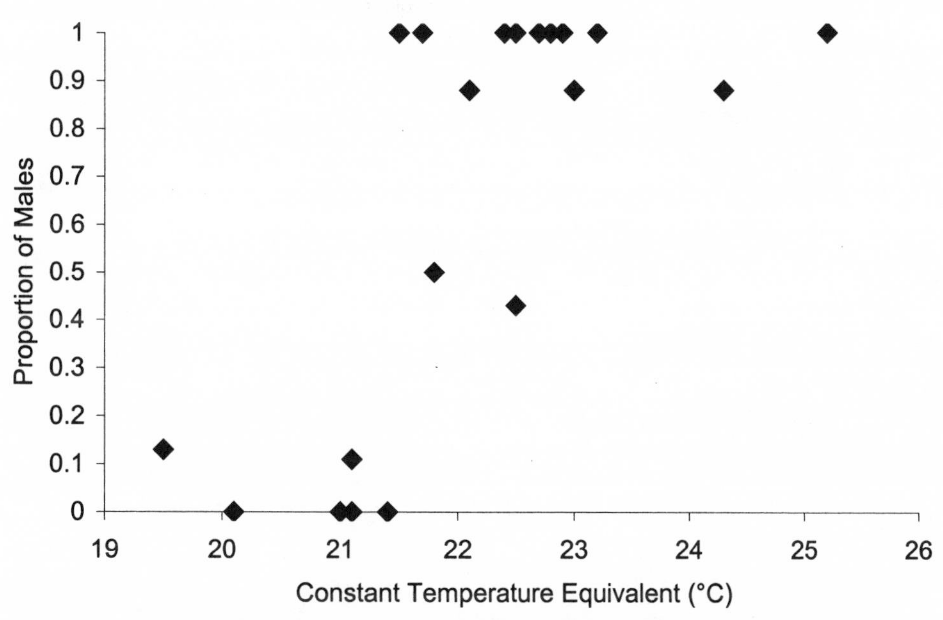

Figure 6.3 The effect of temperature on sex of hatchlings demonstrates that TSD exists in natural nests. The model used to calculate constant temperature equivalents (Georges 1989; Georges et al. 1994) for each nest assumes that hatchling sex depends on the proportion of embryonic development that occurs above the pivotal temperature, not the proportion of time spent above the pivotal temperature.

Future Research Directions

Well-known aspects of TSD for other species of reptiles are yet to be studied in tuatara and could have implications for conservation management. Also, investigation of TSD in tuatara may improve our understanding of the evolution of TSD in reptiles. The authors suggest the following research topics, but acknowledge that many of these studies may not be possible in the short term because they require either sacrifice of embryos (for which it is very difficult to obtain permission) or large numbers of eggs from species/populations that could not provide them.

1. Identification of the thermosensitive period (TSP), with respect to both the developmental stages of embryogenesis during which it occurs and the time of year it occurs in natural nests from different populations.
2. The pattern of TSD over the full range of constant incubation temperatures that may produce fully developed or viable offspring (especially 15–18°C and 23–25°C).
3. Geographic and species variation in pivotal temperatures and TSD patterns (currently in progress; N. Mitchell, pers. comm.).
4. Molecular and endocrine mechanisms by which temperature determines sex in tuatara (molecular studies currently in progress; S. Sarre, pers. comm.).
5. Effects of incubation on phenotype of hatchlings (currently in progress; Nelson 2001).
6. Population sex ratios—especially with respect to naturally incubated hatchling sex ratios.

Acknowledgements—We thank Marsden Fund, WWF–New Zealand, San Diego Zoo, and Victoria University of Wellington for funding; VUW Animal Ethics Committee, New Zealand Department of Conservation, Ministry of Transport, Ngati Koata, Te Atiawa, and the Ngatiwai people for approving our studies; curators of captive-breeding institutions (especially B. Blanchard, B. Goetz, and L. Hazley) and museum curators (Museum of New Zealand Te Papa Tongarewa, Canterbury Museum) for permission to examine specimens; G. Stokes and C. Thorn for histology; the staff of Wellington Public Hospital and B. Gartrell for laparoscopic camera equipment; S. Pledger for statistical advice; A. Georges for use of his temperature models; K. Miller for graphical assistance; and many enthusiastic volunteers for assistance with laparoscopies and in the field.

7

DOMINIQUE CHARDARD,
MAY PENRAD-MOBAYED, AMAND CHESNEL,
CLAUDE PIEAU, AND CHRISTIAN DOURNON

Thermal Sex Reversals in Amphibians

Most amphibian species are gonochoristic. Studies of various aspects of sex determination, including genetics and cytogenetics, the effects and role of sex steroids, and the effects of environmental factors have been carried out in the Anura and the Urodela, but no data are available for the Gymnophiona. Genotypic sex determination (GSD) seems to operate in most if not all species, although approximately 96% of the species examined cytologically did not show morphologically distinguishable sex chromosomes (reviewed in Hayes 1998; Schmid and Steinlein 2001). In these species, the difference between sex chromosomes is too discrete to be observed by classical cytogenetic analysis, possibly because too small a region of these chromosomes is involved in sex determination (Hillis and Green 1990). As in fish and reptiles, the two mechanisms of GSD, male heterogamety (XX/XY) and female heterogamety (ZZ/ZW), exist in amphibians. Female heterogamety is hypothesized to be the ancestral state, and male heterogamety is hypothesized to have evolved several times (Hillis and Green 1990). A shift from one sex-determining mechanism to another is observed in *Rana rugosa*, a species living in Japan: depending on its geographical range, this species exhibits one or the other of the two mechanisms, the sex chromosomes being hetero- or homomorphic (Miura et al. 1998; reviewed in Hayes 1998; Schmid and Steinlein 2001). Unusual systems of sex determination have also been found, such as 0W/00 in the New Zealand frog *Leiopelma hochstetteri*; this system could have originated from a primordial ZW/ZZ system through loss of the Z chromosome (Green 1988).

The different mechanisms of GSD in amphibians are of particular interest for studies on the evolution of sex chromosomes and sex-determining mechanisms. In addition, rearing embryos and/or larvae at abnormally high or low temperatures has been shown to modify gonadal sex differentiation and sex ratio in some anuran and urodelan species, indicating that the action of temperature can be superimposed upon GSD. In amphibians, like in other vertebrates, the anlagen of gonads (genital ridges) are two outpockets of cells on the ventral surface of the mesonephros. They enclose primordial germ cells, the origin of which, mesodermal or endodermal, remains an open question. Gonads are initially undifferentiated with no histologically obvious differences between females and males. Subsequently, in ovarian differentiation, follicles with easily distinguishable oocytes develop early in the cortical part, while a lacuna ("ovarian vesicle") bordered by a flat epithelium develops in the medullary part. In testicular differentiation, the surface epithelium becomes thin, while the future seminiferous tubules develop in the medulla (reviewed in Hayes 1998). In most anuran and urodelan species, these events take place during larval development, so that the identification of phenotypic sex is possible at metamorphosis; these species are "sexually differentiated." However, some species present sexually undifferentiated or semidifferentiated races.

In "sexually undifferentiated races," for example, in some populations of *Rana temporaria* (Witschi 1930), gonads of all animals have an ovarian structure at metamorphosis; testicular differentiation occurs after metamorphosis with a transient structure of ovotestis. In "sexually semidifferentiated" races, such as in the urodele *Hynobius retardatus* (Uchida 1937a,b), phenotypic females with ovaries and intersexes with ovotestes are observed in variable proportions at metamorphosis; intersexes and some phenotypic females then become phenotypic males (reviewed in Dournon et al. 1990). Thus, in undifferentiated and semidifferentiated races, intersexuality may correspond to a transient stage of gonadal differentiation. When investigating the effect of temperature on gonadal sex differentiation, it is important to consider this particularity of some amphibian species in interpreting the experimental results.

Effect of the Rearing Temperature on Gonadal Sex Differentiation

Changes in gonadal differentiation under the effects of the rearing temperature of tadpoles were described for the first time in *R. temporaria* by Witschi (1914). Later, the effects of abnormally high or low temperatures during larval development were again examined in *R. temporaria* (Piquet 1930), and extended to some other anuran and urodelan species (see references in Table 7.1). Masculinizing or feminizing effects were deduced from biased sex ratios and from macroscopic and microcospic examination of gonads at metamorphosis or in recently metamorphosed juveniles. The genetic sex of individuals was unknown except in the more recent studies, dealing with *Pleurodeles* and *Triturus*. Generally, rearing embryos and larvae at ambient temperature (20 ± 4°C) yields a balanced sex ratio (50% males to 50% females). Thus, temperatures ≥ 24°C correspond to a heat treatment, whereas temperatures ≤ 16°C correspond to a cold treatment. The effects of such treatments on the species studied to date are presented in Table 7.1. For each species except one *(Hynobius retardatus)* the mechanism of GSD is indicated, even if it was determined after the experiments were performed.

In the Anura, the observations of Witschi (1914) and Piquet (1930) in *R. temporaria* are concordant: high temperatures (≥ 25°C) are masculinizing, whereas low temperatures (≤ 12°C) are feminizing. However, these results must be taken with prudence. Indeed, as shown by Witschi (1929b, 1930) himself, *R. temporaria* comprises different sexual races; the sexual race of the populations used is unknown. The experiments performed by Witschi (1929a) in *Rana sylvatica* are more convincing. Larvae from this sexually differentiated species were placed at 32 ± 2°C when gonads had begun to differentiate and maintained at this temperature for 15–33 days. At the end of this treatment, 62 out of the 115 individuals were phenotypic males; all the others were intersexes, and their gonads were ovotestes at different steps of ovarian masculinization.

Heat treatment administered for various periods of larval development also had a masculinizing effect in other anuran species, including *Rana japonica* (Yoshikura 1959, 1963), *Rana catesbeiana* (Hsü et al. 1971), and *Bufo bufo* (Piquet 1930). In addition, a possible feminizing effect of cold treatment (at temperatures ≤ 12°C) similar to that observed in *R. temporaria* was suggested in *B. bufo* (Piquet 1930). In the subspecies *Bufo b. formosus*, high temperatures (25 or 30°C) applied all through the larval stage had no significant effect on sex ratio, although some animals had rudimentary or underdeveloped gonads. When high temperatures were applied through the larval and the juvenile stages or only at the juvenile stages, a preponderance of males was obtained. The occurrence of two hermaphrodites among the heat-treated animals was indicative of the sex reversal process (Muto 1961).

In the Urodela, the first study dealing with the effects of temperature on gonadal sex differentiation was performed in a sexually semidifferentiated race of *H. retardatus*. As in the Anura, heat treatment was masculinizing (Uchida 1937b). The more recent studies were carried out in sexually differentiated species: two closely related species of *Pleurodeles*, *P. waltl*, and *P. poireti* (Dournon and Houillon 1984, 1985; Dournon et al. 1984, 1990), and two subspecies of *Triturus cristatus*, *T. c. cristatus*, and *T. c. carnifex* (Wallace et al. 1999; Wallace and Wallace 2000). Since the genetic sex of all or some individuals was identified, they provided evidence of thermal sex reversal, that is, that the temperature-induced sexual differentiation of the gonads did not agree with the sexual genotype.

P. waltl and *P. poireti* display a ZZ/ZW mechanism of GSD. In *P. waltl*, larvae issued from crosses between ZZ standard males and ZW standard females (crosses 1) and from crosses between ZZ standard males and WW thelygenic females (crosses 2) were reared at 30°C or at 32°C. At 30°C, among individuals from crosses 1, 70% were males, 23% females, and 7% intersexes; among individuals from crosses 2 (all with ZW genotype), 44% were males, 46% females, and 10% intersexes. At 32°C, individuals from crosses 1 and 2 were all males, showing that all ZW genotypic females were sex reversed; these "ZW thermoneomales" are fertile (Dournon and Houillon 1984, 1985; Dour-

Table 7.1 Effects of Abnormally High or Low Temperatures Administered during Embryonic and/or Larval Development on Gonadal Sex Differentiation in Amphibians

Species	Sexual Race	GSD	Temperature FT	Temperature MT	Adults	Reference
Anurans						
Rana temporaria	?	XX/XY	≤12°C	≥ 25°C	–	Witschi 1914, Piquet 1930
Rana sylvatica	D	XX/XY	–	≥ 30°C	–	Witschi 1929a
Rana japonica	D	XX/XY	–	≥ 30°C	–	Yoshikura 1959, 1963
Rana catesbeiana	D	XX/XY	–	≥ 30°C	–	Hsü et al. 1971
Bufo bufo	D	ZZ/ZW	≤12°C	≥ 30°C	–	Piquet 1930
Urodeles						
Hynobius retardatus	SD	ND	–	≥ 30°C	–	Uchida 1937b
Pleurodeles waltl	D	ZZ/ZW	–	≥ 30°C	ZW neomales*	Dournon and Houillon 1984
Pleurodeles poireti	D	ZZ/ZW	30°C	–	ZZ neofemales*	Dournon et al. 1984
Triturus c. carnifex	D	XX/XY	13°C		XY neofemales	
				≥ 28°C	XX neomales	Wallace et al. 1999
Triturus c. cristatus	D	XX/XY	≤16°C		XY neofemales	
				28°C	XX neomales	Wallace and Wallace 2000

Note: FT = feminizing temperature, MT = masculinizing temperature, D = differentiated race, SD = semidifferentiated race, ? = unknown race, ND = not determined, * = fertile animals.

non et al. 1990). In *P. poireti,* larvae from a cross between a ZZ standard male and a ZW standard female were reared at 30°C. The sex ratio at metamorphosis was 65.7% females, 22.9% males, and 11.4% intersexes, indicating that several ZZ genotypic males were sex reversed. Thus, high temperatures have opposite effects in *P. waltl* and *P. poireti*, complete or partial sex reversal affecting ZW genotypic females in *P. waltl* and ZZ genotypic males in *P. poireti* (Dournon et al. 1984; reviewed in Dournon et al. 1990). The two species can be crossed to produce hybrid progeny. Unfortunately, due to difficulties in rearing, the effects of heat treatment could not be tested in the hybrids (Dorazi et al. 1995). Likewise, rearing larvae of *Pleurodeles* at low temperatures is difficult, due to their sensitivity to mycosis and epizootis and to the lengthening of development. Nevertheless, in *P. waltl*, a juvenile intersex with both a testis and an ovary was found among animals from standard offspring reared at 15°C between stages 42 and 50 (see Figure 7.3). Whether this animal was a feminized ZZ genotypic male or a masculinized ZW genotypic female was not determined (Dournon et al. 1990).

Studies in *T. c. carnifex* and *T. c. cristatus* showed masculinizing effects at high temperatures and feminizing effects at low temperatures. Both subspecies display an XX/XY mechanism of GSD. Larvae of *T. c. carnifex* reared at 16–26°C show a high survival rate and a 1:1 sex ratio at metamorphosis. Various thermal treatments were carried from the stage "feeding larvae with three fingers" or from "uncleaved eggs" until metamorphosis. In treatments beginning with feeding larvae, the sex ratio was significantly biased in favor of males at 28, 30, and 31°C (67, 74, and 67% males, respectively), and XX neomales were diagnosed in the 30°C trial (see below). In treatments beginning with uncleaved eggs, the sex ratio was biased significantly in favor of females (78%) at 13°C, and XY neofemales were isolated from this trial (Wallace et al. 1999). Similar thermal treatments were performed in *T. c. cristatus*. Trials at 18–24°C showed a 1:1 sex ratio. Treatment of larvae at 28°C resulted in a majority of males (61%), some of which were diagnosed as XX neomales. Trials at and below 16°C resulted in a significant excess of females (68–100%), among which XY neofemales were diagnosed (Wallace and Wallace 2000).

XX neomales of *T. c. carnifex* were reared after metamorphosis until maturity. They failed to fertilize eggs and had miniature testes with premeiotic spermatogonia and spaces presumably left by degenerated spermatocytes. Possibly a fertility factor on the Y chromosome of this species is required for entry into the meiotic stages of spermatogenesis (Wallace et al. 1999). XY neofemales of *T. c. carnifex* also were reared after metamorphosis (Wallace et al. 1999). Apparently they reached maturity (Wallace and Wallace 2000), but no evidence of their fertility was reported.

Cytogenetic and Genetic Proofs of Thermal Sex Reversal

Proofs of sex reversal were provided in *P. waltl*, *P. poireti*, *T. c. carnifex*, and *T. c. cristatus* by different methods including cytogenetic studies, breeding experiments, and searches for gene sex linkage.

Cytogenetic Studies

The mitotic karyotypes of *T. c. carnifex* and *T. c. cristatus* are very similar and display distinct sex chromosomes (fourth pair). Both X and Y chromosomes are submetacentric. The X chromosome carries a faint C band in the middle of the short arm and a subterminal C band on the long arm. The Y chromosome carries these two C bands and an additional one on the long arm. Thus, XX neomales obtained by heat treatment and XY neofemales obtained by cold treatment were identified by examining C-banded chromosome sets from larval tail-tip biopsies in *T. c. carnifex* (Wallace et al. 1999) and *T. c. cristatus* (Wallace and Wallace 2000).

Despite the lack of a detectable heterochromosome in the mitotic karyotype of *P. waltl* and *P. poireti*, sex chromosomes have been identified in both species by examination of lampbrush chromosomes from the oocyte meiotic karyotype of phenotypic females with ZW, ZZ, or WW sexual genotypes. Indeed, lampbrush chromosome loops localized at homologous sites on the two partner chromosomes of a bivalent may show differences in morphological or molecular characteristics (Callan 1986). Thus, in *P. poireti* at ambient temperature, the W chromosome of the bivalent IV has been distinguished by a set of lampbrush loops displaying distinct morphology under a phase contrast microscope. Heteromorphic loops are present in ZW standard females but are not observed in ZZ neofemales resulting from treatment of larvae with estrogen (Lacroix 1970). This specifity was used to identify ZZ thermoneofemales obtained by heat treatment (Dournon et al. 1984; Figure 7.1). In *P. waltl* at ambient temperature, heteromorphic loops in the ZW bivalent are not visible under a phase contrast microscope, but are revealed by in situ protein immunodetection with antibodies (Lacroix et al. 1985) or in situ hybridization with RNA probes (Penrad-Mobayed et al. 1998). Figure 7.2A,B shows heteromorphic RNA-labeled loops on the bivalent IV: they constitute a specific marker for the differential segment of the W chromosome. The RNA-labeled loops were analyzed under various experimental conditions. Results suggest that the RNA sequence interacts with lampbrush loop–associated proteins. This interaction was shown to be thermosensitive: treatment of adult animals for 18 hours at 34°C reduced the labeling signals (Figure 7.2C), whereas treatment for seven days at 8°C enhanced the labeling signals (Figure 7.2D). The identification of such sex-specific RNA-binding proteins is in progress (Penrad-Mobayed et al. 1998, unpubl. results).

Heteromorphic loops on the ZW bivalent, visible under a phase contrast microscope, can be induced in *P. waltl* by treatment at 32°C (Lacroix et al. 1990). Using such induced loops as markers, one trisomic female for bivalent IV (ZZW) was identified. This result pointed to a major role of the W chromosome in the mechanism of female sex determination in this species (Lacroix et al. 1990).

Breeding Experiments

None of the methods characterizing specific transcription loops on the W lampbrush chromosome could be used to identify individuals sex-reversed by a thermal treatment in *P. waltl*, since these animals were males and did not display oocytes. The method chosen was analysis of the progeny. It was also carried out in *P. poireti*. Although time consuming, this method provides an irrefutable demonstration of sex reversal. In addition, it provides data on mating behavior and fertility, which are ultimate stages of sex differentiation. In a number of species, breeding tests have been performed after sex reversal by gonad grafts or hormonal treatments and used to determine which of the males or females were heterogametic (reviewed in Hayes 1998).

The ZW thermoneomales of *P. waltl* and the ZZ thermoneofemales of *P. poireti* are fertile, indicating that the factors controlling spermatogenesis and ovogenesis are autosomal or are not located on the sex-determining region of the W chromosome. When ZW thermoneomales of *P. waltl* were bred with ZW standard females, progenies with 25% ZZ males and 75% ZW and WW females were produced at room temperature as expected. The females with the novel WW genotype were also fertile: when they were crossed with ZZ standard males, the progenies reared at room temperature were 100% females as expected (Dournon and Houillon 1984; reviewed in Dournon et al. 1990). In *P. poireti*, crossing ZZ thermoneofemales with ZZ standard males produced 100% males as expected (Dournon et al. 1984).

Gene Sex Linkage

Peptidase 1 is a dimeric enzyme encoded by two genes, *peptidase 1A* and *peptidase 1B*, located respectively on chromosomes Z and W in *P. waltl*. The polymorphism of its electrophoretic pattern on starch gel plates has been used to identify the ZZ, ZW, and WW sexual genotypes in this species (Dournon et al. 1988). This method also discriminates ZW females and ZZ neofemales in *P. poireti* with, however, different patterns from those obtained in *P. waltl* (Dournon et al. 1984). So far, no other genes that encode enzymes have been localized on the sex chromosomes of the Urodela.

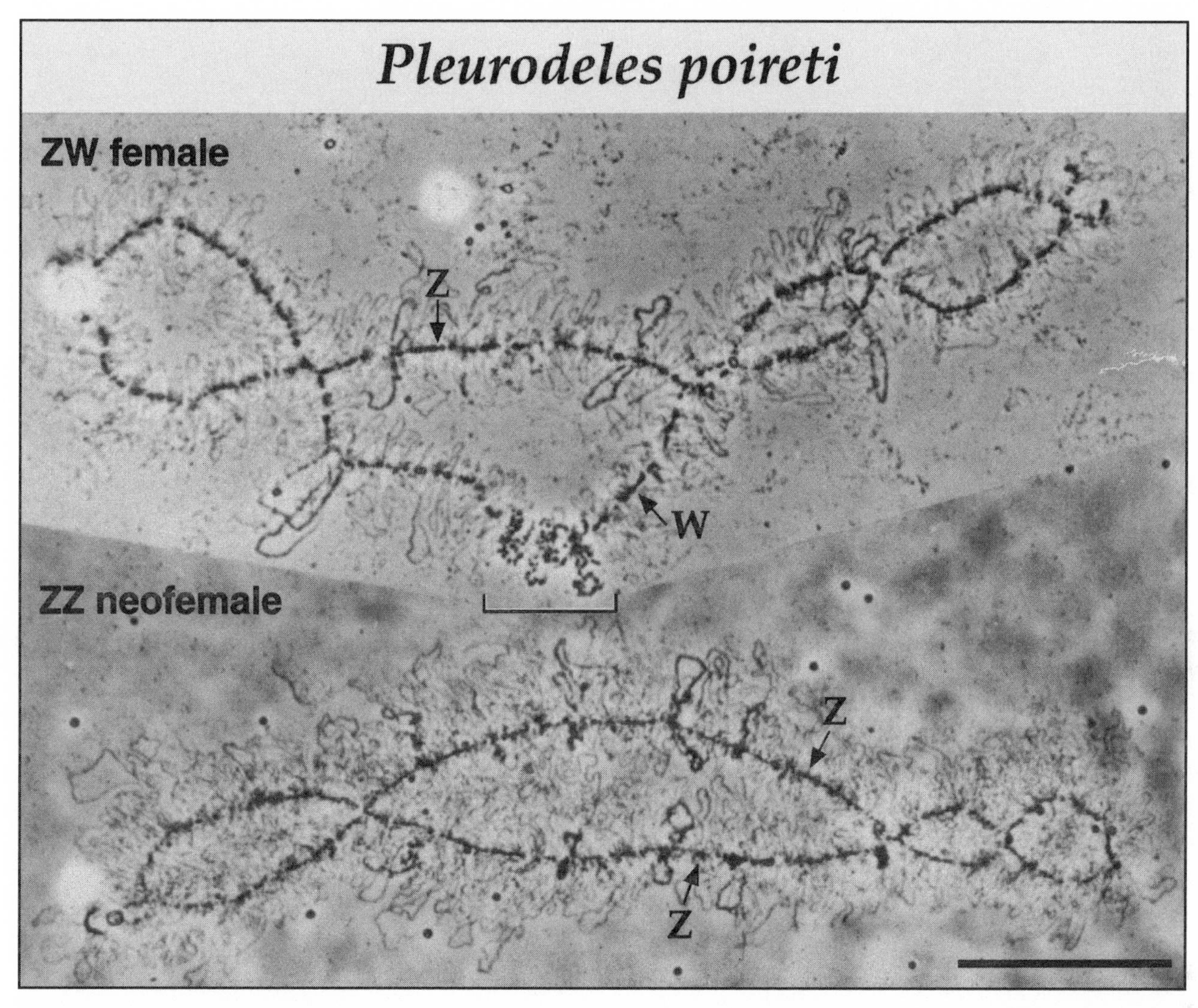

Figure 7.1 Lampbrush sex chromosomes of a standard female and a ZZ neofemale obtained by heat treatment (30°C) during larval development in *Pleurodeles poireti*. The bracket indicates the differential segment of the W chromosome in the standard female. The two Z chromosomes are quite similar in the neofemale. Bar represents 30 μm. [From C. Dournon, F. Guillet, D. Boucher, and J. C. Lacroix, "Cytogenetic and genetic evidence of male sexual inversion by heat treatement in the newt *Pleurodeles poireti*," *Chromosoma* (1984) 90:261–264. Reproduced with permission.]

In contrast, several genes that encode enzymes have been mapped on the sex chromosomes in anurans of the genus *Rana* (reviewed in Schmid and Steinlein 2001; Sumida and Nishioka 2000). The electrophoretic pattern of such enzymes should thus provide a useful tool for demonstrating sex reversal in frogs whose gonadal sex differentiation is sensitive to thermal treatments.

Thermosensitive Period

The thermosensitive period (TSP) is defined by the two stages of development between which a thermal (cold or heat) treatment can induce male or female sex reversal. Before and after this period, thermal treatment does not distort the sex ratio. The period during which heat treatment induces sex reversal of ZW genotypic females has been determined in *P. waltl* by shifting larvae at different stages and for various times from 20 ± 2°C to 30, 31, or 32°C (Dournon and Houillon 1985, unpubl. data). Some results of these experiments are shown in Figure 7.3. All ZW genotypic females were sex reversed when they were reared at 32°C, at least between the larval stages 43 and 54 (based on the developmental stages described by Gallien and Durocher 1957). Rearing at 32°C beginning later than stage 43 or ending earlier than stage 54 did not yield 100% sex reversal. Thus, TSP extends between stages 43 (three fingers formed at forelimbs) and 54 (all fingers formed at hindlimbs) of development in *P. waltl*; it is approximately two months long for both 20 ± 2°C and 32°C rearing temperatures.

Rearing larvae at 30°C for this period of development

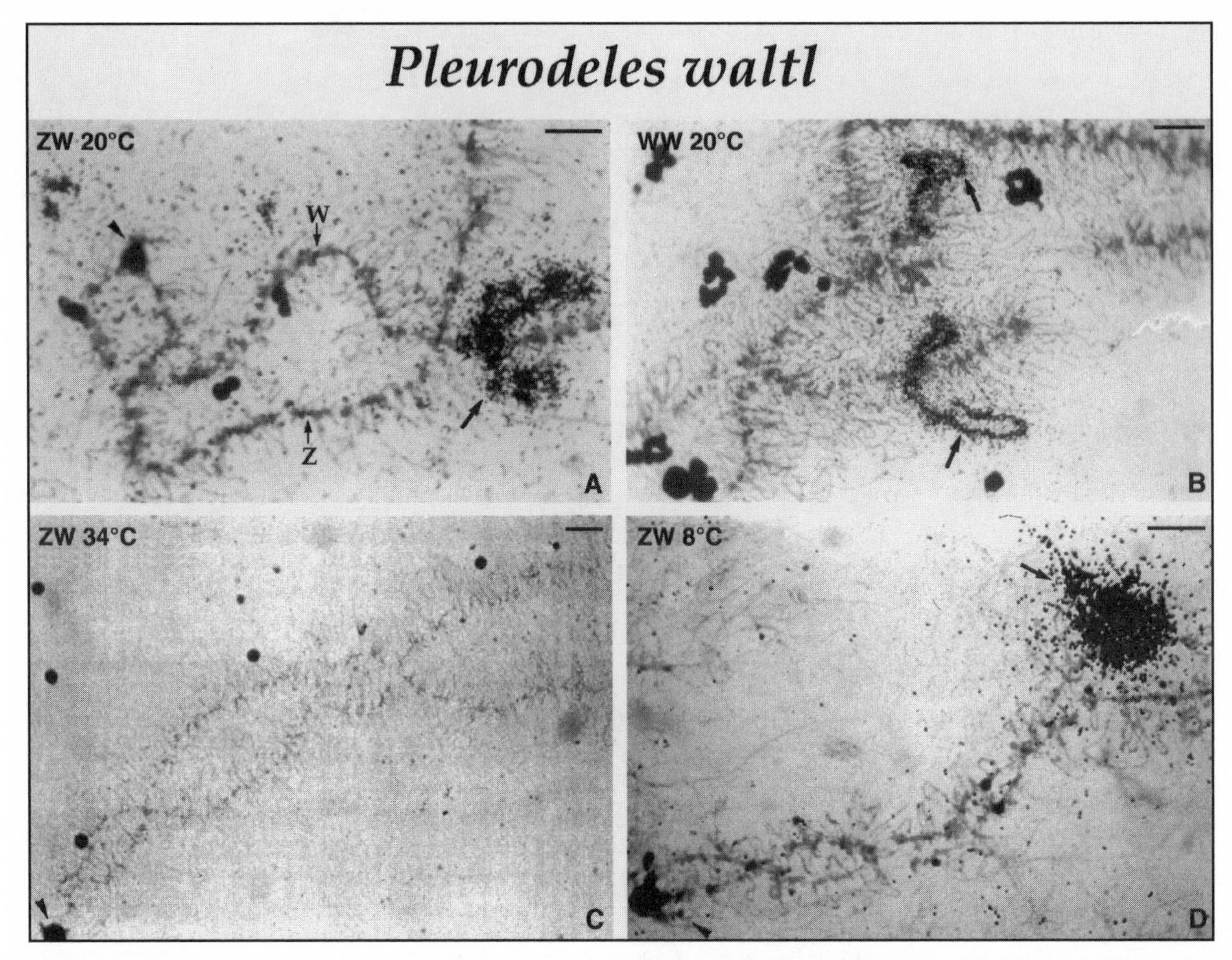

Figure 7.2 In situ hybridization of the ^{35}S-cRNA probe of 159 bp to transcripts of the lampbrush sex chromosomes in control and thermal treated adult females of *Pleurodeles waltl*. (**A**) Control ZW female raised at 20°C: strongly labeled loops (arrows) are observed on one chromosome (W), but not on the other (Z). (**B**) Thelygenic WW female raised at 20°C: labeled loops (arrows) are present on the two chromosomes. (**C**) ZW adult female submitted to heat treatment (18 hours at 34°C): the labeled loops characterizing the W chromosome are absent. (**D**) ZW female submitted to cold treatment (seven days at 8°C): the labeling on the W loops is enhanced. The arrowhead on (A), (C), and (D) points to the sphere characteristic of the sexual bivalent IV. Bars represent 20 μm. [From M. Penrad-Mobayed, N. Moreau, and N. Angelier, "Evidence for specific RNA/protein interactions in the differential segment of the W sex chromosome in the amphibian *Pleurodeles waltl*," *Development, Growth and Differentiation* (1998), 40:147–156. Reproduced with permission.]

also induced sex reversal of genotypic females, but not in all of them. In addition, exposure at 32°C as of stage 43 (or even stage 40; see Figure 7.3) followed by exposure at 30°C as of different stages before the end of the TSP and until metamorphosis did not yield 100% sex reversal. This confirms that 30°C is less efficient than 32°C in inducing female sex reversal. It may be of interest for future research that clones can be obtained by nuclear transplantation in *P. waltl*. Five such ZW clones were reared at 30°C between hatching and metamorphosis: two differentiated as intersexes and three as females (Dournon, unpubl.).

At stage 43 of larval development, the beginning of the TSP, the gonads of *P. waltl* are undifferentiated, and at stage 54, the end of the TSP, they have begun to differentiate into testes or ovaries (see Figure 7.3). Thus, in *P. waltl* gonads appear histologically undifferentiated during the major part of the TSP.

The experiments carried out in other amphibian species did not make it possible to precisely determine the TSP. Treatment at 30°C between stages 42 and 54 of *P. poireti* did not induce all ZZ males to develop as females (Dournon et al. 1984). Experiments in *T. c. cristatus* indicate that heat or cold treatment beginning at the first feeding larval stage needed to be prolonged for most of the larval life to induce female or male sex reversal, respectively. Depending on the rearing temperature chosen, the duration of exposure was between two months (at 28°C) and five months (at 15°C). The TSP in *T. c. cristatus* thus seems similar to that in *P. waltl*,

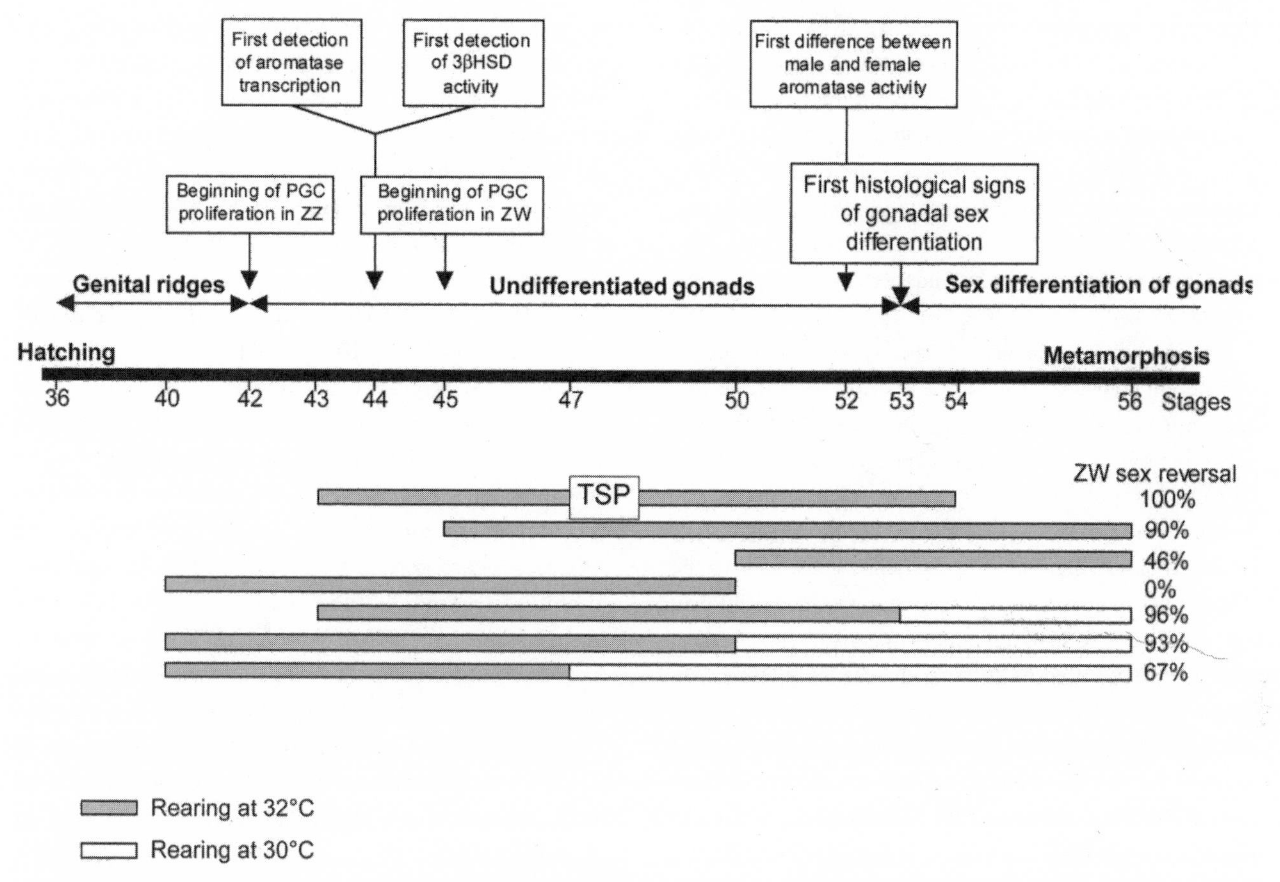

Figure 7.3 Percentages of sex reversal resulting from heat treatments (32°C or 32°C plus 30°C) administered for different times of larval development in ZW females of *Pleurodeles waltl*. Some characteristics of normal gonadal development between hatching and metamorphosis are indicated above the scale of developmental stages. TSP, the thermosensitive period, corresponds here to the minimal duration of exposure at 32°C that resulted in 100% ZW sex reversal. (From Dournon and Houillon 1985 and unpublished results.)

although in no experiment was 100% sex reversal obtained (Wallace and Wallace 2000).

In *T. c. carnifex*, a still longer exposure time, starting before the first feeding larval stage, was required to obtain sex reversal at low temperature (13°C), suggesting that the TSP begins somewhat earlier than in *T. c. cristatus* (Wallace et al. 1999).

Altogether, the experiments in urodelan species show important individual variations in response to thermal treatments; however, in *P. waltl* no significant variation was observed between progenies (Dournon and Houillon 1985).

Comparison with Other Vertebrates: A Common Mechanism for the Action of Temperature?

Thermosensitivity of gonadal sex differentiation has been shown in fish, amphibians, and reptiles. In thermosensitive fish and amphibians, sex chromosomes are homomorphic or present a very discrete heteromorphism. Thus, as shown above in the salamanders, both sex chromosomes have the same size and differ only in their banding patterns *(T. c. cristatus* and *T. c. carnifex)* or the loop patterns in lampbrush chromosomes *(P. poireti* and *P. waltl)*. Variable thermosensitivity between individuals (in fish and amphibians) and between progenies (in some fish species) reflects interactions between temperature and genotype (Conover and Kynard 1981; reviewed in Baroiller and Guiguen 2001). In reptiles, species with temperature-dependent sex determination (TSD) do not display heteromorphic sex chromosomes but present variable responses in incubation at pivotal temperature, potentially reflecting genetic differences between individuals and clutches (Bull 1983; reviewed in Pieau 1996; see also Valenzuela, Chapter 14). It thus appears that in the three classes of vertebrates the action of temperature on gonadal sex differentiation is related with a weak—or the absence of—sex chromosome heteromorphism.

The characteristics of thermosensitivity of gonadal sex differentiation in amphibians appear similar to those in fish, but at first glance differ substantially from those observed in reptiles. In amphibians and most species of fish, there is a wide range of temperature over which the population sex ratios are in agreement with a strict GSD, XX/XY or ZZ/ZW, mechanism (Baroiller and Guiguen 2001; Conover, Chapter 2). Thus, extreme (low or high) temperatures, generally not or rarely encountered in natural habitats, induce sex reversal of one sex. In reptiles, both sexes are obtained within narrower ranges of temperature and male- and female-producing temperatures are normally experienced by the thermosensitive species in the wild (reviewed in Pieau 1996; Chapters 3–6).

In amphibians and fish, gonads appear histologically undifferentiated for most of the thermosensitive period, and begin to differentiate only by the end of this period (reviewed in Baroiller and Guiguen 2001; this chapter). Alternatively, in reptiles gonads are still undifferentiated at the beginning of the TSP but rapidly exhibit histological signs of sexual differentiation; thus, the TSP corresponds to the first steps of gonadal sex differentiation (reviewed in Pieau 1996). However, due to differences in gonadal morphogenesis, the criteria used to distinguish male from female differentiation in fish and amphibians are not the same as those used in reptiles and other amniotes. Considering that germ cells enter the meiotic prophase earlier in the ovary than in the testis, the end of the TSP could correspond to the onset of the meiotic prophase in ovarian differentiation in fish and amphibians as well as in reptiles.

In these animals, a reliable criterion to discriminate between early male or female gonadal differentiation is the stage from which differences in steroid hormone synthesis can be detected. Indeed, estrogens play an important role in differentiation of the ovary. In reptiles, temperature has been expected to be involved in the synthesis of aromatase, the enzyme that converts androgens to estrogens, and different possible mechanisms of the action of temperature have been considered (Pieau 1996; Pieau et al. 2001). Such mechanisms could occur in amphibians. Thus, in *P. waltl*, treatment of larvae with an estrogen led to sex reversal of genotypic males (Gallien 1951), whereas treatment with an aromatase inhibitor led to sex reversal of genotypic females (Chardard and Dournon 1999). Moreover, estrogenic treatment was shown to counteract the masculinizing effect of high temperature (Zaborski 1986). At 20 ± 2°C, aromatase activity in gonad-mesonephric complexes was significantly higher in ZW females than in ZZ males already at stage 52. It then strongly increased in differentiating ovaries whereas it remained low in differentiating testes. At 32°C, aromatase activity in ZW neomales was similar to that in normal ZZ males. The effects of temperature shifts from 20 ± 2°C to 32°C for 48 hours were examined at different stages. Gonadal aromatase activity decreased when larvae were shifted at stage 53 (before the end of the TSP), but not at stages 55 and 56 (after the TSP), strongly suggesting that, as in reptiles, the temperature acts on regulation of aromatase synthesis during the TSP in *P. waltl* (Chardard et al. 1995). In accordance with this hypothesis, the onset of *aromatase* gene transcription has been shown to occur between stages 43 and 44, that is, at the beginning of the TSP (Kuntz et al. 2003). Moreover, the stage from which gonadal aromatase activity is no longer inhibited by heat treatment corresponds to the end of the TSP (Chardard et al. 1995).

The main problems that remain to be resolved are therefore to determine how the *aromatase* gene is regulated in amphibian gonads during the thermosensitive period and how temperature intervenes in this regulation. It can be assumed that, at temperatures normally experienced by the species, sex chromosomes are implicated in the regulation of aromatase synthesis: the Y chromosome could encode a factor leading to repression of aromatase synthesis, whereas the W chromosome could encode a factor leading to activation of aromatase synthesis. Thus, in *T. c. cristatus* and *T. c. carnifex*, which both display a XX/XY mechanism of GSD, it has been postulated that a sex-determining gene on the Y chromosome is indirectly implicated in the reduction of aromatase activity (Wallace and Wallace 2000). Extreme temperatures could act on such determining genes or on other genes directly or indirectly involved in the regulation of aromatase synthesis.

Unfortunately, little is known about the genes involved in amphibian sex determination, although homologues of genes involved in mammalian sex determination have been identified. *Wt1*, a factor implicated in the formation and differentiation of mammalian gonads, was cloned in the newt *Cynops pyrrhogaster:* mRNA expression was shown to be higher in testes than in ovaries in the adult, but was not examined during gonadal sex differentiation (Nakayama et al. 1998). *Sox* (SRY-related HMG box) genes were identified in *Xenopus laevis* (Miyata et al. 1996; Hiraoka et al. 1997) as well as in *P. waltl* (Chardard et al. 1993); they did not show a sex-specific expression pattern. Two *Sox9* mRNA isoforms were isolated in *Rana rugosa* and found to be equally expressed in both differentiating testes and ovaries (Takase et al. 2000), a pattern different from the one found in reptiles, birds, and mammals, in which *Sox9* displays a testicular-specific expression. The expression of three other genes, *Dmrt1, Dax1,* and *Sf1*, was also examined in *R. rugosa*. A *Dmrt1* homologue was isolated from a testis cDNA library, and its expression was examined during gonadal differentiation and in sex-reversed gonads. *Dmrt1* transcripts were detected in differentiating testes but not in differentiating ovaries. Moreover, *Dmrt1* was also clearly expressed in gonads of sex-reversed genotypic (XX) females treated with testosterone, showing that this gene, as in other vertebrates, is implicated in testicular differentiation of *R. rugosa* (Shibata et al. 2002). By contrast, the expression of *Dax1* during gonadal development was shown to increase in testes and to decrease and disappear in ovaries, a pattern opposite to that observed in mammals (Sugita et al. 2001). Additionally, SF1 protein was expressed in both gonads during the first steps of their differentiation (Kawano et al. 2001). A more complete study of SF1 protein expression was performed in *Rana catesbeiana*, a frog whose gonadal sex differentiation is sensitive to heat treatment (Hsü et al. 1971; Table 7.1). Using a quantitative Western blot analysis system, the expression of SF1 protein at the time of sexual differentiation of the gonads was shown to increase in ovaries and decrease in testes (Mayer et al. 2002). This expression pattern is similar to that found in the alligator, *A. mississippiensis*, a reptilian species that exhibits TSD (Western and Sinclair 2001). This result is of particular interest, since SF1, among other roles, is a transcriptional activator of *aromatase* and could play a pivotal role in gonadal sex differentiation (Pieau et al. 2001). However, several other factors are problably implicated in the regulation of *aromatase* gene transcription. While the molecular target of temperature is still unknown, it can be hypothesized that one gene involved in sex determination is thermosensitive, or its expression is dependent on thermosensitive proteins. In mammals, the heat shock protein HSP70 is expected to stabilize a multiprotein transcriptional complex through an interaction with SOX9 (Marshall and Harley 2001) and WT1 (Maheswaran et al. 1998) in the pathway of AMH induction. Heat shock proteins could also interact with steroid hormone pathways, as they are known to bind their receptors. In *P. waltl*, western blot analysis has shown that a high level of HSP70 expression is induced and maintained during all the TSP when larvae are reared at 32°C, suggesting that this protein could play a role in temperature-induced sex reversal (D. Chardard, A. Chesnel, S. Flament, and M. Penrad-Mobayed, unpubl. data).

Conclusion

To date, a very low number of amphibian species (5 out of 4,381 anuran and 4 out of 429 urodelan species) have been examined for the effects of temperature on gonadal sex differentiation. It appears that all of them display a balanced sex ratio at ambient temperatures, reflecting a GSD mechanism of sex determination. Abnormally high or low rearing temperatures of larvae induce distorted sex ratios, but only in *P. waltl* were individuals all of the same sex obtained. Recent genetic and cytogenetic studies carried out in three salamanders have provided evidence of sex reversal. Heat treatment (temperatures of 25–34°C) generally induces sex reversal of genotypic females, whereas cold treatment (temperatures of 12 to 16°C) induces sex reversal of genotypic males. However, in *P. poireti* heat treatment induces sex reversal of genotypic males.

The extreme temperatures used in the laboratory are rarely experienced by amphibian species in their natural habitats. However, a few cases of spontaneous sex reversal, such as XY neofemales in *Hyla japonica*, have been reported (reviewed in Wallace et al. 1999). Moreover, a comparison of the temperatures producing sex reversal in the laboratory to the temperatures found in the wild suggests that natural sex reversal could occur in *T. cristatus* (Wallace and Wallace 2000). Thus, considering that very few species have been examined, thermal sex reversals in amphibians could be more widespread than currently believed.

The mechanism by which temperature acts on gonadal sex differentiation is unknown. However data in amphibians strongly suggest that thermal treatments, directly or indirectly, repress or activate aromatase synthesis and thus estrogen synthesis, as shown in reptiles and fish. A common fundamental mechanism of action of temperature with various modalities would thus exist in the three classes of vertebrates.

Whether temperature sensivity is an ancestral mechanism of gonadal sex differentiation, conserved by reptiles and progressively lost by amphibians, or an additional mechanism to genotypic sex determination (GSD), gained by fish and amphibians and completed in reptiles, is a question that arises from observations of temperature-induced sex reversal. As GSD evolved repeatedly, environmental sex determination has been considered ancestral in vertebrates (Bull 1983). It now remains to be determined if temperature acts on genes controlling GSD in amphibians and reveals a potential TSD-like mechanism similar to that naturally observed in reptiles.

The elucidation of the mechanism of temperature-induced sex reversal in amphibians, as well as in fish, would greatly help our understanding of the evolution of sex-determining mechanisms in vertebrates.

Acknowledgments—We thank Dr. Mireille Dorizzi for her help with the manuscript.

Part 2

Thermal Effects, Ecology, and Interactions

8

MARSHALL D. MCCUE

General Effects of Temperature on Animal Biology

Thermal Limits

Temperature is simply a measure of the average kinetic energy in a body. Although this term is occasionally confused with heat, temperature is but one of three components of the total heat in a body (heat = specific heat × mass × temperature). Since living cells are chiefly composed of water, the body temperatures (T_b) of most organisms are bounded by water's transition temperatures (i.e., melting and boiling points). The T_bs of most active organisms occur within a relatively narrow thermal range (5–50°C); however, several types of thermal specialists (i.e., many bacteria) are able to survive at more extreme temperatures.

The highest body temperatures (T_b) achieved by terrestrial vertebrates are documented in some desert-dwelling lizards that are able to achieve T_bs of approximately 50°C for short periods (Cole 1943; Mayhew 1964). The highest T_bs reported among aquatic vertebrates (approximately 45°C for limited periods) are reported in estuarine fishes (Rajaguru 2002). Active animals such as polar and deep-sea fishes commonly inhabit temperatures ranging between −5 and 0 °C (Montgomery and MacDonald 1990). However, the lowest T_bs among terrestrial vertebrates are found in hibernating ectotherms. Some studies have documented body temperatures of hibernating reptiles and amphibians as low as −10 °C (Packard and Packard 1990, 1993).

In this chapter, the general effects of temperature on animal biology are discussed. The first section reviews thermoregulatory terminology and measurement of thermal sensitivity. Later sections review the effects of temperature on animals on two time scales (acute and chronic exposure). Acute exposure to a specific temperature induces a variety of reversible physiological and biochemical changes in an organism. Chronic exposure to specific temperatures can result in a variety of compensatory effects that may or may not be reversible.

Thermoregulatory Terminology

The distribution of environmental temperatures available to animals, particularly terrestrial animals, is frequently complex and is composed of a continuum of thermal microenvironments (Scott et al. 1982). Furthermore, thermal conditions in these microenvironments are continually changing over time, and all animals are constantly challenged by this thermal heterogeneity. Within this ecological diversity, animals vary widely with regard to physiological thermal sensitivity, behavioral thermal selectivity, and primary modes of thermoregulation.

Many terms categorize animals by their major patterns of thermal sensitivity and thermoregulation (Cowles 1962). It should be noted that these patterns are frequently species specific (Bogert 1949) and are temporally dynamic (Christian and Tracy 1983). The terms *homeotherm* and *heterotherm* (Table 8.1) differentiate between the degrees of thermal fluctuation among animal species. While all animals are het-

Table 8.1 Definitions of Thermoregulatory and Temperature Terms Used in This Chapter

Term	Definition
Acclimation	A compensatory shift in a physiological parameter occurring in the laboratory
Acclimatization	A compensatory shift in a physiological parameter occurring under natural conditions
Activity range	Range of temperatures at which an animal is able to carry out routine activities
Controlled freezing	Allowing selected tissues to freeze in a manner that is not harmful to the whole animal
Ectotherm	An animal that obtains the majority of its body heat from its environment
Endotherm	An animal that obtains the majority of its body heat from metabolic processes
Eurytherm	An animal that can tolerate a wide range of environmental temperatures
Heliotherm	An ectotherm that obtains the majority of its body heat from solar radiation
Heterotherm	An animal that demonstrates spatially or temporally variegated body temperature
Homeotherm	An animal that maintains constant body temperature
Lower lethal temperature	Temperature at which an animal will succumb due of a lack of body heat
Maximum critical temperature	Temperature at which an animal begins to lose physiological control
Mean selected temperature	The mean body temperature an animal voluntarily selects given a thermal gradient
Metabolic scope	Factorial increase in metabolism resulting from increased temperature, activity, etc.
Poikilotherm	An animal whose body temperature fluctuates with that of its environment
Stenotherm	An animal that can tolerate a narrow range of environmental temperatures
Supercooling	Process of allowing the body temperature to drop below its melting point
Thigmotherm	An ectotherm that obtains its body heat through direct contact with a heated surface
Upper lethal temperature	Temperature at which an animal will succumb due of an excess of body heat
Voluntary temperature	Temperatures to which an animal will voluntarily subject itself given a thermal gradient

erothermic to some degree, the term heterotherm does not specify whether thermal variability is spatial or temporal. As a result, the term *poikilotherm* (Table 8.1) is more commonly used in place of heterotherm.

The terms *endothermy* and *ectothermy* differentiate animals according to their principal source of thermal energy (Table 8.1). Because of their precise definitions, these terms replaced the terms *warm-blooded* and *cold-blooded* in scientific literature. Most animals are ectothermic and obtain the majority of their heat from the environment. However, thermal conditions among environments can be enormously variable, and the fluctuations of T_b among ectothermic animals inhabiting these environments vary.

The terms *stenotherm* and *eurytherm* are used to classify the natural breadth of thermal regimes of ectothermic animals into two categories (Table 8.1). Terrestrial ectotherms are frequently eurythermic due to diel and seasonal fluctuations in atmospheric conditions, whereas the thermal inertia in large bodies of water results in stenothermy in many aquatic animals. Many of these stenothermic species can be very sensitive to temperature changes. For example, polar fishes are often so specialized that they quickly succumb to temperatures above 5°C (Di Prisco and Giardina 1996).

Terrestrial animals have two primary modes of absorbing external thermal energy, *thigmothermy* and *heliothermy* (Table 8.1). All ectotherms are thigmothermic to some degree. *Heliothermy* is most frequently employed by animals inhabiting environments that do not offer sufficient opportunities to heat convectively (Navas 1997). Heliothermic lizards are able to maintain T_bs over 15°C despite ambient temperatures (T_a) of 0°C (Van Damme et al. 1990) or achieve T_bs 30°C above T_a (Cole 1943; Pearson 1977). Several terms are also employed to refer to specific and critical thermal limits of an animal (Table 8.1). The values of these limits vary widely among species but follow a general pattern (Figure 8.1).

Thermal Sensitivity

The thermal sensitivity of any biological process can be described by its temperature coefficient (Q_{10}). Thermal sensitivity is calculated using the following equation,

$$Q_{10} = (R_2/R_1)^{[10/(T2 - T1)]}$$

where R is any physiological rate and T is temperature (°C). By definition, a Q_{10} of 1 indicates complete thermal independence, and a Q_{10} of 2 indicates a doubling of a rate over a 10-degree interval. Since Q_{10}s are not always linear responses, it is important to describe the specific thermal range over which Q_{10} is calculated.

Physiological responses typically have Q_{10}s unique to a

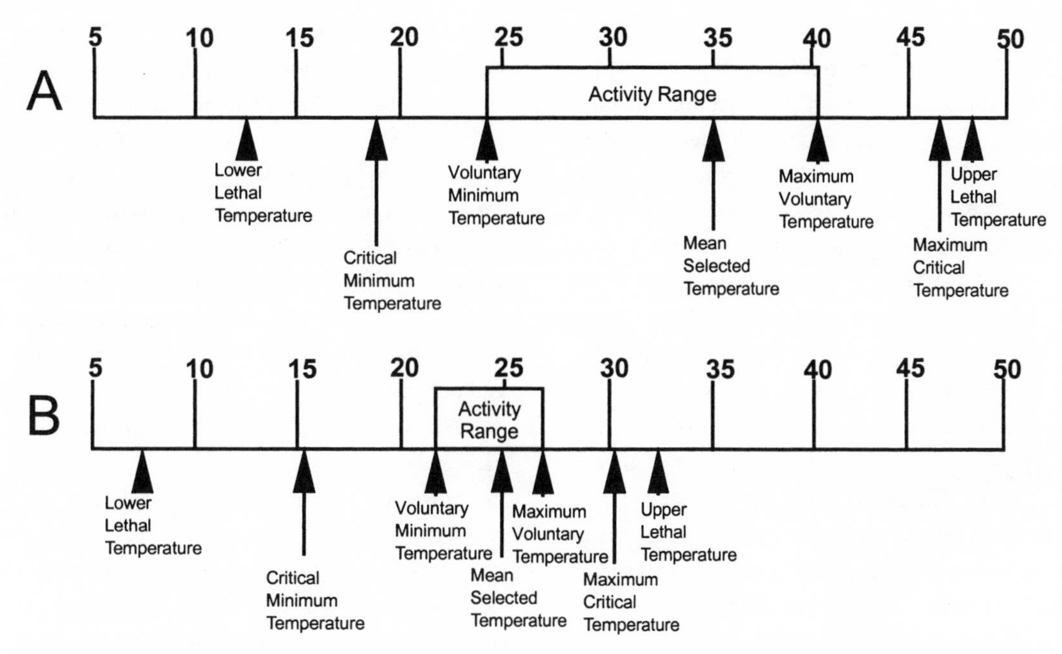

Figure 8.1 Diagrams comparing thermal parameters of two terrestrial animals. (**A**) A eurythermic lizard, *Uta stansburiana*. (**B**) A stenothermic toad, *Bufo marinus*. [Adapted from data presented in Brattstrom (1963, 1965).]

particular range of temperatures. Typically, thermal sensitivities are lowest within the activity range of an animal and increase at extreme temperatures. Different physiological responses may also differ in thermal sensitivity. Such disparities among Q_{10}s of closely related processes can be deleterious. The direction of thermal change may also affect Q_{10}. For example, Birchard and Packard (1997) report that the heart rates of young turtles are much more thermally sensitive when warming than when cooling over an identical thermal range.

Although different physiological systems within an animal may demonstrate unique thermal sensitivities, comparisons of thermal sensitivities among analogous tissues in endotherms and ectotherms reveal only small differences (Berner 1999; Else and Hulbert 1981; Paladino 1985). Although this observation may seem reasonable since the fundamental metabolic pathways of ectotherms and endotherms are identical, the similarities in thermal tolerance between endotherm and ectotherm tissues are generally overlooked for three reasons. (1) Endothermic homeotherms demonstrate overall metabolic rates that are 5–15 times greater than those of similar sized ectotherms (Brand et al. 1991; Pough 1980) due to qualitative and quantitative differences among their mitochondria (Akhmerov 1986; Else and Hulbert 1981). (2) The relative amount of highly metabolically active tissues (i.e., brain, heart, liver, kidney) is greater in endotherms (Brand et al. 1991; Else and Hulbert 1981). (3) Endothermic homeotherms rarely experience (or tolerate) the thermal variability required for testing their sensitivity to broad thermal ranges (Cassuto 1971).

Acute Responses to Thermal Exposure

Acute exposure to particular temperatures elicits a diversity of physiological and behavioral responses in animals, many of which (e.g., locomotion and food processing) are directly connected to fitness. Predicting the acute effects of temperature on these fitness responses can be complex because temperatures influence suites of physiological responses, each of which may have a different thermal sensitivity. For example, the interaction between temperature and locomotion can take many forms and involve several biological parameters including neural transmission, sprint speed, aerobic or anaerobic scope, endurance, and postexercise recovery. The variables most frequently reported (i.e., locomotion, digestion) are the simplest to measure.

Locomotion

Most reports agree that sprint performance in ectothermic animals increases with increasing temperature over a broad thermal range. However studies of frogs report that jumping distance is relatively temperature insensitive (Knowles and Weigl 1990). This phenomenon may be related to the differential recruitment of aerobic and anaerobic muscle tissues under different thermal conditions (Rome 1990); but it may also be explained by the fact that anaerobic metabolism is much less temperature dependent (demonstrating Q_{10}s of approximately 1.2–1.3) than aerobic metabolism (Bennett 1980; Bennett and Licht 1972).

Increased locomotor speed could be related to increased neural transmission at warmer temperatures. This may also help explain behavior in herptiles under predatory threat. It is generally found that warmer individuals flee, whereas cooler individuals tend to remain stationary and become defensive (Crowley and Pietruszka 1983; Hertz et al. 1982). It should be noted that temperature can also influence locomotor parameters other than maximal sprint speed. Depending on the particular species, temperature can influence aerobic scope, endurance, and postexercise recovery, each of which can have significant fitness components (Bennett and John-Alder 1984; Huey 1979).

Food Processing

The rate of food processing is well known to increase with temperature in ectotherms (Avery et al. 1993; Skoczylas 1970; Van Damme et al. 1991; Waldschmidt et al. 1986). However, like locomotion, its rate is dependent on several components including rates of food intake, mechanical digestion, chemical digestion, and gastric clearance. While appetite is generally increased (Greenwald 1974; Lang 1979) and prey-handling times are generally reduced with increased temperature (Avery et al. 1982; Van Damme et al. 1991), much less is known about the specific mechanisms accounting for accelerated food processing with increased temperature (Greenwald and Kanter 1979).

Other Physiological Effects

Many acute effects of temperature can be explained by the effect of temperature on the most fundamental levels of biological organization. Several subcellular and molecular changes immediately follow changes in temperature (Di Prisco and Giardina 1996). Similar trends result from thermal acclimation (Feder 1983). Such changes lead to increased biochemical processing that results an increase in overall metabolic rate. The index of metabolic rates under different conditions (i.e., active vs. resting, warm vs. cool) is termed *metabolic scope* (Table 8.1).

Temperature alone is responsible for a metabolic scope of 100 over the temperature range of 10–40°C in the turtle *Pseudemys scripta* (Gatten 1974). However the thermal effect on metabolic rate can also be confounded with changes in activity level. For example, in the same turtle species, Gatten (1974) found that the metabolic scope between resting rates at 10°C and active rates at 40°C approached 1,000. Bartholomew and Tucker (1963) demonstrated that lizard metabolic rates were least temperature sensitive within their activity range; this is representative of most animals. For example, Q_{10} values for resting metabolism across the activity range generally fall between 2 and 3 (Andrews and Pough 1985; Gatten 1974; John-Alder et al. 1989; Rao and Bullock 1954).

The solubility of most biologically significant gases (i.e., carbon dioxide, oxygen) is generally decreased at warm temperatures, which can lead to reduced capacity for respiratory gas exchange (Johnston et al. 1996). This relationship contrasts to the ubiquitous increase in whole-animal metabolism associated with increased temperatures and may limit oxygen uptake at high temperatures.

The propensity for evaporative water loss (EWL) also increases with temperature, and can become a serious liability for many animals by tapping into water stores (Bently and Schmidt-Nielsen 1966; Shoemaker et al. 1989). Furthermore, animals experiencing osmotic stress typically demonstrate lower tolerances to thermal extremes (Beuchat et al. 1984; Preest and Pough 1989). However, EWL is also the primary means of lowering T_b, and thereby relieving heat stress, in warm terrestrial climates.

Blood parameters (i.e., viscosity, pH, oxygen-carrying capacity) are highly influenced by temperature. Acute increases in blood viscosity may lead to additional cardiac workload and reduced blood flow to peripheral capillary beds (Snyder 1971), which can minimize heat loss at the outer surface of some reptiles. This effect has been reported in a variety of animals (Glidewell et al. 1981; Grigg and Alchin 1976; Smith 1976; Weathers and White 1971).

Maintaining Homeostasis

Although enzymes are well known to denature at extreme temperatures, the catalytic effect of many enzymes begins to deteriorate at temperatures much lower than those responsible for denaturation (Somero 1995; Somero et al. 1996), due to small, heat-induced conformational changes in the enzymes. Enzymes found in animals can vary widely in their degree of thermal stability, and those produced by animals inhabiting extreme habitats (e.g., desert environments) tend to be the most thermally stable (Hochachka and Somero 2002).

Thermal stability can be increased at various levels

of protein organization (Jaenicke 1991; Kumar et al. 2000; Somero 1995), and also in genetic material (McDonald 1999), albeit these changes often reduce their biological activity.

Animals demonstrate complex methods for minimizing the effects of temperature on vital molecules. For example, thermally induced conformational changes in proteins and enzymes can be minimized with the help of specialized chaperone molecules termed stress/shock proteins, or other types of thermally protective proteins (Feder 1999; Hochachka and Somero 2002; Somero 1995).

Cool Temperatures

Acute exposure to low temperatures poses a number of molecular and physiological challenges to animals. Many of these biological responses are simply the converse of the effects caused by high temperatures. For example, at certain temperatures, relatively demanding activities like digestion (Harwood 1979; Skoczylas 1970) or locomotion can be reduced or even halted (Emshwiller and Gleeson 1997; John-Alder et al. 1989; Rome 1990; Van Damme et al. 1991).

Cool temperatures pose several unique obstacles such as behavioral torpor and reduced physiological performance. However, when T_bs approach 0°C, ice formation becomes a threat. This can prove deleterious to living cells by increasing cytosol osmolarity or puncturing cell membranes. Animals use several methods to prevent intracellular ice crystal formation, including the production of antifreezing agents such as glycoproteins and glucose (Churchill and Storey 1992a,b; Storey and Storey 1984), *controlled freezing*, and *supercooling* (Churchill and Storey 1992a; Packard and Packard 1990, 1993; Storey and Storey 1984).

Chronic Responses to Thermal Exposure

Acclimation and Acclimatization

Most animals are able to make some type of compensatory physiological adjustments when exposed to warm or cool temperatures for extended periods. *Acclimation* and *acclimatization* (Table 8.1) occur in varying degrees among animals (Ragland et al. 1981) and are believed to have significant genetic components (Bennett 1990; Wilson and Echternacht 1987). The ability of animals to acclimate has been hypothesized to account for many large-scale patterns in distribution of animals (Brattstrom 1968; Hutchison 1961; John-Alder et al. 1989; Navas 1996; Wilson and Echternacht 1987).

Many amphibians are reported to be incapable of thermal acclimation in some aspects of biological performance such as locomotion (Else and Bennett 1987; Knowles and Weigl 1990; Rome 1983; Whitehead et al. 1989). However, since many locomotor activities are primarily powered through anaerobic metabolism in amphibians (Bennett and Licht 1974), thermal acclimation of aerobic processes may be widespread among amphibians. For example, thermal acclimation influenced aerobic, but not anaerobic performance in salamanders (Feder 1978), and differential thermal exposure induces differential levels of aerobic enzyme expression in toads and salamanders (Feder 1983).

Thermal acclimation is usually quantified by comparing the frequency or velocity of a physiological character at two acclimation temperatures. Depending on the species and the temperature, the degree of compensation typically follows a continuum from complete to partial; however, acclimatory responses may involve overcompensation or reverse compensation (Figure 8.2a). The time course of acclimatory responses is rarely linear and tends to occur at a logarithmic rate. Furthermore, thermal acclimation to warm temperatures has been found to occur faster in salamanders than acclimation to cool temperatures (Hutchison 1961).

Symptoms of thermal acclimation can affect animals on several levels of biological organization. Animals acclimated to a specific thermal regime typically exhibit a corresponding shift in thermal preferences (Crawshaw 1984; Geiser et al. 1992; Murrish and Vance 1968; O'Steen 1998). Critical thermal limits are also frequently influenced by acclimation (Figure 8.2b). One of the best-studied symptoms of thermal acclimation is the compensatory shift in resting metabolic rate. Generally, resting metabolic rates, measured at a given body temperature, are elevated in cool acclimated species and reduced for warm acclimated individuals. However, the use of standard metabolic rate as an effective measure of thermal acclimation has been questioned, as it is simply an ultimate product of subordinate thermally sensitive processes (Clarke 1991).

Blood properties (i.e., viscosity, pH) are among the most plastic characters with regard to thermal acclimation. To counter chronic changes in viscosity, hematocrit is often reduced in warm-acclimated and increased in cool-acclimated animals (Snyder 1971). Changes in muscle structure are also influenced by chronic thermal exposure. These phenomena are well studied, particularly in fishes (Johnston et al. 1996).

The mechanisms that underlie thermal acclimation are limited in number, and they most frequently involve enzymatic changes. In short, these mechanisms can modulate either the "quantity" or "quality" of catalytic enzymes. The former method is simpler and involves up- or downregula-

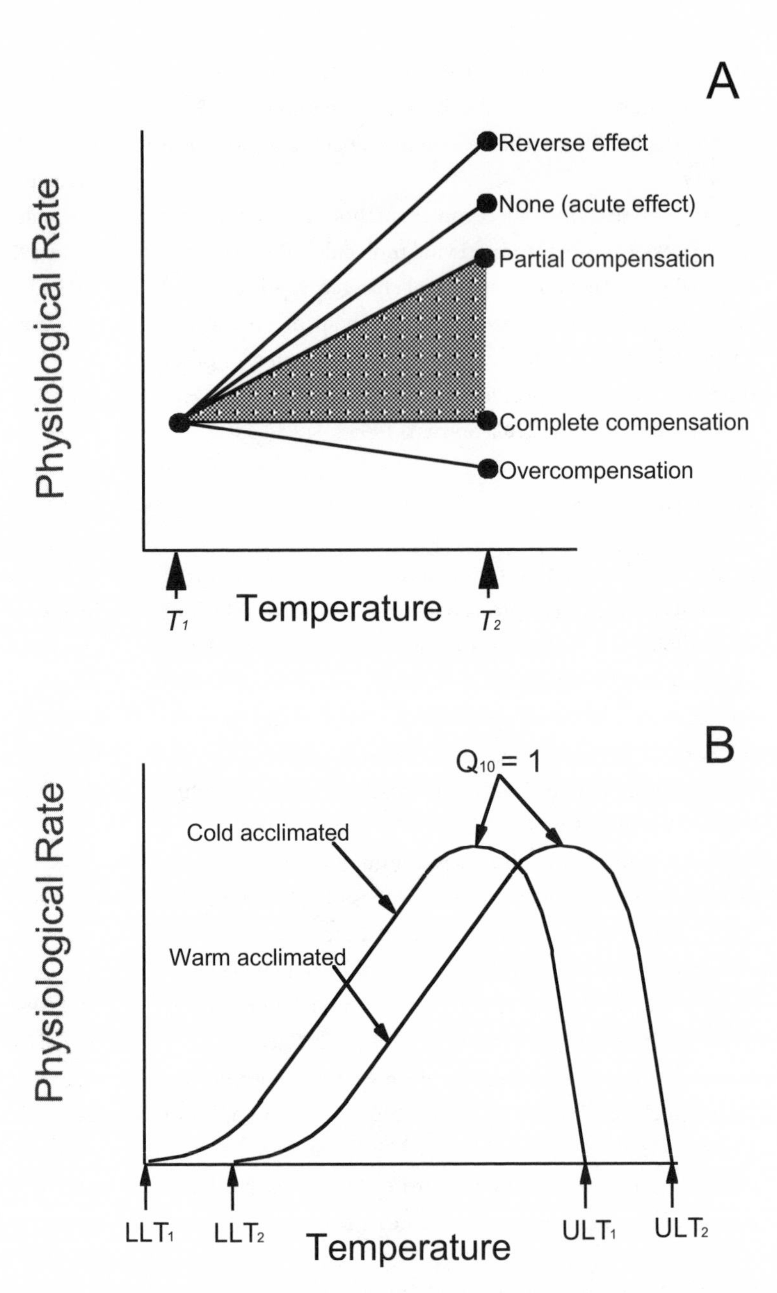

Figure 8.2 Diagrams illustrating the effects of thermal acclimation on a rate processes. (**A**) Illustrates the variety of physiological responses of any parameter to thermal acclimation. Although most acclimatory responses are compensatory and fall into the shaded section, reverse compensation and overcompensation are occasionally reported. (**B**) Illustrates compensatory shifts in thermal tolerances. In this case, lower and upper lethal temperatures (LLT, ULT) are shifted following thermal acclimation. Also, notice that the zone of thermal insensitivity (and presumably thermal preference) is shifted.

tion of genes coding for specific enzymes. The latter method involves changing the relative expression of specific isozymes that have differential thermal sensitivity. Variations in the quantity and quality of enzymes are believed to play a significant role in determining thermal breadth and influencing biogeography (Lin and Somero 1995; Tsugawa 1980).

Temperature, Development, and Phenotype

Many environmental variables affect development in animals. Such external cues may cause developmental switches via differential gene expression in the developing animals (Huey and Berrigan 1996). The resulting alterations in phenotype are termed *polyphenisms*. Polyphenisms are a specific type of phenotypic modulation involving individuals that express differential developmental modifications. It should be noted that these differences are distinct from polymorphisms in that the latter typically involve genetic differences. Although many factors cause polyphenisms, the following section focuses on thermally induced phenotypic plasticity in animals (see also Georges et al., Chapter 9; Rhen and Lang, Chapter 10).

Thermally induced polyphenisms are most commonly observed in oviparous animals for several reasons. First, oviparous species are more likely to experience broad ranging thermal variability during early development than viviparous or ovoviviparous species (Deeming and Ferguson 1991b). Second, the logistics involved in subjecting developing young to a specific thermal regime is far simpler in these species than in live-bearers. Finally, in oviparous species, the effects of a changing thermal environment on the developing embryo can be clearly differentiated from maternal effects. Because of this, scientific literature describing the effects of temperature focuses almost exclusively on oviparous animals.

Embryonic Stages

The most frequently documented developmental effect of temperature is on incubation period. In general, rates of embryonic development increase with temperature, and warmer temperatures are reported to reduce incubation period in several animals (Deeming and Ferguson 1991b; Leshem et al. 1991; Shine et al. 1997a). One study documents an incubation period reduced by a factor of six in an amphibian (Volpe 1957).

The large number of observations of animals maintaining elevated T_b while gravid further demonstrates the importance of increased temperature during development. This behavior is well documented in snakes (Graves and Duvall 1993; Sievert and Andreadis 1999). Even species that do not demonstrate increased mean T_bs while gravid appear to thermoregulate more precisely (Beuchat 1986; Osgood 1970). One lizard species has been reported to reduce its T_b when gravid (Beuchat and Ellner 1987). This observation illustrates the potential tradeoff between the optimal developmental temperatures and those preferred by adults. The importance of specific thermal requirements during early development can also be seen in female pythons. These animals incur tremendous costs while "shivering" to thermally provision their developing young; this behavior can increase their metabolic rates by 20-fold, even at room temperature (Van Mierop and Barnard 1978).

As documented elsewhere in this volume (see Chapters 2–7 and 9), sex is thermally dependent in a variety of vertebrates. However, other less dramatic changes in animal biology can be influenced by developmental temperature (see also Rhen and Lang, Chapter 10). Developmental temperature has been linked to thermal preference in turtles (O'Steen 1998), relative tail length in lizards (Shine et al. 1997a), and swimming behavior in fishes (Johnston et al. 2001). The temporal windows where developmental events are most likely to be influenced are termed thermosensitive periods (TSPs) (Lang and Andrews 1994). The most drastic of these developmental periods occur in the earliest embryonic stages (Deeming and Ferguson 1991b). Thermosensitive periods are unique to each aspect of development. During these periods, temperature alters patterns of tissue differentiation (Johnston et al. 1996). As a result, it may be important to minimize extraneous thermal variability during developmental processes.

Juvenile Stages

The scientific literature is replete with examples of thermally expedited growth rates. However, growth rate (usually measured by change in dry mass) is not a single variable; it is influenced by innumerable behavioral, physiological, and morphological variables that may be individually influenced by temperature. For example, expedited growth at increased temperatures may be linked to increased appetite (Greenwald 1974; Lang 1979), increased prey capture and handling ability (Avery et al. 1982; Van Damme et al. 1991), and/or increased foraging efficiency (Avery et al. 1982; Ayers and Shine 1997; Greenwald 1974). Nevertheless, warm temperatures are most often reported to be positively correlated with growth rates (Brett 1971; Elliott 1982; Lillywhite et al. 1973; Mrosovsky 1980; Van Damme et al. 1991). It has been noted that when "normal" developmental rates are not concomitant with "normal" growth rates, juveniles animals may become large and underdeveloped or undergo arrested development and remain small (Burger 1990; Johnston et al. 1996).

It should be noted that warmer temperatures do not always lead to increased growth rates. In some cases warm temperatures increase metabolic costs by an amount actually sufficient to retard growth (Porter and Tracy 1974). For example, O'Steen (1998) and Spencer et al. (2001) found that cool-incubated turtles compensated by growing faster than warm-incubated turtles, and Rhen and Lang (1995) found that warm-incubated lizards grew more slowly. Letcher and Bengtson (1993) and Johnston et al. (2001) report that fish given equivalent rations grow slower when maintained at warmer temperatures. However, the latter examples are likely associated with increased swimming behavior that also occurs at warmer temperatures (Crawshaw 1984).

Summary

Temperature influences every level of biological organization, from molecular interactions to large-scale evolutionary patterns. The effects of temperature on animals can best be seen on two temporal scales (acute and chronic).

Acute responses involve the immediate effects of temperature on cellular structures and biochemical processes. As a result, most animals make some attempt (often behaviorally) to avoid extreme temperatures and minimize acute changes in T_b. Acclimatory responses may take hours to weeks to occur, and they generally involve changes in tissue composition and enzyme expression. Animals vary widely in their ability to thermoacclimate, and the ability to compensate to specific temperatures is frequently correlated with behavioral and biogeographic patterns.

Acknowledgments—Thanks to A. F. Bennett and T. J. Bradley for helpful comments on this manuscript. Support provided by NSF grant IBN 0091308 awarded to A. F. Bennett and J. W. Hicks and NSF-GRF awarded to M.D. McCue.

9

ARTHUR GEORGES, SEAN DOODY,
KERRY BEGGS, AND JEANNE YOUNG

Thermal Models of TSD under Laboratory and Field Conditions

Recent studies have demonstrated a remarkable range of interactions between environmental conditions and developmental attributes and outcomes in reptilian eggs. Rate of embryonic development and length of incubation period (Ewert 1985), yolk reserves remaining at hatching (Allsteadt and Lang 1995a), hatchling size and morphology (Osgood 1978), coloration (Murray et al. 1990; Etchberger et al. 1993), posthatching behavior (Lang 1987; Burger 1991; Janzen 1993; Shine and Harlow 1996), posthatching growth rates (Joanen et al. 1987; Rhen and Lang 1995), and offspring sex (Bull 1980; Ewert and Nelson 1991; Janzen and Paukstis 1991a) may all be directly influenced by the incubation environment. These results have been established primarily in laboratory experiments using constant temperatures. Much less focus has been placed on reproducing, in the laboratory, the thermal regimes that prevail in reptile nests (but see Paukstis et al. 1984; Packard et al. 1991; Georges et al. 1994). Yet daily temperature fluctuations, variable weather conditions, seasonal trends, thermal gradients within nests, and stochastic events such as rainfall, which temporarily depress nest temperatures, can all be expected to complicate the relationship between nest temperature and developmental outcomes. Developmental times of insect eggs and larvae may be affected by daily fluctuations in temperature, quite independent of the effects of average temperature (Hagstrum and Hagstrum 1970), and there is at least one instance where this is so for a reptile species (Shine and Harlow 1996). Daily fluctuations in temperature also have an impact on phenotypic attributes, including sex (Georges et al. 1994). In this chapter, the evidence in support of an independent influence of variability in temperature on developmental times and offspring sex ratios in reptiles is reviewed. The consequences for translating the results of laboratory experiments into a field context, the context in which sex-determination traits evolved, are explored. In particular, the potential of degree-hour approaches for predicting offspring sex ratios is explored, and these approaches are extended to cover cases where there is a nonlinear response of developmental rate to changes in incubation temperature. An improved algorithm for calculating the daily constant-temperature equivalent (CTE) for a natural nest is presented.

Key Challenges

On the basis of early laboratory work, reptile nests with mean daily temperatures above the temperature-dependent sex determination (TSD) pivotal temperature, established in the laboratory, were expected to yield one sex, and nests with mean temperatures below the threshold were expected to yield the other. Results of early field studies were in broad agreement with laboratory studies—females typically emerge from hot, exposed turtle nests, males from cool, shaded nests (Bull and Vogt 1979; Morreale et al. 1982;

Wilhoft et al. 1983) —but it was soon clear that mean nest temperature was not the best predictor of hatchling sex. For example, predominantly female hatchlings of the turtle *Emys orbicularis* emerged from nests that had a longer daily exposure to temperatures below than to temperatures above the pivotal temperature of 28.5°C (Pieau 1982), the opposite of what was expected. The mean temperature in these nests was considerably lower than the pivotal temperature, and on this basis alone should have produced only male hatchlings. In another study, hatchling sex ratios of *Chrysemys picta* were most closely related to time spent between 20.0 and 27.5°C, the upper and lower threshold temperatures, and not to mean temperature in natural nests (Schwartzkopf and Brooks 1985). In yet another study, both mean temperature and variance in temperature were required to account for sex ratio differences among nests of map turtles in the genus *Graptemys* (Bull 1985). A single mean nest temperature was inadequate as a threshold for natural nests, because the mean temperature that best discriminated between male and female nests decreased as temperatures fluctuated more widely each day.

By way of explanation, several authors have noted that because embryonic developmental rates are greater at higher temperatures than at lower temperatures (within limits), more development will occur at temperatures above the mean than below it (Bull and Vogt 1981; Pieau 1982; Mrosovsky et al. 1984a; Bull 1985). An embryo incubating under a daily sinusoidal cycle of temperature will spend 50% of its time at temperatures above the mean, but much more than 50% of development will occur during that time. It seems that the outcome of sexual differentiation depends more on the relative proportion of development taking place above and below the pivotal temperature than on the relative time spent above and below the pivotal temperature.

The first key challenge for those working on reptile sex determination in field nests was to understand precisely how mean temperature and daily variability in temperature interact to influence sexual outcomes. The relationship between temperature and sex in field nests required a concise formulation in order to translate the results of constant-temperature experiments in the laboratory to a field context.

The second key challenge derives from the need to model the relationship between developmental rate and incubation temperature so as to be able to estimate the period in field nests when temperature exerts its influence on sex. Sex is irreversibly influenced by temperature only during a thermosensitive period (TSP), typically the middle third of incubation (Yntema 1979; Bull and Vogt 1981; Pieau and Dorizzi 1981; Yntema and Mrosovsky 1982; Ferguson and Joanen 1983), and it is defined in terms of embryonic stages (Wibbels et al. 1991b). This TSP is readily identified under constant conditions, because it coincides with the middle third of incubation in both time and embryonic stage of development. However, when nest temperatures are subject to wide diel fluctuations, seasonal shifts, and abrupt changes brought about by rainfall, the middle third of development, the progression of embryonic stage, and duration of incubation become uncoupled. Estimating the middle third of incubation in terms of embryonic stage becomes problematic in field nests. Eggs have been sampled and embryos examined, either destructively (Schwartzkopf and Brooks 1985) or by candling (Beggs et al. 2000), to overcome these problems, but this may conflict with study objectives when clutch sizes are small or the study species is of conservation concern. If we are to either use temperature traces alone to estimate the TSP noninvasively or use incubation duration as a surrogate for estimating offspring sex ratios (Marcovaldi et al. 1997; Mrosovsky et al. 1999), then we require detailed knowledge of the relationship between developmental rate and incubation temperature for all temperatures experienced within the natural nests.

Obtaining such detailed knowledge has long been a focus of study in entomology (Wagner et al. 1984; Liu et al. 1995), although there are conflicting reports on the effects of fluctuating temperatures on insect development (Eubank et al. 1973). Similar conflicting results are available in the more limited literature on reptile development under fluctuating regimes. For example, Georges et al. (1994) found that fluctuations of up to ± 8°C about a mean of 26°C had no effect on incubation period for the marine turtle *Caretta caretta*, whereas fluctuations of ± 9.75°C about a mean of 23°C substantially reduced the incubation period of the Alpine skink *Bassiana duperreyi* compared to that at a constant 23°C (Shine and Harlow 1996).

The third key challenge faced by those working on sex determination in the field is to predict the likely sex ratios that result from the thermal regime experienced during the TSP, once it has been identified. Temperatures vary during the TSP on two scales—there is periodic variation on a daily scale driven by the cycle of day and night, and these periodic fluctuations track aperiodic variations on a broader temporal scale as the TSP progresses. Variation on the broader scale is caused by chance rainfall events, air temperature variation driven by weather, and seasonal trends in temperature. The daily periodic variation inflates the effective nest temperatures (Georges 1989; Georges et al. 1994) and both sources of variation can cause temperatures to move between the male-producing and female-producing conditions during the TSP, which complicates prediction of sex rations from nest temperature measurements. The influence of daily temperature fluctuations (Georges et al. 1994) and the influence of variations on a broader time scale during the TSP (Valenzuela 2001b) may well need to be considered independently.

So there are three key practical issues for the field biologist wishing to draw upon the extensive research on TSD in the laboratory. How do we accommodate the interaction between mean nest temperature and daily variation in temperature, both of which are known to influence incubation duration and sexual outcomes? How do we noninvasively identify the period of incubation during which the embryos are influenced by the thermal environment of the nest, which in the field no longer corresponds to the middle third of development in time? Finally, how do we predict the outcome of sexual differentiation under the complex thermal conditions during the TSP, which involve aperiodic variation overlaid by periodic daily fluctuations, especially when conditions involve both the male and female domains?

Daily Temperature Variation

The Degree-Hour Approach

Early developmental models, based on linearity in the response of developmental rate to changes in temperature and no hysteresis in the action of temperature on developmental rate, were coupled with the temperature summation rule of de Candolle (1855) and Reibisch (1902) to develop the notion of degree-hours widely used in the applied biological sciences under a variety of names (cumulative temperature units, time·temperature equivalents, thermal units, cumulative heat units). Under a degree-hour model, developmental rate increases linearly as temperature (T) increases from a developmental zero (T_0). No development occurs when temperatures drop below the developmental zero.

$$\frac{ds}{dt} = A(T - T_0) \qquad \text{for } T > T_0 \qquad [1]$$

$$\frac{ds}{dt} = 0 \qquad \text{for } T \leq T_0$$

where A is the rate of increase. The equation is constrained by the biologically realistic assumption that growth cannot be reversed ($A \geq 0$). This is a degree-hour model because developmental rate is simply proportional to temperature, when temperature is measured with respect to the developmental zero (i.e., $T' = T - T_0$) (Georges et al. 1994). The developmental zero can be estimated by regressing developmental rate against temperature, where developmental rate is obtained directly from embryos of sacrificed eggs incubated at a range of temperatures spanning the TSP (Georges et al. 1994), or it can be estimated by regressing the inverse of incubation period against temperature (Georges 1989; Demuth 2001)

Georges (1989) proposed that under fluctuating temperature regimes, female turtles will be produced if more than half of embryonic development occurs at temperatures above the pivotal temperature each day, and males will be produced if more than half of daily embryonic development occurs below the pivotal temperature. If the development that occurs above the pivotal temperature is the same as that which occurs below it, both sexes will be produced. The key statistic for predicting sex ratios from natural nests under this proposition is not the mean temperature or its variance, but rather the temperature above and below which half of development occurs, calculated on a daily basis. This statistic is referred to as the constant-temperature equivalent (CTE) (Georges et al. 1994). A temperature regime that fluctuates about a stationary mean with constant variance will be equivalent, in terms of phenotypic outcomes such as hatchling sex ratios, to a constant-temperature incubator set at the value of the CTE (Georges et al. 1994). The CTE can also be calculated on a daily basis for an assessment of the likely influence that day will have on offspring sex. The CTE statistic can in theory be used across a range of underlying models of development with temperature, but under linear assumptions, it is convenient to consider the CTE as the temperature above and below which half of the degree-hours occur.

A general approach to calculating the CTE when full temperature traces are available draws upon the observation that the CTE is a form of developmental median. The point temperatures obtained from the data logger are interpolated using a procedure such as a cubic spline to yield temperatures that are spaced at equal but arbitrarily small intervals for the day in question (PROC EXPAND, SAS Institute 1988). The CTE for that day is the median temperature, where the contribution of each temperature to that median is weighted by the corresponding value of developmental rate, obtained from the degree-hour model. This is readily achieved using most statistical packages, such as SAS (SAS Institute 1988).

If the daily temperature trace is modeled with a continuous equation, based perhaps on fewer daily measurements or the maximum and minimum only, then the process of

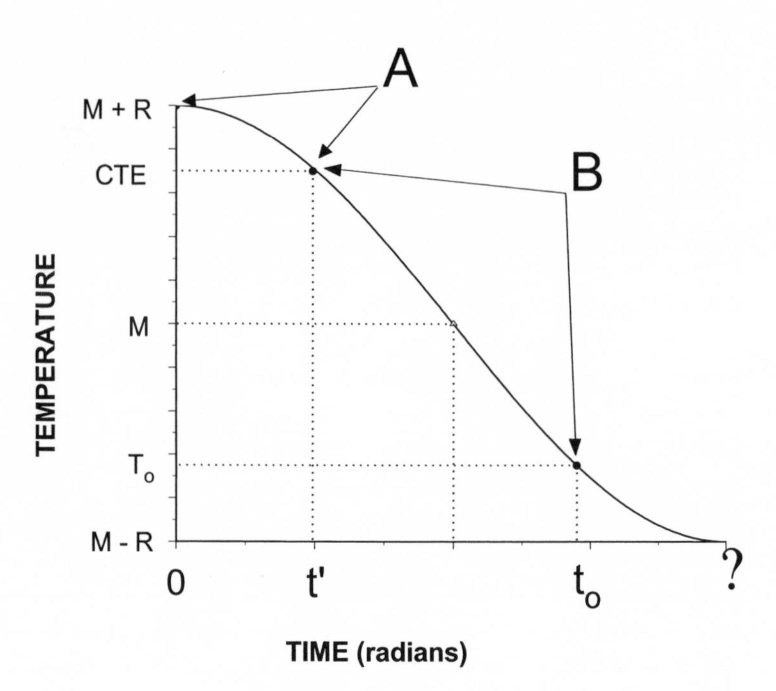

Figure 9.1 Segment of a sinusoidal cycle defining the parameters used in the derivation of the CTE statistic. *M*, mean nest temperature; *R*, amplitude of the daily cycle in temperature; *CTE*, temperature above and below which half of development occurs; t', time at which the *CTE* is achieved (in radians); T_0, temperature at which development ceases; t_0, time at which development ceases (in radians). The *CTE* is obtained by integrating developmental rate along segments A and B of the temperature cycle, setting the two equal, and solving for temperature *T* corresponding to *CTE*.

summation becomes integration. Assuming a daily sinusoidal cycle with mean nest temperature *M* and range 2*R* (Figure 9.1),

$$T = R \cdot Cos(t) + M \qquad [2]$$

it is possible to solve for the temperature above and below which half of development occurs (CTE). The solution can be obtained by integrating developmental rates as they vary along the diel temperature cycle such that

$$\int_0^{t'} \frac{ds}{dt}\, dt = \int_{t'}^{t_0} \frac{ds}{ds}\, dt \qquad [3]$$

where t' is the time at which the CTE is achieved and t_0 is the time at which temperatures drop to the developmental zero T_0, with time measured from the daily peak in temperature (Figure 9.1). A solution to this integral equation is given by

$$\begin{aligned} CTE &= R \cdot Cos(t') + M && R \geq 0 \\ t' &= \frac{t_0}{2} + \frac{R}{2(M - T_0)} Sin(t_0) - \frac{R}{M - T_0} Sin(t') && \\ t_0 &= Cos^{-1}\left[\frac{T_0 - M}{R}\right] && \text{for } T_0 > M - R \\ t_0 &= \pi && \text{for } T_0 \leq M - R \end{aligned} \qquad [4]$$

where the second equation above is solved by application of the standard bisection method. This method relies on the intermediate value theorem, which states that any continuous function that takes on a negative value at one end of an interval and a positive value at the other must pass through zero at least once in between (Hille 1964, 155). The function $f(t')$ is continuous over the interval and

$$f(0) = \frac{t_0}{2} + \frac{R}{2(M - T_0)} Sin(t_0) = -f(t_0) \qquad [5]$$

so when the function evaluates positive at one end of the interval, it evaluates as negative at the other end. Furthermore, if a function is monotonic, decreasing or increasing, or has at most a single stationary point, it will pass through zero once only, that is, the solution will be unique. The function above is monotonic decreasing for $M > T_0$ and monotonic increasing for $M < T_0$ over this interval, provided development occurs at some time during each day, so any solution is unique. Convergence of the bisection method is slow but guaranteed. Note that this method is an improvement on that recommended by Georges (1989), because it converges to a solution for the CTE for all scenarios where temperatures drop below the developmental zero each day. A SAS program for undertaking this analysis is available from the senior author.

Whether we assume a sinusoidal cycle, or measure the daily temperature trajectory directly, the CTE can be calculated for each day, and trends in the CTE during the TSP can be examined for an assessment on the likely outcome of sexual differentiation. For turtles with a single pivotal temperature (TSD Ia), female hatchlings will emerge if the

CTE consistently exceeds the pivotal temperature for sex determination during the TSP, and male hatchlings will emerge if the CTE is consistently less than the pivotal temperature. Both sexes will emerge if the CTE consistently falls on the pivotal temperature or *may* emerge if the CTE moves through or oscillates about the pivotal temperature during the themosensitive period. This assessment would typically be done for temperature traces from the top, core, and bottom of nests, as daily temperature fluctuations are dampened with depth and can vary greatly in magnitude within single nests (Georges 1992; Demuth 2001).

Support for the Degree-Hour Model and CTE Statistic

Strong support for the degree-hour model and CTE statistic stemmed from a reanalysis of the data on map turtles in the genus *Graptemys*. Bull (1985) compared the sex ratios of nests that differed with respect to mean temperature and variance in temperature; he found that nests producing females had higher means or higher variances than nests producing males. A straight line with a negative slope best discriminated between male and female nests, in contrast to the vertical line that would be expected if mean temperature alone determined hatchling sex. Reanalysis of these data using the degree-hour model generated a sloping line as the expected boundary between male-producing nests and female-producing nests (CTE on the pivotal temperature) that was remarkably close to the line produced empirically by Bull (Georges 1989). The effect noted by Bull could be explained entirely by the degree-hour model and use of the CTE statistic.

Specific experiments to address the influence of daily variation in temperature on offspring sex were even more convincing. The marine turtle *Caretta caretta* has a pivotal temperature of around 28.5°C, with 100% males produced under constant-temperature regimes of 26 and 27°C (Georges et al. 1994). When eggs of *Caretta caretta* are incubated at a mean temperature of 26°C, but with progressively increasing daily fluctuations in temperature from ± 0°C to ± 8°C, offspring sex ratios moved progressively from 100% male to 100% female (Georges et al. 1994). Daily fluctuations in temperature alone were demonstrated to influence offspring sex, independent of mean temperature. Furthermore, the sex ratios emerging from the fluctuating temperature regimes were in very close agreement with those predicted from the degree-hour model and the CTE statistic.

Other studies have been equivocal in their support for this approach. Souza and Vogt (1994) found that mean temperature in conjunction with variance and the number of hours at the pivotal temperature were the two indices that best described the sex ratio in *Podocnemis unifilis*. They rejected the CTE approach because all CTEs for their nests were found to lie below the pivotal temperature, yet many such nests produced females. Reanalysis of their data (mean and variance for each nest taken from their Figure 1B and reworked as per Georges 1989) with the degree-hour model and CTE shows that at least nine nests, including all nests with a mean temperature greater than 31°C, have a CTE exceeding the pivotal temperature of 32°C. One nest had a CTE of 33.5°C. It is difficult to tell from the paper which of the data points, in their Figure 1B, Souza and Vogt used to assess the various models of sex determination, but it appears that they may have made a mistake in the computation of the CTE.

Valenzuela et al. (1997) regarded the CTE approach as inappropriate for *Podocnemis expansa*. The daily and overall mean temperatures observed in their study were well below the pivotal temperature and should therefore, according to their interpretation of the CTE model, produce only males. Females were produced. In fact, the primary prediction of the CTE approach is that in many cases where the mean temperature is well below the pivotal temperature, females will be produced, which is in qualitative agreement with the outcomes described for *Podocnemis expansa*. Valenzuela et al. (1997) also make the observation that daily variances in temperature are typically not constant in natural nests, which they regard as a violation of the assumptions of the CTE model. However, we would argue that the CTE should be calculated separately for each day of the TSP and that trends in the CTE should be examined in relation to the pivotal temperature in the same way as one would examine trends in mean temperature if that were thought to be influential. There is no need to assume equal variances across days.

The degree-hour and CTE approach appears to have worked well in a study of the effects of constant and fluctuating incubation temperatures on sex determination in the tortoise *Gopherus polyphemus* (Demuth 2001), though the number of natural nests involved was limited. The CTEs correlated well with offpring sex ($R^2 = 0.76$), much better than did mean temperature ($R^2 = 0.139$), and Demuth used CTEs extensively in analyzing and interpreting his results.

We conclude from these observations and experiments that the degree-hour model and the CTE statistic are of value in predicting offspring sex ratios of some species when temperatures fluctuate on a daily cycle, provided we can assume linearity in the response of developmental rate to temperature change. Linearity might seem a rather unreal-

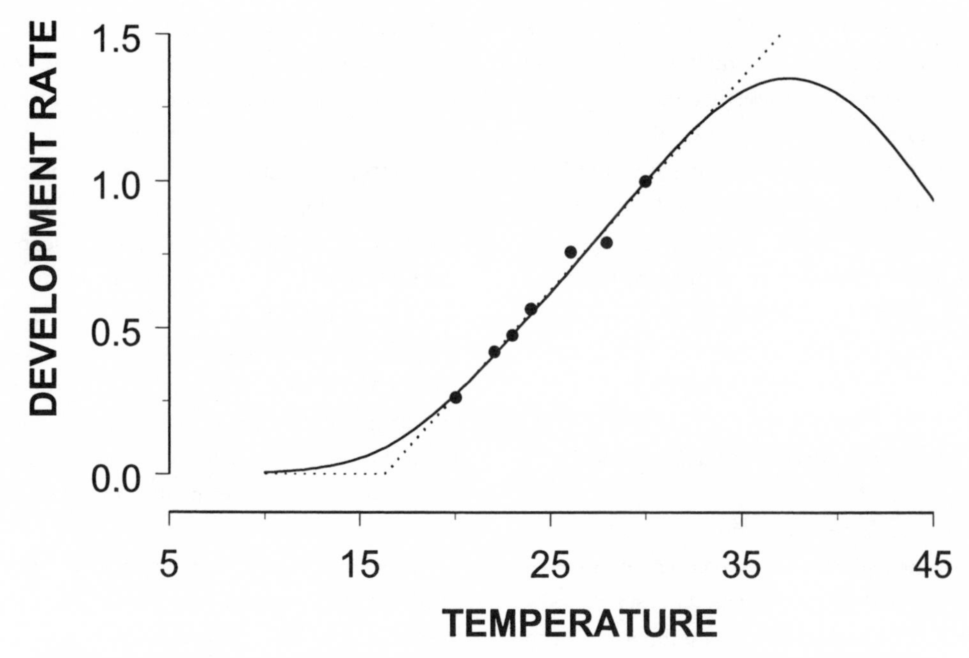

Figure 9.2 The Sharpe-DeMichelle model (solid line) and linear degree-hours model (broken line) applied to developmental rates of the temperate-zone lizard *Bassiana duperreyi*. Data are calculated from Figure 2 of Shine and Harlow (1996). Development rate is incremental change in head width, expressed as a percentage of final hatchling head width, per day. The linear model was fitted by least squares regression ($ds/dt = 0.07251\ T - 1.1818$). The Sharpe-DeMichelle model was optimized at lower temperatures using development rates for 23.0 ± 3.75°C and 23.0 ± 9.75°C (Shine and Harlow 1996). The curvature at higher temperatures is not supported by data and is shown for illustration only. Parameters of the Sharpe-DeMichelle curve are $RHO_{25} = 0.7$, $H_A = 17199.9$, $T_L = 291.47$, $T_H = 311.71$, $H_L = -64401$, and $H_H = 41965$. Note the close agreement between the linear model and the Sharpe-DeMichelle model at intermediate temperatures.

istic assumption, and it certainly will not be true of all species in all circumstances, but there are indications that it is a reasonable assumption for a wide range of natural scenarios. These indications derive from an unlikely source, the study of nonlinear responses of developmental rate to change in temperature.

Why Expect Linearity?

The most widely accepted model of poikilotherm development is that of Sharpe and DeMichele (1977). They extended the work of Eyring (1935), Johnson and Lewin (1946), and Hultin (1955) to formulate a biophysical model that describes the nonlinear response of developmental rate to incubation temperature at both high and low temperatures, as well as a linear response at intermediate temperatures (Figure 9.2). At low temperatures, developmental rate does not decrease linearly to the developmental zero, but rather decelerates to approach zero almost asymptotically. At high temperatures, developmental rate does not increase linearly without limit. Rather, it achieves a maximum, establishing an optimum temperature for development (at least in terms of rates). It then drops precipitously as temperatures rise above the optimum and enzymatic deactivation comes to dominate over the tendency for exponential increase with temperature (Figure 9.2).

Despite its overall curvilinearity, a triumph of the Sharpe-DeMichelle model is its explanation of a strong linear response at intermediate temperatures (Figure 9.2), establishing the validity of the degree-hour concept and the CTE approach in the midtemperature region. At low temperatures, developmental rates are greater than anticipated on the basis of a solely exponential relationship (the Eyring equation) because of progressive activation of enzymes with increasing temperature. At high temperatures, developmental rates are lower than would be expected because of the progressive deactivation of enzymes with increasing temperature. These two compensating effects cause a linearization of the Eyring equation at midrange temperatures (Sharpe and DeMichele 1977).

The parameters of the Sharpe-DeMichelle model are presumably subject to selection, and for many species, an advantage of extending the midrange linear region would be to ensure coincident development of eggs at different depths within a nest. This is because mean temperature varies little with depth (Collis-George et al. 1968, 14), and mean temperature under linear assumptions is the prime determinant of developmental rate. In shallow-nesting species, eggs experience roughly the same mean temperatures, but quite different thermal ranges, each day (Georges 1989; Demuth 2001). In the presence of a nonlinear response of developmental rate to change in temperature, differing daily thermal ranges among eggs would lead to hatching asynchrony. A linear response across the range of temperatures experienced by embryos in natural nests is therefore of likely advantage, and the linearization of the response of developmental rate to midrange temperatures, as described by Sharpe and DeMichelle, would be subject to positive selection. It also means that for many species, and under many circumstances in the field, the degree-hour model and the CTE will yield good estimates of sex ratios for nests.

Nonlinear Models and the CTE Statistic

Notwithstanding the broad applicability of the degree-hour model, there are circumstances where it will fail. Linear relationships apply to good approximation within the limits of constant temperatures that support successful development for some species (Georges et al. 1994; Shine and Harlow 1996), but not all (Muth 1980; Yadava 1980). For these latter species, the degree-hour approach is not appropriate. Also, for many shallow-nesting species, nest temperatures will commonly vary to extremes, well beyond the intermediate linear region of the Sharpe-DeMichelle model, for some period of each day. In many poikilotherms, temperatures that result in embryo death when held constant throughout incubation can readily support development, provided the embryo is exposed to those temperatures for only a limited period each day (Dallwitz and Higgins 1992; Morales-Ramos and Cate 1993; Demuth 2001; Valenzuela 2001b). For example, temperatures in nests of *Carettochelys insculpta* varied from 18 to 45°C, well beyond the range of constant temperatures that will support successful incubation (26–34°C) (Georges and Doody, unpubl. data). These brief daily exposures to more extreme temperatures were tolerated and supported development to successful hatching. Again, because the relationship between developmental rate and temperature cannot be expected to be linear at extremes, and because many embryos are experiencing these extremes briefly each day, the CTE statistic calculated using the degree-hour approach can be expected to perform poorly in predicting the outcome of sexual differentiation for embryos in such nests.

The approach to take under these circumstances involves first estimating the parameters of the nonlinear relationship between instantaneous developmental rate and incubation temperature within and beyond the range of temperatures that support successful development when held constant, then applying summation to estimate the proportion of development that occurs above and below the pivotal temperature each day to estimate the CTE.

A computational formula is available for the Sharpe-DeMichelle model (Schoolfield et al. 1981), but the model is more demanding of data than the simpler degree-hour model. Estimating its parameters will usually require estimates of developmental rate for temperatures outside the range of the constant temperatures that support successful development, that is, beyond the constant-temperature survival thresholds (Georges et al. 2005). The necessary data typically cannot be obtained by constant-temperature experiments alone.

Survival Thresholds

Three temperature zones can be identified with respect to reptile development (Figure 9.3). There is a central zone defined by those constant temperatures that support successful incubation (between T_2 and T_3 of Figure 9.3, often referred to as the critical thermal minimum and the critical thermal maximum, respectively (Ewert 1979)). There are the two absolute lethal limits (T_1 and T_4), beyond which even brief daily exposure to temperature extremes causes embryonic death. There are the sublethal temperature zones ($T_1 < T < T_2$ and $T_3 < T < T_4$) where the embryo can tolerate exposures of short duration each day, and where a positive correlation exists at each temperature between mortality and duration of daily exposure.

The position of these zones in relation to the Sharpe-DeMichele curve or Dallwitz-Higgins (1992) alternative (Figure 9.3) will determine how successful constant-temperature data are in supporting parameter estimates for the models. For example, constant temperatures will support incubation of eggs of the marine turtle *Caretta caretta* (Georges et al. 1994) and the lizard *Bassiana duperreyi* (Shine and Harlow 1996) only in the linear region of the Sharpe-DeMichele model. Data in support of estimates of parameters governing curvilinearity at high and low temperatures cannot be obtained from constant-temperature experiments on these species. In contrast, high-temperature inhibition is clearly evident in constant-temperature experiments conducted on eggs of the lizards *Dipsosaurus dorsalis* (Muth 1980) and *Physignathus lesueurii* (Harlow 2001). At the other extreme, higher developmental rates occur at 22°C in the eggs of the freshwater turtle *Kachuga dhongoka* than would be expected

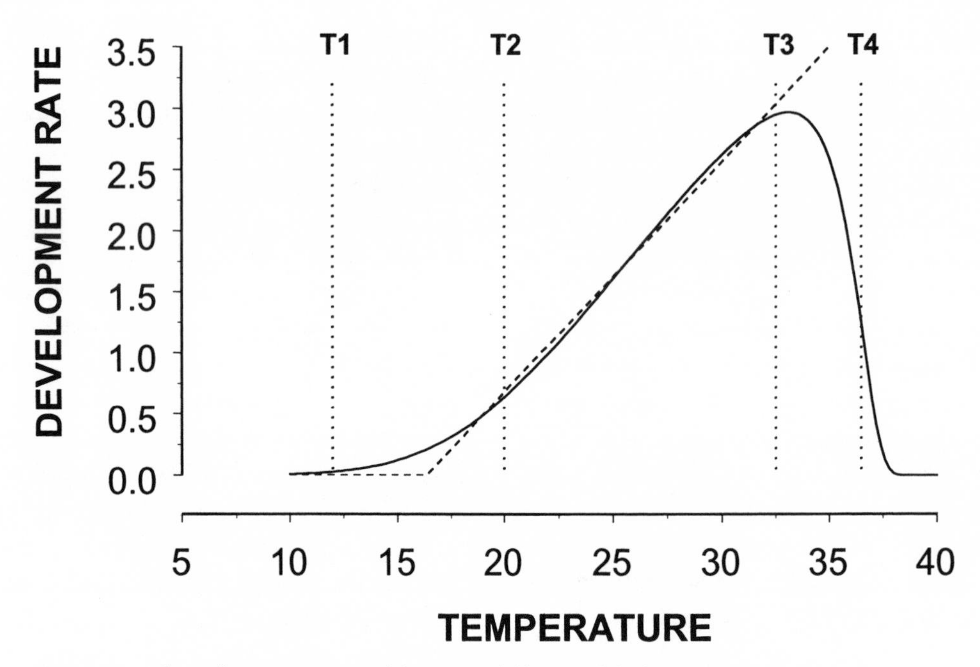

Figure 9.3 The Dallwitz-Higgins model (1992) (solid line) and the linear degree-hours model (broken line) applied to hypothetical data. Developmental rate is incremental change in head width, expressed as a percentage of final hatchling head width, per day. Note the close agreement between the linear model and the Dallwitz-Higgins model at intermediate temperatures, similar to that of the Sharpe-DeMichelle model of Figure 9.2. T_1 and T_4 are the lower and upper absolute lethal limits, outside which even brief exposure causes embryo death or gross abnormality. T_2 and T_3 are the constant-temperature lethal limits, outside which a temperature held constant throughout incubation will cause embryo death or gross abnormality. Temperatures in the sublethal ranges T_1-T_2 and T_3-T_4 will support embryonic development, provided exposure is for a part of each day only, but the duration of exposure that can be tolerated will decline as one moves to extremes.

from consideration of the linear relationship between developmental rate and temperature in the range 25–33°C (reworked from the data of Yadava 1980). For these species, estimation of high or low temperature parameters (respectively) of the curvilinear models may be possible from constant-temperature experiments alone.

In most cases, parameters of the Sharpe-DeMichelle model will have to be estimated by the method of Dallwitz and Higgins (1992). They use data from both constant and fluctuating temperature regimes to estimate the developmental rate function, across a greater range of temperatures than just those constant temperatures that support development. Initial values for the model's parameters are selected, and developmental rate is integrated over each of several temperature traces, generated in the laboratory or taken from field nests, to obtain estimates of the development that should accompany each trace. Each such estimate is then matched to observed development, and the parameters of the model are adjusted iteratively using nonlinear techniques (e.g., PROC NLIN of SAS) to minimize the squared deviations between the two. In this way, we get estimates of the instantaneous developmental rate for temperatures well outside the range of temperatures that support development when held constant.

SAS programs are available in the literature to fit the full Sharpe-DeMichelle model from constant-temperature data (Wagner et al. 1984), or to fit restricted, four-parameter submodels using fluctuating temperature data (Hagstrum and Milliken 1991). The restricted four-parameter models are useful where high-temperature inhibition or low-temperature nonlinearity occur, but not both. Dallwitz and Higgins (1992) provide a FORTRAN program to fit an alternative five-parameter model from uncontrolled temperature traces and accompanying incubation periods. A SAS program for fitting the full six-parameter Sharpe-DeMichelle model using constant and fluctuating temperature traces matched to known developmental increments has not been published, but it is available from the senior author.

Once the Sharpe-DeMichelle function has been determined for the species in question, the CTE can be calculated for each day of incubation as with the degree-hour model. It is the daily median temperature, where temperature is weighted by its corresponding developmental rate under the Sharpe-DeMichelle or Dallwitz-Higgins models. A program for computing the daily CTE for a given temperature trace based on these underlying models is also available from the senior author.

Identifying the Thermosensitive Period

Progress of incubation in a natural nest can be estimated by integrating developmental rate along the temperature trace recorded at the same depth as and in the vicinity of the egg. In practice, temperatures taken sequentially using a data logger are interpolated using cubic splines (PROC EXPAND, SAS Institute 1988) to yield temperatures that are evenly spaced at equal but arbitrarily small intervals. The amount of development, S, to occur over time increments $t = 1$ to $t = t'$ is then calculated as

$$S = \sum_{t=1}^{t=t'} \frac{ds}{dt} \Delta t \qquad [6]$$

where ds/dt is developmental rate as a function of temperature, as specified by the linear degree-hour, Sharpe-DeMichelle, or Dallwitz-Higgins models described above. It is convenient to express development (S) as the percentage of total development and developmental rate as a percentage of total development per day. The TSP will begin when $S = 33.3$ and end when $S = 66.7$, and incubation will be complete when $S = 100$. Note that the TSP will correspond to the middle third of incubation in terms of development (Figure 9.4). It will not correspond to the middle third of incubation in terms of timing or duration.

It is important to note that daily fluctuations in temperature should have no influence on developmental times, over and above that of the daily mean, provided developmental rate and temperature are approximately linearly related over the range of temperatures experienced by the eggs (Georges et al. 1994). Thus, in cases where the degree-hour model is appropriate and temperatures remain above the developmental zero, mean temperature and duration of incubation can be used to estimate the progress of development.

Predicting Sex

Once the TSP is identified, the CTE statistic provides a major advance over previous approaches to predicting offspring sex ratios from temperature traces because it promises to unambiguously predict sex in cases where the CTE remains below or above the pivotal temperature for all of the TSP. This may well be the case for most nests. Where it will potentially fail is when the CTE crosses the pivotal temperature during the TSP. Mixed-sex nests will not necessarily result when this occurs because switch experiments indicate that irreversible masculinization may occur from a pulse of cool temperatures of only a few days, even if the majority of the TSP is spent in the female-producing domain (Bull and Vogt 1981). How much of the TSP must fall within the female-producing domain to produce females and how much must fall in the male-producing domain to produce males is a question that has not been resolved satisfactorily. Are the embryos presumptive females until a cool pulse of sufficient duration during the TSP switches their developmental trajectory to male? Does the efficacy of a cool pulse depend on the magnitude of the temperature shift as well as its duration? What about two pulses—are their effects cumulative? If so, can degree-hour approaches applied across the TSP provide good predictions of offspring sex? How can the CTE modeling, which applies to individual days, be factored into these calculations?

A number of authors have used correlative approaches to gain insight to the factors other than mean temperature that operate during the TSP to influence offspring sex ratios. Hatchling sex ratios in natural nests of *Chrysemys picta* were most closely correlated to time spent between 20.0°C and 27.5°C, the upper and lower pivotal temperatures (Schwartzkopf and Brooks 1985). A linear regression involving mean temperature and variance best predicted sex ratios of *Podocnemis unifilis* (Souza and Vogt 1994). However, the most thorough field study of this question was undertaken on *Podocnemis expansa* (Valenzuela et al. 1997; Valenzuela 2001b). Valenzuela (2001b) developed a statistical model to account for the effects of heterogenous daily fluctuations of natural nest temperatures on development and sex ratios using such indices as mean temperature, degree-hours (cumulative temperature units), and time spent below the low constant-temperature survival threshold. Degree-hours and time spent below the survival threshold explained significant variations in sex ratio when laboratory and field data were combined (Valenzuela 2001b). In a previous study in which only partial thermal profiles were available, the number of hours greater than 31°C, during a two-day period only, provided the best prediction of offspring sex ratios in field nests (Valenzuela et al. 1997). However, this statistical result had little biological relevance as the two-day period may not have been representative of a single developmental stage or the entire TSP (Valenzuela et al. 1997; Valenzuela 2001b).

These correlative studies establish the deficiency of

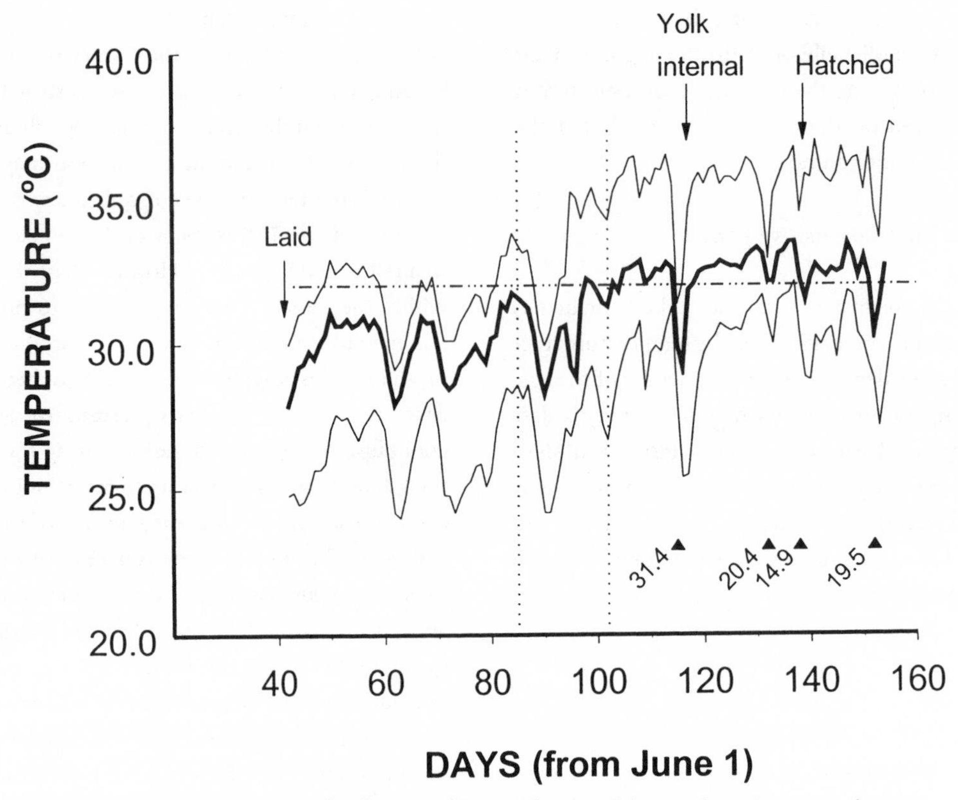

Figure 9.4 A temperature trace for the core of a nest of *Carettochelys insculpta* (Ci98–001) showing traces for the maximum and minimum daily temperatures (thin lines) and the constant-temperature equivalent (CTE) (thick line) for the Sharpe-DeMichelle model. Time is in days from June 1, 1998, and temperature is in °C. The known points of laying and hatching are marked. The threshold for sex determination (32°C; Young et al., in press) is shown as a horizontal broken line, and the thermosensitive period lies between the two vertical broken lines. Note that the thermosensitive period does not correspond to the middle third of incubation, either in position or duration, owing to the nonstationary trend in temperatures with season. The point of yolk internalization is estimated from the model, and hatching is presumed to be possible after that point. Heavy rainfall or inundation stimulates hatching in this species (Georges 1992; Webb et al. 1986), and four rainfall events of 31.4, 20.4, 14.9, and 19.5 mm are shown. The nest produced male offspring.

mean temperature alone in predicting offspring sex. Those combining a temperature with a duration of exposure, such as the model derived by Valenzuela, provide indirect support for a degree-hour approach, provide a potentially useful approach for predicting offspring sex ratios averaged over a number of nests, and provide hope that accurate prediction of offspring sex from particular nests with a thermal regime that spans the pivotal temperature may one day be possible.

Discussion

There are a number of practical implications stemming from knowledge that it is the proportion of development at a temperature that is important for sex determination and not simply duration of exposure to a temperature, implications that become clearer from the modeling presented above. While the pivotal temperature for sex determination established in the laboratory is a fairly well-defined concept, its definition in a field context is less well defined. The boundary between male-producing and female-producing conditions depends on both mean nest temperature and the magnitude of daily fluctuations in temperature (Bull 1985; Georges et al. 1994). The CTE provides a means of evaluating a nest against the pivotal temperature established under constant conditions in the laboratory, whether it is calculated under the degree-hour model (Georges et al. 1994) or the more complex nonlinear scenarios (as yet untested). The CTE, as a single value extracted from the complex daily thermal regime of a natural nest, promises to be of considerable practical utility, especially if coupled with the degree-hour approaches applied to the aperiodic variation in temperature across the TSP (Valenzuela 2001b).

Regimes for monitoring temperatures in reptile nests for study of sex determination will need to consider daily variation in temperature if the nests are shallow. Spot temperatures each day are of little value. Temperatures monitored only in the core of the nest fail to recognize the contribution of temperature fluctuations on offspring sex, and that there are likely to be strong gradients in the magnitude of those fluctuations with depths above 30 cm (Georges 1992; Demuth 2001). Ideally, each nest should be fitted with three probes, one immediately above the top egg, one in the core of the nest, and one at the bottom of the nest chamber.

The foundation for using incubation period as a surrogate for temperature in predicting offspring sex is weak, especially for species that have shallow nests. Under the linear model describing the relationship between developmental rate and incubation temperature, and probably under all but extreme cases of nonlinearity, incubation period will be closely linked to mean nest temperature. Sex, on the other hand, will be influenced by both mean temperature and variability in temperature. In this sense, the average developmental rate reflected by incubation period and the thermal influences important for sex are uncoupled, and incubation period will not necessarily be a good basis for predicting sex ratio, especially in the case of individual nests.

Finally, the approaches above are of value not only in reconciling the results of laboratory and field studies, but also in reconciling disparate results of laboratory studies where cyclic temperature regimes are applied (Paukstis et al. 1984; Packard et al. 1991; Georges et al. 1994).

Clearly, the relationship between incubation temperature and development, and understanding how the form of this relationship influences the interplay between the thermal environment of a nest and the outcome of sexual differentiation is critical for reconciling laboratory and field data. From the insights derived from the degree-hour approaches, in general, and the formulation of the CTE statistic, in particular, we now understand why mean nest temperature is a poor indicator of hatchling sex ratios. We can understand why indicators such as hours spent above the threshold, hours above 30°C, degree-hours above 31°C, and other related indices perform better than the mean in predicting sex in empirical approaches such as multiple regression, but have a theoretical foundation for preferring the CTE if the assessment is to be made on a day-by-day basis.

The degree-hour approaches and their nonlinear counterparts, coupled with the CTE statistic, provide us with a general framework for integrating experiments using constant temperatures with those in the field or laboratory using fluctuating regimes, but there is still work to be done. These approaches adequately integrate the effects of average temperature and periodic daily variation in temperature, but we have yet to find a satisfactory model for integrating the effects aperiodic variation in temperature across the TSP with those of average temperature. The nonlinear approaches outlined in this paper have yet to be tested in studies of reptile sex determination, either in the laboratory or in the field. Nor have we applied these ideas to species with dual thresholds: Are degree-hours above the high pivotal temperature in turtles with dual thresholds equivalent in their feminizing effect to degree-hours below the low pivotal temperature?

If we look beyond the practical application of these models to ecological implications, the models discussed in this paper yield important insights. They explain why mixed sex ratios occur in more nests than would be expected from the very narrow pivotal temperature range of many species, even in the absence of gradients in mean temperature with depth (Georges 1992; Demuth 2001). The models provide us with more scope for exploring how reptiles with TSD might respond to climatic change, latitudinal variation in climate, or other disturbances to the incubation environment, because they identify a range of additional parameters that shallow-nesting species can manipulate in order to compensate for climatic change or variation with latitude. We have focused on the outcome of sexual differentiation as the phenotypic trait of interest, but the approaches used to model sex may apply equally well to other phenotypic traits, especially those without a posthatching metabolic component, such as coloration or morphometrics. This latter aspect would be a fruitful area for further investigation in both reptiles (Shine and Harlow 1996; Doody 1999) and insects.

10

TURK RHEN AND JEFFREY W. LANG

Phenotypic Effects of Incubation Temperature in Reptiles

Developmental temperatures of embryos vary in both time and space and determine sex in many lower vertebrates. In this chapter, the effects of embryonic temperature on traits other than gonadal sex are discussed. With examples from diverse model systems, the widespread and long-term effects of incubation temperature in reptiles, primarily those with temperature-dependent sex determination (TSD) are highlighted, but the pervasive effects of developmental temperatures on phenotypes in species with genotypic sex determination (GSD) are also illustrated. From the available literature, it is clear that there is a gap in our understanding of the genetic, molecular, physiological, and developmental mechanisms that underlie the phenotypic effects of incubation temperature. Only by linking these mechanisms to the fitness effects of developmental temperature in representative species can we fully elucidate the ecology and evolution of the modes and patterns of sex determination that are evident in living reptiles. Such data could also help clarify the evolution of viviparity, which may be related to the beneficial effects of retaining eggs at maternal body temperatures (Shine 1995).

Given that incubation temperature determines gonadal sex and influences many other traits, it is critical to acknowledge the potentially confounding effects of incubation temperature and gonadal sex on phenotype in species with TSD (Figure 10.1). In particular, temperature could have indirect effects that are mediated by sex-specific patterns of hormone production after temperature determines gonadal sex. Alternatively, temperature could influence phenotype by mechanisms independent of its effect on sex determination and the subsequent production of sex-specific hormones. Finally, there may be important interactions between incubation temperature and gonadal sex. This review is therefore centered on the basic tenet that incubation temperature and gonadal sex are independent variables that influence phenotype, but ones that can interact in complex ways and in patterns that are species specific.

Two basic approaches have been taken to tease apart temperature versus sex effects in TSD species. In certain studies, knowledge of the physiological mechanism of TSD has been used to separate temperature and sex effects experimentally. Specifically, exogenous estrogens have been administered to embryos to produce females at temperatures that normally produce only males, and an aromatase inhibitor has been administered to produce males at a temperature that normally produces a female-biased sex ratio. Aromatase is the key enzyme that converts androgens into estrogens. This approach is particularly useful when the transitional range of temperatures is narrow and most incubation temperatures produce exclusively one sex or the other. In contrast, some investigators have used an alternate approach with species in which males and females are produced over a broad range of incubation temperatures, albeit in different proportions. There are advantages and disadvantages to both approaches, but the general picture acquired from these investigations is that the thermal

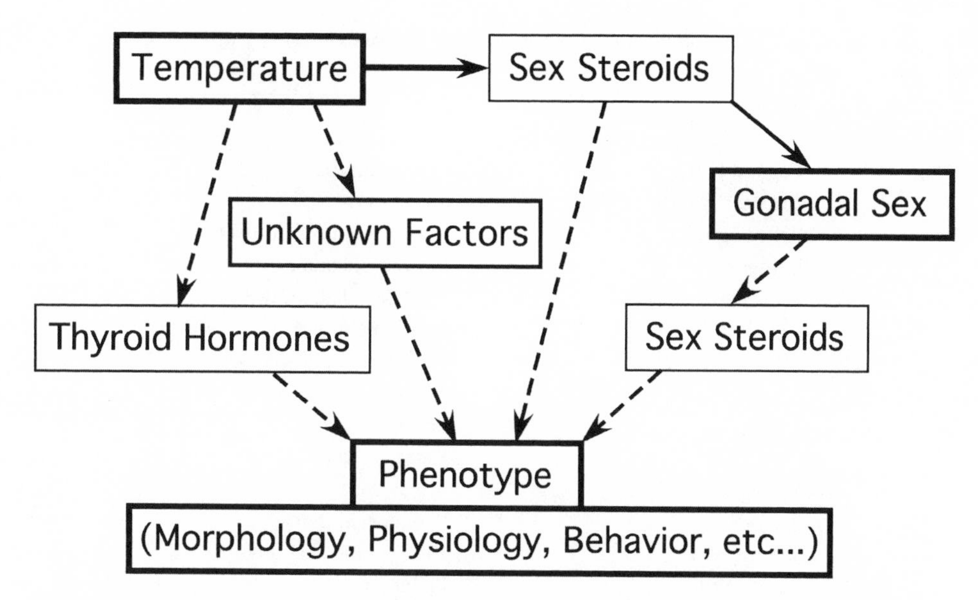

Figure 10.1 Phenotypic effects of incubation temperature may be mediated by its effect on sex steroid metabolism in embryos, by its effect on sex determination and the subsequent production of sex steroids, or by mechanisms independent of sex steroids, including thyroid hormones or other as yet unknown mechanisms.

regime during incubation produces (1) phenotypic effects that are independent of its effect on sex determination, (2) phenotypic effects that are mediated by its effect on gonadal sex, and (3) possibly significant interactions between the effects of temperature and sex. In other studies of TSD species, researchers have not attempted to control for sex versus temperature effects, but nonetheless their findings indicate that one factor or the other strongly influences phenotype. Further evidence for these conclusions is provided by studies that demonstrate temperature effects, sex effects, and interactions between these factors in reptiles with GSD.

Temperature Effects in GSD Species

Sex is determined by genotype (GSD) in all snakes studied to date. Nevertheless, temperature during development is known to alter neonate morphology, physiology, behavior, and survival in diverse species, including egg-laying and live-bearing species (Table 10.1). Additional support for direct phenotypic effects of incubation temperature is provided by similar findings in a number of lizards in which sex is determined genotypically. Specific traits such as body size, growth rate, running speed, and thermal preference are strongly affected by incubation regimes regardless of gonadal sex. Developmental temperatures in egg-laying as well as live-bearing lizards have been shown to influence various aspects of neonate morphology, physiology, and behavior (Table 10.1). Temperature effects on traits such as hatchling morphology, growth, and locomotor performance have been observed in turtles with GSD (Table 10.1). Together, these studies indicate that incubation temperature can influence phenotype independently of any effect on sex determination. Another important point drawn from these studies is that the phenotypic effects of incubation temperature may differ in sex-specific and/or species-specific ways.

Incubation Temperature Effects on Natural Males and Females in TSD Species

A number of studies have used TSD reptiles in which males and females are naturally produced over a range of incubation temperatures. In these studies, temperature and sex effects were not confounded, allowing comparisons of temperature effects within each sex as well as comparisons between males and females produced at the same temperatures. For example, both incubation temperature and sex influence hatchling traits in the frillneck lizard (Harlow and Shine 1999). While a low temperature (26°C) produced only females, intermediate (29°C) and high temperatures (32°C) produced slightly male-biased sex ratios. Males from the high temperature had smaller SVL (snout-vent length) and tail length and weighed less than males from the intermediate temperature. Males had longer tails than females at the intermediate and high temperatures, but did not differ from

Table 10.1 Incubation Temperature Effects on Phenotypic Characteristics in Snakes, Lizards, and Turtles with Genotypic Sex Determination (GSD)

Characteristic	Snakes	Lizards	Turtles
Hatchling morphology	Osgood 1978 Gutzke and Packard 1987b Reichling and Gutzke 1996 Burger et al. 1987 Shine et al. 1997a Ji et al. 1999 Ji and Du 2001a Ji and Du 2001b	Van Damme et al. 1992 Overall 1994 Phillips and Packard 1994 Elphick and Shine 1998 Elphick and Shine 1999 Ji and Braña 1999 Qualls and Andrews 1999 Andrews et al. 2000 Booth et al. 2000 Flatt et al. 2001 Ji et al. 2002 Radder et al. 2002	Janzen 1993 Booth 1998 Du and Ji 2001 Greenbaum and Carr 2001 –
Locomotor performance	Shine et al. 1997a Burger 1990 Burger 1991 Burger 1998a	Van Damme et al. 1992 Elphick and Shine 1998 Elphick and Shine 1999 Qualls and Andrews 1999 Andrews et al. 2000 Vanhooydonck et al. 2001	Janzen 1993 Doody 1999
Antipredator behavior	Shine et al. 1997b Burger 1990 Burger 1998a Burger 1998b	Elphick and Shine 1998 Flatt et al. 2001	–
Thermal responses	Burger 1998a Blouin-Demers et al. 2000	Qualls and Andrews 1999 Booth et al. 2000 Flatt et al. 2001 Ji et al. 2002 Radder et al. 2002 Vanhooydonck et al. 2001 Shine 1995 Qualls and Shine 1996 Blumberg et al. 2002	–
Feeding responses	Shine et al. 1997a Burger 1990 Burger 1991	–	–
Growth	–	Van Damme et al. 1992 Andrews et al. 2000 Alberts et al. 1997	Booth 1999
Ecdysis	Burger et al. 1987	–	–
Sex ratios	Burger and Zappalorti 1988 Dunlap and Lang 1990	–	–

females in SVL or mass. Temperature effects were more complex in females. Whereas tail length decreased with increasing temperature, mean SVL did not differ between the low and intermediate temperatures but was smallest at the high temperature. Female mass increased from the low to the intermediate temperature and decreased again at the high temperature.

The leopard gecko has proven to be another excellent model species because both sexes are produced across a range of temperatures (Viets et al. 1993). Incubation of leopard gecko eggs at 26°C produces only females, incubation at 30°C produces a female-biased sex ratio (~30% males), incubation at 32.5°C produces a male-biased sex ratio (~70% males), and incubation at 34°C again produces mostly females (~5% males). Crews and colleagues have extensively studied temperature and sex effects from hatching through reproductive maturity in leopard geckos (Crews et al. 1998). The general findings are that both incubation temperature and gonadal sex affect growth during juvenile development, as well as numerous features after reproductive maturity, including body size and shape, sexual and aggressive behavior, circulating levels of sex steroids, and the metabolic capacity of certain brain nuclei. Our discussion will focus on temperature effects evident in adulthood in this species.

Body size is sexually dimorphic in leopard geckos: males are generally larger than females produced at the same temperature. Incubation temperature also has an effect on adult size (Tousignant and Crews 1995). Females from the male-biased temperature (i.e., 32.5°C) are larger than females from the low (i.e., 26°C) and high (i.e., 34°C) temperatures. However, females from different temperatures also display differences in body shape (Crews et al. 1998). Females from the male-biased and high temperatures have higher ratios of body weight to SVL than females from the low temperature, whereas females from the female-biased temperature (i.e., 30°C) display intermediate weight to SVL ratios. On average, males from the male-biased temperature are larger than males from the female-biased temperature.

Sex differences in reproductive and aggressive behavior have been examined at female- and male-biased temperatures and appear to follow the general pattern found in mammals (Rhen and Crews 1999, 2000). Females from either temperature do not scent mark or display male-typical sexual behavior even when treated with androgens that induce these behaviors in males. Sex differences in responsiveness to hormones in adulthood are most likely organized by differences in sex steroid levels during ontogeny (Sakata et al. 1998). Although incubation temperature influences certain behaviors in females, we will concentrate on temperature effects in males. Males from the female-biased temperature are more sexually active and less aggressive towards females than are males from the male-biased temperature (Flores et al. 1994). Incubation temperature also influences endocrine physiology in adult males: the ratio of androgen to estrogen levels is lower in males from the female-biased temperature than in males from the male-biased temperature (Tousignant and Crews 1995; Crews et al. 1996b). Nevertheless, temperature-induced differences in certain behaviors do not depend upon temperature-induced differences in circulating levels of steroids. When treated with the same levels of testosterone or dihydrotestosterone in adulthood, castrated males from the male-biased temperature scent mark more than do castrated males from the female-biased temperature (Rhen and Crews 1999). Conversely, and across the same hormone treatments, males from the female-biased temperature mount and attempt to copulate with females more than do males from the male-biased temperature. Taken together, these findings indicate that temperature during embryogenesis has a permanent effect on the organization of the male leopard gecko brain.

In accord with this conjecture, temperature effects on brain morphology and physiology correlate with temperature effects on the sexual and agonistic behavior displayed by adult males (Crews et al. 1996b; Coomber et al. 1997). For example, more sexually active males from the female-biased temperature have greater metabolic capacity in the preoptic area (POA), dorsoventricular ridge, and torus semicircularis than less sexually active males from the male-biased temperature. Metabolic capacity was measured by quantitative cytochrome oxidase histochemistry. Temperature-induced difference in metabolic capacity in the POA might be expected because this nucleus plays an important role in regulating male sexual behavior in vertebrates. Temperature during embryonic development also affects metabolic capacity in other brain regions known to be involved in the regulation of agonistic behavior (Crews et al. 1996b; Coomber et al. 1997). Temperature-induced correlations in metabolic capacity among various nuclei might therefore reflect differences in functional relationships among brain regions critical for the display of social behaviors (Sakata et al. 2000). In summary, incubation temperature and gonadal sex interact to influence growth and endocrine physiology, behavior, and metabolic capacity in several hypothalamic nuclei. These results demonstrate that temperature has general effects, as well as sex-specific effects, on various traits in the leopard gecko.

Incubation Temperature Effects on Sex-Reversed Males and Females in TSD Species

Both sexes are produced across a narrow range of constant incubation temperatures in many TSD reptiles, which precludes a clear view of temperature versus sex effects. For example, the viable range of incubation temperatures spans approximately 10°C in painted turtles, *Chrysemys picta*, from Minnesota, but the transitional range of temperatures that produce males and females occurs over just 1.5°C (Rhen and Lang 1998). In another population of painted turtles, incubation temperature (or sex) has a weak effect on hatchling mass (Janzen and Morjan 2002). Egg size explained much of the variation in initial hatchling size, as it does in other oviparous reptiles. Egg size was still correlated with body size of turtles at one year of age, but this effect was much smaller and was superseded by the effect of temperature (or sex). More importantly, temperature (or sex) had a strong effect on posthatching growth: turtles incubated at 30°C (i.e., females) are approximately 30% heavier than turtles incubated at 26°C (i.e., males) at one year of age (Janzen and Morjan 2002). Such observations illustrate the normally confounded nature of temperature and sex effects and highlight the need for their dissociation.

The authors have used hormone manipulations during embryonic development to separate temperature effects from sex effects in the common snapping turtle, *Chelydra serpentina*. In this species, the TSD pattern is generally female-male-female (TSD II) from low to high temperatures (Yntema 1979; Ewert et al. 1994). In the Minnesota population the authors studied, predominantly males are produced from 21.5 to 22.5°C (~90% males), 23–27°C produces exclusively males, mixed sex ratios occur from 27.5 to 29.5°C, and all females are produced above 29.5°C (Rhen and Lang 1998). Viability is greatly reduced in eggs incubated at constant temperatures lower than 21.5°C or higher than 31.5°C. Therefore hormone manipulations at three temperatures symmetrically distributed across the viable range were used: eggs were incubated at two temperatures (i.e., 24 and 26.5°C) that normally produce only males and a third (i.e., 29°C) that produces mostly females. Eggs were treated during the thermosensitive period with 17β-estradiol, a potent aromatase inhibitor, or a vehicle control, or they were not treated at all (Rhen and Lang 1994).

In agreement with studies in other TSD reptiles, gonadal sex was reversed by hormonal manipulations but not by treatment with the vehicle control: 17β-estradiol induced differentiation of ovaries at male-producing temperatures, and the aromatase inhibitor induced differentiation of testes at the female-producing temperature. Further study showed that sex-reversed turtles have sex steroid levels consistent with their gonadal phenotype and indistinguishable from temperature-induced males and females (Rhen et al. 1996). Males had higher levels of testosterone than females, both before and after treatment with follicle stimulating hormone (FSH). Whereas testosterone levels increased approximately 11-fold after FSH challenge in males, testosterone levels remained low in females. In contrast, levels of 17β-estradiol after FSH challenge were higher in females than in males, but showed no difference before FSH challenge.

Although manipulation of embryonic sex steroid levels clearly reversed the gonadal and physiological sex of snapping turtles, neither hormone manipulations nor gonadal sex influenced posthatching growth. Embryonic temperature, however, had very strong effects on growth rate after hatching. These results provide persuasive evidence that sex steroids do not mediate temperature effects on growth physiology in this species. Temperatures that normally yield only males produce faster growing turtles than the temperature that produces mostly females (Rhen and Lang 1995). This study was conducted in a thermal gradient in which turtles had access to a wide range of ambient temperatures. Consequently, a subset of turtles was held at constant cool or warm ambient temperatures to determine whether differences in growth were due to a temperature effect on growth potential per se or to differences in thermoregulatory behavior (Rhen and Lang 1999a). As expected, juvenile turtles grew more slowly in a cool (19°C) than in a warm (28°C) environment. Nevertheless, incubation temperature effects on growth persisted even when turtles were not able to behaviorally thermoregulate. In a parallel study, turtles from different incubation temperatures displayed different patterns of temperature choice in a thermal gradient. In general, there was an inverse relationship between embryonic temperature and juvenile temperature choice. It was tentatively concluded that embryonic temperature regulates postembryonic growth in snapping turtles in two ways: (1) via a behavioral mechanism (i.e., through altered choice of ambient temperatures) and (2) via a mechanism independent of its effect on behavioral thermoregulation (i.e., through an intrinsic effect on growth physiology) (Rhen and Lang 1999a).

In further studies, temperature and sex effects on hatchling size, energy reserves, and use of those reserves during a 30-day fasting period were also examined (Rhen and Lang 1999b). Residual yolk mass increased with increasing incubation temperature, but was not affected by gonadal sex. Residual yolk had virtually disappeared in turtles that had been fasted for 30 days after hatching. While incubation temperature influenced abdominal fat mass in hatchlings, gonadal sex did not. Interestingly, both incubation temperature and sex influenced fat mass after 30 days of fasting. Incubation temperature did not affect total hatchling mass initially, but this changed so that the temperature effect was significant after 30 days of fasting. Turtles from the lowest incubation temperature lost more mass than turtles from the intermediate temperature. Likewise, turtles from the intermediate temperature lost more mass than turtles from the high incubation temperature. Moreover, controlling for these temperature effects, females lost more mass than males when fasted for 30 days. These data strongly suggest that some traits were influenced directly by temperature, whereas others were independently affected by sex (i.e., possibly mediated by differences in circulating levels of sex steroids).

Incubation Temperature (or Sex) Effects in TSD Species

There has been work in many TSD reptiles in which temperature effects versus sex differences were not explicitly addressed or compared. Although potential temperature and sex effects are intertwined in these studies, significant differences indicate that one factor or the other influences phenotype. In fact, numerous studies of turtles have documented that incubation environments (or sex) affect hatchling characteristics, including body size, thermal and metabolic responses, and growth (Table 10.2). These studies serve to reinforce and/or complement the findings just discussed. Incubation temperature (or sex) has clear effects on post-hatching traits in crocodilians, a group in which all species so far studied exhibit TSD (Lang and Andrews 1994). Hatchling mass and dimensions vary with incubation temperature, but results have often been confounded by other factors, notably clutch of origin and temperature-by-clutch interactions (e.g., in alligators, reviewed by Allsteadt and Lang 1995a). These investigations provide excellent background information for designing experiments to separate and determine the mechanisms responsible for incubation temperature and/or sex effects on phenotype (Table 10.2). As in GSD species, patterns of incubation temperature (or sex) effects do not generalize across all reptiles, and they often appear to be group and/or species specific.

Clutch Effects and Interactions between Clutch and Incubation Temperature

Clutch effects and interactions between clutch and incubation temperature are nearly ubiquitous for the traits discussed above, as well as for sex ratio itself (Bull et al. 1982a; Janzen 1992; Rhen and Lang 1994, 1998). Assuming that a clutch of eggs is sired by a single male, the variation evident among clutches includes additive, dominance, and epistatic genetic effects as well as maternal effects (Falconer 1989). Given the long generation time in reptiles, it would be very difficult to analyze the relative contribution of genetic mechanisms versus maternal effects using breeding or selection experiments. To further complicate the issue, recent reports indicate that multiple paternity, in some instances associated with long-term sperm storage, occurs in representative snakes, lizards, and turtles, thereby qualifying the assumption that a clutch of eggs is sired by a single male in analyses of putative genetic effects.

Nevertheless, genetic and maternal effects have extremely important implications for the evolution of temperature-dependent reaction norms. For instance, responses to natural and sexual selection would be expected to differ for a temperature-sensitive trait when the genetic basis of plasticity varies: that is, different responses would occur if additive genetic effects were present than would occur in genotype-by-temperature (G×T) interactions. Maternal effects, due to variation in yolk steroids, for example, would also modify evolutionary responses. Alternative experimental methods will therefore be required to address questions about the mechanistic basis for clutch effects and clutch-by-temperature (C×T) interactions. Manipulations of circulating steroid levels in females and monitoring of hormonal changes in yolking follicles and eggs will be an important way of dissecting potential maternal and genetic effects on offspring phenotype (Bowden et al. 2002; Janzen et al. 2002). Molecular techniques like differential display polymerase chain reaction (PCR) or subtractive hybridization should be used to identify genes that are differentially regulated by incubation temperature and that display differences in patterns of regulation among clutches or between populations.

Potential Mechanisms for Temperature Effects on Phenotype

While it is clear that temperature influences developmental, behavioral, physiological, and morphological traits in reptiles, it is uncertain how these effects are mediated at a molecular level. In some TSD species, studies suggest that temperature determines sex by altering sex steroid metabolism during embryogenesis (Crews 1996; Pieau et al. 1999a). In these species, it is thought that temperature modulates the level and/or activity of aromatase, the enzyme that converts androgens to estrogens. Estrogens produced in a temperature-dependent manner are thought to play a central role in ovarian differentiation in egg-laying vertebrates, including TSD reptiles (Lance 1997). Because sex determination is a threshold trait, individuals with estrogen levels above a specific threshold develop as females. In contrast, individuals with estrogen levels below the threshold develop as males. This model predicts that individuals of the same sex from different temperatures are exposed to different hormonal environments. Consequently, temperature-

Table 10.2 Incubation Temperature (or Sex) Effects on Phenotypic Characteristics in Turtles and Crocodilians with Temperature-Dependent Sex Determination (TSD)

Characteristic	Turtles	Crocodilians
Hatchling morphology	Gutzke et al. 1987	Hutton 1987
	Packard et al. 1988	Joanen et al. 1987
	Packard et al. 1989	Manolis et al. 1987
	Souza and Vogt 1994	Webb et al. 1987
	Rhen and Lang 1995	Whitehead and Seymour 1990
	Booth and Astill 2001	Allsteadt and Lang 1995a, b
	Hewavisenthi and Parmenter 2001	Congdon et al. 1995
	Janzen and Morjan 2002	
	Steyermark and Spotila 2001	
	Reece et al. 2002	
Pigmentation	Etchberger et al. 1993	Murray et al. 1990
Egg dynamics	Bilinski et al. 2001	Elf et al. 2001
	Elf et al. 2001	Conley et al. 1997
	Elf et al. 2002c	
Locomotor performance	Demuth 2001	–
Metabolic rate	St. Clair 1995	–
	O'Connell 1998	
	O'Steen and Janzen 1999	
	Steyermark and Spotila 2000	
Thermal responses	Rhen and Lang 1999a	Lang 1987
	O'Steen 1998	
Antipredator behavior	O'Connell 1998	–
Feeding responses	O'Connell 1998	–
Growth	Rhen and Lang 1995	Joanen et al. 1987
	O'Connell 1998	Webb and Cooper-Preston 1989
	O'Steen 1998	
	Ryan 1990	
	Brooks et al. 1991	
	McKnight and Gutzke 1993	
	Bobyn and Brooks 1994	
	Spotila et al. 1994	
	Roosenburg 1996	
	Roosenburg and Kelley 1996	
	Foley 1998	
	Demuth 2001	
	Freedberg et al. 2001	
	Janzen and Morjan 2002	

induced variation in sex steroid levels during embryogenesis could have pleiotropic effects on many tissues in each sex.

There is evidence showing that temperature influences levels of sex steroids in embryonic red-eared slider turtles and that these temperature effects persist for at least six weeks after hatching (White and Thomas 1992a,b; Rhen et al. 1999). Indirect evidence in American alligators also suggests that incubation temperature alters circulating levels of sex steroids during embryonic development. The overall size, shape, and level of vascularization of the genitalia are affected by incubation temperature in a graded fashion in hatchling alligators (Allsteadt and Lang 1995b). Phallus dimensions increase with increasing temperature within each sex. Sex differences in genital morphology also occur at temperatures that produce both males and females, suggesting that both temperature and gonadal sex influence circulating levels of sex steroids in alligator embryos.

Whereas temperature effects on certain traits may be mediated by temperature-induced variation is sex steroid levels in some species, studies in snapping turtles indicate that temperature can have effects mediated via alternative mechanisms. In experiments described earlier, snapping turtle embryos were treated with exogenous estrogen or a potent aromatase inhibitor at the beginning of the thermosensitive period (TSP). These manipulations reversed the gonadal sex of turtles, yet did not alter the effect of incubation temperature on a whole suite of other traits. In a study that partly controlled for sex effects, O'Steen and Janzen (1999) found that thyroid hormone levels and resting metabolic rate decreased with increasing incubation temperature in hatchlings. Moreover, treatment of embryos with exogenous triiodothyronine raised resting metabolic rate in hatchlings. Consistent with these temperature-induced differences in metabolic rate, individuals from different temperatures, as well as males and females, lost weight and utilized fat stores at different rates when fasted (Rhen and Lang 1999b). In another study of juvenile snapping turtles, the incubation temperature effect on metabolic rate was reversed (Steyermark and Spotila 2000), but this result may reflect differences in the age at which metabolic rate was measured and/or differences in the incubation regime. Given that thyroid hormones have well-characterized effects on metabolism, growth, and thermoregulatory behavior, it is possible that temperature effects on thyroid hormone physiology are responsible for sex-steroid–independent effects of incubation temperature in snapping turtles.

Despite evidence that temperature can influence circulating levels of sex steroids and thyroid hormones, it remains unclear whether temperature influences only the physiology of the gonads and thyroid or whether the entire neuroendocrine system is affected (i.e., the hypothalamic-pituitary-gonadal axis and the hypothalamic-pituitary-thyroid axis). The finding that temperature influences the morphology and metabolic capacity of certain hypothalamic nuclei in adult leopard geckos makes it conceivable that temperature alters the development and homeostatic interactions of entire organ systems. Consequently, one must use an integrative approach to elucidate temperature effects from the molecular to the organismal level.

Despite some common features in the process of sex determination in the TSD reptiles studied to date, the mechanistic details involving networks of genes and regulatory signals appear to vary, for example, between sea turtles and alligators (Merchant-Larios 2001). Work in the Olive Ridley sea turtle, *Lepidochelys olivacea,* suggests that incubation temperature can modulate steroid levels and metabolism in the embryonic brain (Salame-Mendez et al. 1998). The diencephalon, which includes the hypothalamus, had a higher content of testosterone and ability to transform testosterone into 17β-estradiol at a female-producing temperature than at a male-producing temperature. Aromatase activity in the brain also differed between male- and female-producing temperatures in the slider turtle (Willingham et al. 2000), but not in the alligator (Milnes et al. 2002). These results reveal additional species differences in temperature effects and underscore the need for caution when proposing mechanisms that may underlie the diverse phenotypic effects produced by incubation temperature.

Identifying Common Features and Future Directions

In this chapter, many examples of phenotypic traits altered by embryonic incubation temperature in snakes, lizards, turtles, and crocodilians have been carefully documented and described. In general, incubation temperature directly or indirectly affects body size and shape, performance, activity levels, metabolism, growth, reproduction, behavior, and/or survival in reptiles, regardless of their mode of sex determination. Key features of thermally sensitive traits are identified in Table 10.3. This general framework should help guide future experiments designed to both elucidate the proximate mechanisms underlying temperature effects, and contribute to a better understanding of their ultimate consequences. It is important to note that these features also characterize the process of temperature-dependent sex determination (TSD).

Reaction norms for specific traits can be established by incubating eggs at constant temperatures across the viable range of incubation temperatures in the same way that the reaction norm for sex determination is established. In TSD species, the thermosensitive period (TSP) for sex determination is determined by shifting eggs between male- and female-producing incubation temperatures at critical times during development (e.g., Lang and Andrews 1994). One can determine the TSP for other phenotypic traits in a similar manner; that is, by exposing eggs to variable thermal regimes during development (Lang 1987). Information on reaction norms and TSPs of development will be critical for understanding how natural thermal regimes experienced by embryos influence the development of particular phenotypes. With this knowledge, it should also be possible to establish the molecular basis for the development of thermosensitive traits.

Recent studies of the Australian skink *Bassiana duperreyi* are especially instructive in this regard. In this GSD lizard, Shine and Elphick (2000) examined the effects of short-term weather fluctuations on nest temperatures and demonstrated how variable incubation temperature affected the

Table 10.3 Common Features of Thermosensitive Traits

The Phenotypic Effects of Developmental Temperature Are	
Thermally induced	Responses not due to correlated environmental factors, e.g., moisture or oxygen availability
Continuous	Responses that typically exhibit linear or curvilinear change as temperature increases or decreases, e.g., not discrete or categorical features like sex
Time dependent	Responses that are induced by exposure during a specific period or "window" of embryonic development
Long term	Responses that persist in juveniles and/or in adults
Species specific	Responses that vary between related species and among diverse lineages
Sex specific	Responses that may affect only one sex, both sexes similarly, or influence the sexes differentially
Trait specific	Responses that affect particular morphological, physiological, and/or behavioral characters
Natural	Responses that occur in nests, and are sensitive to fluctuating and constant thermal regimes
Sources of variation	Of major significance when compared with variation from nonenvironmental sources, e.g., clutch effects of genetic and maternal origin

development of phenotypic traits in hatchling lizards. In a series of experiments designed to mimic the thermal regimes in natural nests, eggs were exposed to diel cycles of temperature (± 5°C) that ranged from cool (17°C) to warm (22°C) to hot (27°C). Eggs were then shifted to warm or hot conditions for brief periods that simulated transient periods of hot weather. It was found that the timing of exposure, as well as the temperatures experienced during that period, modified a range of phenotypic traits in hatchling lizards, in direct proportion to the thermal exposure associated with each treatment.

These results parallel findings on the thermal induction of sex in TSD species (e.g., Lang and Andrews 1994; Wibbels et al. 1994). In turtles and crocodilians, the sex-determining effect of temperature is dependent on both the temperature level and its duration during the middle third of development (see also Georges et al., Chapter 9). Temperature exposure acts in a cumulative fashion to influence sex determination in TSD species. In the skink described above, early exposure to hot temperatures affected hatchling phenotypes most strongly and produced "better" hatchlings, relative to exposures to warm temperatures later in development. The match between the reaction norms for sex ratio and for other traits in response to incubation temperature is of special interest for understanding the adaptive significance of TSD (Charnov and Bull 1977; Shine 1999; Valenzuela, Chapter 14).

Conclusions

While a review of the literature indicates that the phenotypic effects of incubation temperature are widespread in reptiles, incubation temperature does not always have a detectable effect on phenotype. Given that embryonic temperature determines sex and influences a number of morphological, physiological, and behavioral characteristics in TSD species, we have stressed the importance of the potentially confounded effects of temperature and sex on phenotype. Such a perspective is not only essential for understanding how incubation temperature has its effects, but also for addressing the evolutionary and ecological significance of TSD (Rhen and Lang 1995; Shine 1999). Some model systems illustrate that temperature can have effects that are simply mediated by its effect on sex determination (and the subsequent production of sex-specific hormones). Temperature also appears to have more subtle effects that are mediated by temperature-induced differences in sex steroid levels within each gender. In other cases, notably reptile species with GSD, incubation temperature has been shown to influence phenotype independently of its effect on sex determination. Temperature-induced variation in thyroid hormone levels may represent an alternative mechanism by which temperature influences a variety of traits. Finally, there are important interactions between incubation temperature and gonadal sex in a number of TSD species. Consequently, it is critical to realize that incubation regimes have species-, sex-, and trait-specific effects on phenotype. At present, a large gap exists in our understanding of the genetic, molecular, physiological, and developmental mechanisms that underlie the phenotypic effects of incubation temperature. Future work should be directed at elucidating these mechanisms. Such findings are an important first step for understanding the ecology and evolution of TSD and GSD in vertebrates.

11

ALLEN R. PLACE AND VALENTINE A. LANCE

The Temperature-Dependent Sex Determination Drama

Same Cast, Different Stars

Amniote vertebrates exhibit a profound dichotomy in their mode of sex determination and differentiation. Placental mammals exhibit a female phenotype in the absence of the hormonal signals emanating from the fetal testes, whereas egg-laying vertebrates exhibit a male phenotype in the absence of embryonic gonads, or if estrogen synthesis by the embryonic ovary is blocked by an CYP19 (aromatase) inhibitor (e.g., Elbrecht and Smith 1992). Thus a male phenotype is imposed on the neutral female condition or a female phenotype is imposed on the neutral male condition. What the default condition in egg-laying mammals (prototherians) is in the absence of embryonic gonads is unanswerable without available research animals. It is not clear if there is a neutral or default condition in anamniote vertebrates (amphibians and fishes). It is clear that hormones are necessary for male differentiation in mammals and for female differentiation in birds and reptiles (Lance 1997). What is not clear is that despite this difference many of the genes involved in gonadal differentiation are common to all vertebrates (see Table 11.1). Exponentially growing research on sex determination followed the discovery of *Sry* by Sinclair et al. (1990) (reviewed in Merchant-Larios and Moreno-Mendoza 2001; Morrish and Sinclair 2002; Parker and Schimmer 2002; Pieau et al. 1999a, 2001; Swain and Lovell-Badge 1999). A large number of new genes (see Table 11.1) seemingly involved in the gonadal differentiation process have been identified, and some are expressed in egg-laying vertebrates. The molecular basis for sex determination and sex differentiation in mammals is becoming clearer as more genes are discovered in the developmental cascade, but the molecular mechanisms involved in temperature-dependent sex determination (TSD) remain poorly understood. The rapid pace of research in this area gives one cause for hope.

The processes of sex determination and sex differentiation are far more complicated than first envisioned. In the late 1980s, it was thought that a single gene on the Y chromosome accounted for the difference between a male and a female gonad. It is now apparent that each phase in the process, from migration of the primordial germ cells (PGCs) to the final steps of gonadal differentiation involves numerous genes, whose products form complex webs of protein-protein and protein-DNA interactions. Many of the genes perform many different roles and interact with many other genes in a cell-specific and developmental stage–specific manner that is slowly beginning to be unraveled. Many of these interactions are still not understood.

In order to understand TSD, it is necessary to know how the process of sex determination unfolds in mammals, where genetic factors alone determine sex. In the following review the authors briefly cover what is known on the molecular mechanisms involved in sex determination and differentiation in mammals, contrast this with the mechanisms in birds, and then discuss the similarities and differences in

Table 11.1 Known Members of the Mammalian Sex Determination/Differentiation Network Described and Referenced in the Text.

Gene	Human Chromosome Location	Gene Family	Putative Function	Gonadal Cell Lineage
Sf1(NR5A1)	9q33	Nuclear receptor	Transcription factor	Steroidogenic cell precursors
Wt1	11p13	Nuclear receptor	Transcription factor	Sertoli cell lineage
Sry	Yp11	HMG protein	Transcription factor	Pre–Sertoli cell lineage
Sox9	17q24	HMG protein	Transcription factor	Pre–Sertoli cell lineage
Dax1(NR0B1)	Xp21.3	Nuclear receptor	Transcription factor	Pre–Sertoli cell lineage
Dmrt1	9p24	DM-domain protein	Transcription factor	Sertoli cells and primordial germ cells
Amh	19q13	Transforming growth factor (TGF) β	Growth factor	Pre–Sertoli cell lineage
Wnt4	1p35	Wnt	Growth factor	Interstitial cell lineage
Dhh	3p21	Hedgehog	Signaling molecule	Sertoli cell lineage
Foxl2	3q23	Winged/forkhead	Transcription factor	Pre–granulosa lineage
Gata 4	8p23	GATA	Transcription factor	Germinal epithelium, followed b expression in Sertoli cells and interstitial cells including Leydig cells
Atrx	Xq13.3	SWI/SNF (helicase superfamily)	Chromatin remodeling family	Unknown
Lhx9	1q31	Lim homeobox	Transcription factor	Germinal epithelium
Emx1	2p13	Emx homeogenes	Transcription factor	Germinal epithelium
Fgf9	13q12	Fibroblast growth factors	Growth factor	Expressed within testicular cords
Vnn1	6q23	Glycosyl phosphatidylinositol anchored cell surface molecule	Cell migration and homing	Expressed within testicular cords
Bmp4/Bmp8b	14q22/1p34	Transforming growth factor (TGF) β	Growth factor	Extraembryonic ectoderm
Lim1	10q23	Lim homeobox	Transcription factor	Coelomic epithelium
M33	17q25 (?)	Polycomb	Homeotic transcription repressors	
Pn1	2q33–35	Protease nexin-1	Serine protease inhibitor	Expressed within testicular cords
Pod-1/Capsulin	6q23	Basic helix-loop-helix	Transcription factor	Sertoli cells
KIAA0800	3p21	Unknown	Unknown	Same expressoin as *Sox9*
Mfge8	15q26	Integrin-binding protein family	Soluble integrin-binding protein	Unique cell lineage from coelomic epithelium
Testatin/cystatin C	20p11(?)	Related to cystatins	Small secreted protease inhibitors	Pre–Sertolic cells
3β-Hsd	16p11	Steriod dehydrogenase	Sex steroid precursors	Leydig cells
Cyp17 (17α-hydroxylase)	10q24	Cytochrome P450	Production of androgens	Leydig cells
Cyp11A1	15q24	Cytochrome P450	Sex steroid precursors	Leydig cells
Cyp19 (aromatase)				Granulosa cells

these mechanisms when they are compared with those in egg-laying vertebrates that show TSD. Many of the gene functions are revealed by the phenotype of natural or experimental mutants that cannot be described in detail here. Instead, the main conclusions derived from such observations are highlighted.

Sex the Mammalian Way

Mammalian gonads arise in both sexes from bilateral genital ridges capable of developing as ovaries or testes, that is, a bipotential gonad (reviewed in Capel 1998, 2000; Koopman 2001a; Lovell-Badge et al. 2002; Merchant-Larios and

Moreno-Mendoza 2001; Parker and Schimmer 2002; Swain and Lovell-Badge 1999). In mice, the genital ridge forms on embryonic day (E) 9.5. At approximately E10.5, the mammalian sex-determining switch, *Sry* (sex-determining region Y chromosome)(Koopman et al. 1991), turns on in the XY gonads and sets in motion the testis developmental pathway—differentiation of supporting cell precursors into Sertoli cells rather than granulosa cells (Albrecht and Eicher 2001; Burgoyne et al. 1988). Apparently these terminally differentiated Sertoli cells signal ALL subsequent development of the male gonad (Jost and Magre 1988; Magre and Jost 1984, 1991), including cell proliferation (Schmahl et al. 2000), testes-specific migration of mesonephric cells into the gonad (Buehr et al. 1993; Martineau et al. 1997; Merchant-Larios and Moreno-Mendoza 1998), and differentiation of the other cell lineages (Byskov 1986; Jost et al. 1973). This SRY-initiated differentiation leads to the development of two distinct gonadal compartments: (1) testis cords containing germ cells (arrested in mitosis) and Sertoli cells surrounded by peritubular myoid cells (precursors of seminiferous tubules) and (2) an interstitial region containing steroidogenic cell precursors (Leydig cells). The germ cells go into meiotic arrest in the fetal ovary.

The internal genitalia derive from the genitourinary tract and are identical in both sexes at the indifferent stage. Male and female embryos have two identical sets of paired ducts: (1) the Müllerian (paramesonephric) ducts and (2) the Wolffian (mesonephric) ducts. Once the male developmental pathway is activated, the Müllerian ducts degenerate, and the Wolffian ducts develop into seminal vesicles, epididymis, and vas deferens. In the absence of testicular hormones, the Wolffian ducts regress, and the Müllerian ducts develop into the oviducts, Fallopian tubes, uterus, and upper vagina. As predicted by Jost (Jost 1953; Jost et al. 1973), the mediator of Müllerian duct regression in males is a glycoprotein hormone, Müllerian-inhibiting substance (MIS, also know as anti-Müllerian hormone or AMH), which is produced by Sertoli cells within the testicular cords. Testicular androgens, synthesized by Leydig cells in the interstitial region, cause male differentiation of the Wolffian ducts and external genitalia.

The external genitalia, like the internal genitalia, are also derived from structures common to both sexes, including the genital tubercle, the urethral folds, the urethral groove, and the genital swellings. Again, androgens are critical for virilization, although full virilization requires conversion of testosterone to dihydrotestosterone by 5α-reductase.

The gonads are unique, unlike the genital ducts and urogenital sinus, in that they are formed from cells originating in tissues outside the organ. Prior to differentiation, all somatic cells in the urogenital ridge appear to be derived from a cortical epithelial layer and are similar in presumptive males and presumptive females.

The primordial germ cells (PGCs) originate in the proximal epiblast, migrate through the hindgut and dorsal mesentery, and colonize the urogenital ridges. In mice, the specification of PGCs involves the interaction of at least three bone morphogenetic proteins (BMPs), BMP2, BMP4, and BMP8B, members of the transforming growth factor β superfamily (Kierszenbaum and Tres 2001). BMP binding to DNA signals to *Smad-A* and *Smad-B* transcription factors, which then activate transcription of specific target genes. Both BMP4 and BMP8B are derived from the extraembryonic ectoderm. Along the migration pathway, the *c-kit* receptor (a tryosine kinase) and its ligand (*c-kit* ligand) are expressed. The *c-kit* receptor is expressed in PGCs from E7.5 to E13.5. Once in the urogenital ridge, the PGCs begin expressing *ErbB3* (another tyrosine kinase receptor) and its coreceptor *ErbB2*. *Neuregulin-B,* a ligand for *ErbB3* and *ErbB2*, is expressed mainly in the developing nervous system, but also in the urogenital ridge, and is presumably necessary for PGC proliferation (Kierszenbaum and Tres 2001).

During the colonization of the urogenital ridge, PGCs express an ortholog of the *Drosophila* gene *Vasa*, a member of the ATP-dependent RNA helices in the DEAD-box family proteins. Its function is unknown, but it may regulate the translation of specific mRNAs by interacting with translation initiation factor 2. *Integrins* $\alpha 3\beta 1$ and $\alpha 6\beta 1$ are also expressed during the migration phase of PGCs and are necessary for their colonization of the urogenital ridge. After colonization, PGCs undergo a wave of apoptosis around E13.5, regulated by the balance of expression for *Bcl-x* (a cell survival factor) and *Bax* (a cell death factor). As the PGCs proliferate in the urogenital ridge, they form aggregations mediated by expression of *E-cadherin*, a cell adhesion molecule. The complexity of PGC development and the significance of somatic cell–PGC interaction are stressed by this brief summary (Kierszenbaum and Tres 2001). It is likely that a similar complexity is involved in migration and colonization of PGCs in egg-laying vertebrates, but very few of the genes discussed above have yet been identified.

Distinctive functional attributes of the somatic cell population of the extraembryonic site along the migration pathway and in the urogenital ridge contribute to PGC development. Sertoli cells that surround the germ cells in the testis cord seem to prevent the male germ cells from entering meiosis, whereas in the ovary, meitotic germ cells are essential for formation and maintenance of the ovarian follicles (McLaren 1995).

Six genes essential for bipotential gonad formation have been identified: *Sf1, Wt1, Lhx9, Emx2, M33,* and *Lim1* (Birk et al. 2000; Katoh-Fukui et al. 1998; Kreidberg et al. 1993; Luo et al. 1994; Miyamoto et al. 1997; Shawlot and Behringer 1995), and their role is detailed elsewhere (Merchant-Larios and Moreno-Mendoza 2001; Parker and Schimmer 2002). Two of these transcription factors *(Sf1* and *Wt1)* are described in more detail.

SF1 (steroidogenic factor 1; also known as *Ad4BP*; *Nr5a1*) is a member of the subfamily of nuclear receptors, the orphan receptors, for which no clear activating ligand has been found. SF1 is expressed during embryonic development in mice regions associated with endocrine function such as gonads, adrenals, pituitary, and hypothalamus.

The Wilm's tumor–associated gene *(Wt1)* is also important in early gonad and kidney development. In mammals, *Wt1* is comprised of 10 exons with four major mRNAs generated by two different splicing events. There are at least 16 possible forms of the protein (Reddy and Licht 1996; Sharma et al. 1994). Isoforms that are otherwise identical, except for the presence or absence of three amino acids (lysine, K; threonine, T; and serine, S: KTS) have different affinities for DNA. WT1 and LHX9 function as direct activators of the *Sf1* gene, and only the −KTS form of WT1 can bind to and transactivate the *Sf1* promoter (Wilhelm and Englert 2002). Additionally, the +KTS isoform colocalizes with proteins involved in RNA splicing (Caricasole et al. 1996; Charlieu et al. 1995; Davies et al. 1998; Larsson et al. 1995). During embryogenesis, WT1 is expressed throughout the intermediate mesoderm at E9.5 and later in the gonad and the differentiating metanephric mesenchyme.

The transcription factor, SOX9 (*Sry*-like HMG box), is the earliest known marker of differentiating Sertoli cells and is therefore postulated as an immediate downstream target of *Sry*. SOX9 is expressed in both male and female mouse genital ridges when the gonad first develops, but immediately after SRY at E10.5, it is upregulated in the male gonad and turned off in the female gonad. By E12.5, SOX9 is expressed in Sertoli cells, where it persists throughout life, whereas it is not seen at all in the ovary (Kent et al. 1996). The key factor regulating AMH expression in mammals is SOX9 (Arango et al. 1999; Vidal et al. 2001; Clarkson and Harley 2002). SOX proteins bind a specific DNA sequence using a high-mobility group (HMG) domain that mediates interactions with other transcription factor proteins and contains signals for nuclear import. As transcription factors are formed in the cytoplasm, they must be transferred from the cytoplasm to the nucleus, assisted by importin proteins, in order to function. Besides up-regulation of SOX9 expression in males, the subcellular localization of SOX9 protein resulting from a nuclear import/export equilibrium could be a regulatory switch that represses (in females) or triggers (in males) male-specific sexual differentiation (Gasca et al. 2002). Moreover, SOX9 interacts with heat-shock proteins (Marshall and Harley 2001) and with splicing factors involved in pre-mRNA splicing (Ohe et al. 2002), but the interaction with specific DNA sequences is not fully understood.

As the gonads emerge, the signaling molecule WNT4 (an ovary-determining factor) is expressed in the mesenchyme and mesonephros of the indifferent gonad in both sexes. Activation of steroidogenesis in developing testes requires downregulation of *Wnt4* expression and, correspondingly, ovary-specific expression of *Wtn4* suppresses Leydig cell development and activation of the steroidogenic enzymes (Vainio et al. 1999). Leydig cells seem to originate from mesonephric precursors that migrate into the genital ridge before peak expression of *Sry*. Leydig cells differentiate after Sertoli cells; this differentiation is probably controlled by Sertoli cell–secreted factors.

Dax1 (dosage-sensitive sex reversal; *Nr0B1*) is an orphan nuclear receptor that acts as a suppressor of *Sf1 (Nr5A1)*, and as an anti-*Sry* factor during gonadal sex differentiation (Yu et al. 1998), thus it is an "anti-testes" gene, rather than a ovary-determining gene (Meeks et al. 2003). DAX1 co-occurs with SF1 in many tissues, including the gonads (Ikeda et al. 1996; Kawabe et al. 1999), interacts with it, and represses its transcriptional activity (Crawford et al. 1998; Gurates et al. 2002; Kawabe et al. 1999; Tremblay and Viger 2001b). *Dax1* gene transcription is activated by β-catenin, a key signal-transducing protein in the *Wnt* pathway, acting in synergy with SF1 (Mizusaki et al. 2003). *Wnt4* signaling mediates the increased expression of *Dax1* as the ovary becomes sexually differentiated. *Dax1* also plays a crucial role in testis differentiation by regulating the development of peritubular myoid cells and the formation of intact testis cords (Meeks et al. 2003).

In testes, WT1 and SF1, cooperatively bind to the promoter and activate transcription of the *Amh* gene, initiating degeneration of the Müllerian ducts. In the ovaries, on the other hand, the orphan nuclear receptor DAX1 binds to SF1, and inhibits transactivation by WT1/ SF1 and thereby suppresses the induction of MIS expression and permits retention of the Müllerian ducts. Additionally, WT1 itself upregulates *Dax1* transcription. The LIM-only coactivator FHL2 and WT1 functionally interact (Du et al. 2002), and importantly, FHL2 potentiates the synergistic induction of *Mis* gene expression by *Wt1*/ *Sf1*. Moreover, FHL2 coactivates transactivation of the *Dax1* promoter by *Wt1*. The ability to modulate both *Dax1* and *Mis* expression might allow FHL2 to act in the molecular fine-tuning of *Wt1*-dependent control mechanisms in the reproductive organs, perhaps through interactions with *Sox9*.

Fibroblast growth factors (*Fgfs*), have been shown to be involved in the embryogenesis of several organs, including lungs, limbs, and anterior pituitary. FGF9 appears to act downstream of *Sry* to stimulate mesenchymal proliferation, mesonephric cell migration, and Sertoli cell differentiation in the embryonic testis.

Proper cell-cell signaling is required for normal patterning and development during embryogenesis. In mammals, three hedgehog genes and their products have been described: Desert hedgehog *(Dhh),* Indian hedgehog, and Sonic hedgehog. *Dhh* is expressed in Sertoli cells of the undifferentiated male gonad at E10.5. The receptor for *Dhh* is *patched (Ptc),* which is expressed in the male, but not the female, gonad of mice at E11.5, after onset of *Dhh* expression. After formation of the seminiferous cords, expression of PTC is limited to Leydig cells and to the pertitubular cells (Pierucci-Alves et al. 2001). Although formation of testis cords and development of other cell types normally take place in a tightly regulated sequence, each of these events can occur independently of the others.

Dmrt1 belongs to a family of known or putative transcription factors sharing a conserved DNA-binding motif, the DM domain (Raymond et al. 1999a). *Dmrt1* is conserved across vertebrates and is expressed in the embryonic gonads of fishes, reptiles, birds, and mammals (Kettlewell et al. 2000; Moniot et al. 2000; Raymond et al. 1998; Smith et al. 1999a). In all these groups, expression is higher in developing male than in developing female gonads. In the mouse from E10.5, *Dmrt1* is expressed exclusively in the bipotential gonads of both sexes. From E12.5, *Dmrt1* transcripts are localized to the Sertoli and germ cells of the testis and in a spectacled pattern in the ovary. No downstream targets have yet been found for *Dmrt1*.

The zinc-finger transcription factors GATA2 and GATA4 (members of the GATA-binding proteins family) are implicated in the development and function of the gonads. So far the target genes for GATA4 include *Mis* (Tremblay and Viger 1999; Viger et al. 1998; Watanabe et al. 2000), *inhibin/activin β-B-subunits* (Feng et al. 2000), *Sf1* (Tremblay and Viger 2001a), *StAR* (Steroidogenic acute regulatory protein) (Silverman et al. 1999; Tremblay and Viger 2001a), and *Cyp19/aromatase* (Jin et al. 2000; Tremblay and Viger 2001a). GATA2 expression is restricted to germ cells in XX gonads (Siggers et al. 2002), suggesting that it plays a role in ovarian germ cell development.

The normal in vivo function of GATA factors requires physical interaction with multitype zinc-finger proteins of the FOG (Friend of GATA) family, FOG1, FOG2, xFOG, and USH (Cantor and Orkin 2001). FOG1 and FOG2 can either activate or repress the action of GATA4 (Lu et al. 1999; Svensson et al. 1999, 2000). GATA4 and FOG2 expression are evident in the bipotential urogenital ridge and their expression persists in the fetal mouse ovary. In contrast, FOG2 expression is lost in fetal Sertoli cells along with the formation of the testicular cords (Anttonen et al. 2003). FOG2 diminishes the transactivation of *Mis* promoter by GATA4 in cultured postnatal rat Sertoli cells (Tremblay et al. 2001) and postnatal granulosa cells (Anttonen et al. 2003) endogenously expressing FOG2. *Foxl2* is a member of the winged helix/forkhead family of transcription factors. Forkhead proteins are found in all eukaryotes and serve important functions in the establishment of the body axis and development of tissues from all three germ layers in animals. Mouse *Foxl2* is expressed almost exclusively in ovarian follicular cells and the developing eyelid.

In mice, CYP19/aromatase is not expressed until late in ovarian development (Greco and Payne 1994), so ovarian development is largely independent of estrogen synthesis in mammals.

A summary of the gene action at early steps in mammalian gonadal development is presented in Figure 11.1.

Sex on the Wing

Gonadal sex differentiation in birds differs markedly from that in mammals, but many of the same genes are expressed during gonadal differentiation in both groups, although sometimes in different sequences. Female birds are heterogametic (designated ZW), and males are homogametic (ZZ). Avian sex chromosomes are not homologous to mammalian X and Y but share many features (Ellegren 2000). However, the Z chromosome is homologous to a large part of human chromosome 9, including 9p, a region associated with female sex reversal and gonadal dysgenesis in XY humans. Two copies of one or several genes within this region are required for normal male development. Interestingly, *Dmrt1* maps to this region in humans and has been implicated as a strong testes-determining gene in mammals (Raymond et

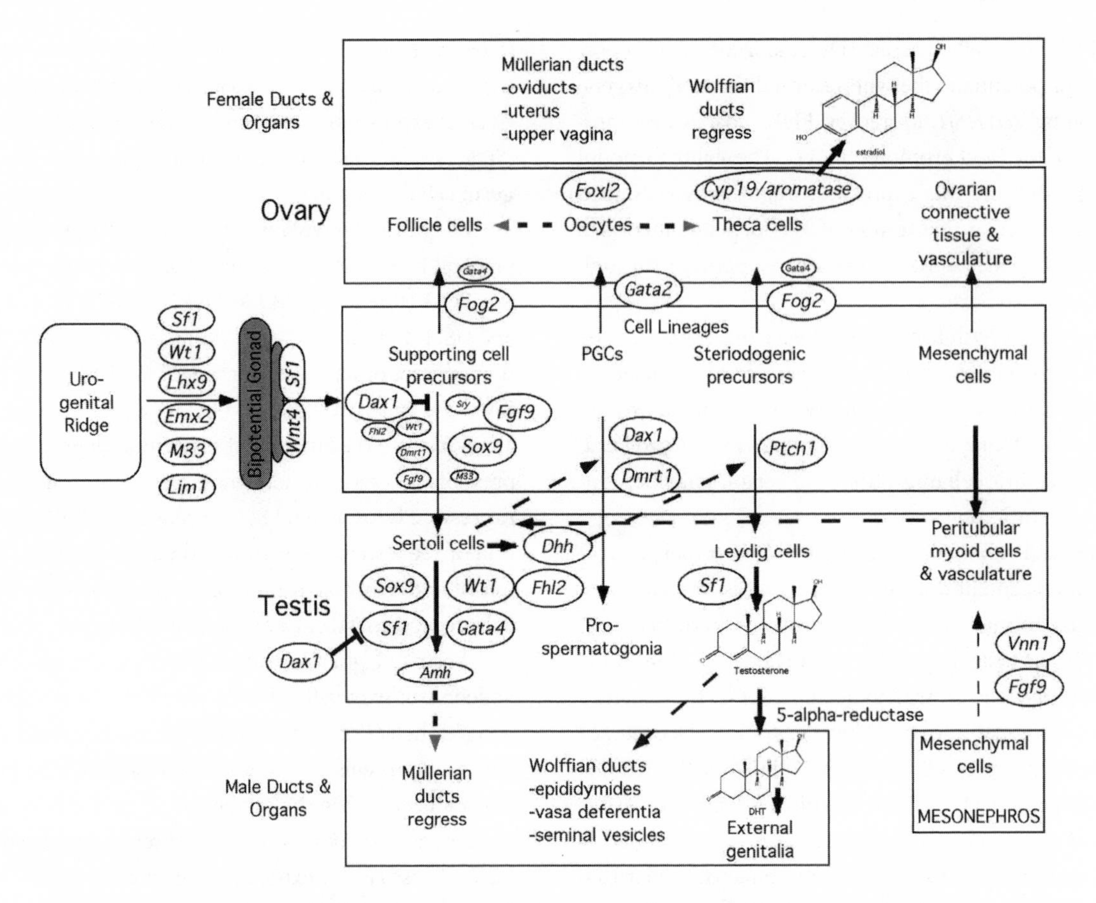

Figure 11.1 Gene action and cellular interactions during mammalian gonadal induction and early differentiation. Pathways of cellular differentiation and/or migration are indicated by thin black arrows, biosynthetic pathways by thick black arrows, and hormonal signaling or cellular migration by dashed arrows. (Modified from Koopman 2001b and Scherer 2002.)

al. 1999b, 2000). The avian homolog of *Dmrt1* maps to the Z chromosome, suggesting that it might also be involved in sexual differentiation in birds, perhaps through Z chromosome dosage (Nanda et al. 2000). What also maps to the Z chromosome is the receptor for hedgehog signaling, *patched (Ptc),* whose importance in testes cord formation in mice has already been described. *Dmrt1* is expressed in the genital ridge, from the time it starts to form, as well as in the Wollfian ducts, which become the male-specific reproductive structures. Moreover, *Dmrt1* expression is higher in male embryos than in female embryos (Raymond et al. 1999a; Smith et al. 1999a) but can be upregulated in ZW chick embryos treated with an CYP19/aromatase inhibitor and undergoing female-to-male sex reversal (Smith et al. 2003). The single Z in females is slightly methylated, whereas the two male Z chromosomes are hypermethylated in a region (MHM) adjacent to the *Dmrt1* locus (Teranishi et al. 2001). The MHM region is transcribed in females but is silenced in males by the hypermethylation, which is controlled by the W chromosome (Teranishi et al. 2001).

The W chromosome (like the mammalian Y) has a high heterochromatin content and contains few genes with homologs on the nonrecombining parts of the Z (Ellegren 2000). Among them is an altered form of a protein kinase C inhibitor *(PkciW),* widely conserved among nonratite birds and expressed in developing ovaries and various adult tissues of female birds (Hori et al. 2000; O'Neill et al. 2000). *PkciW* mRNA is destabilized by ASF/SF2 splicing factors (Lemaire et al. 2002), which mediate the sex-specific splicing of *doublesex* mRNA in *Drosophila* (Lynch and Maniatis 1996). A homolog of *PkciW* that differed in structure from its partner on the W chromosome by being more similar to the mammalian protein HINT (Hori et al. 2000) was found on the Z chromosome *(PkciZ).* PKCIW has a dramatically altered nucleotide binding site, and it might function by a dominant-negative mechanism through heterodimerization with PCKIZ, thereby inducing female development (Hori et al. 2000; O'Neill et al. 2000). An *in silico* experiment suggests that an extra dose of *PkciZ* may overcome the feminizing influence of *PkciW* and lead to the intersexes seen in triploids (Pace and Brenner 2003).

At present, the mechanism of avian sex determination is unclear. It may involve a dominant ovary determinant carried on the W sex chromosome (perhaps *PckiW*), or it may depend on the Z-chromosome dosage (copies of *PckiZ*) (Ellegren 2000), or both. Dosage compensation of sex (Z) linked genes was considered nonexistent but there is expression equalization for many Z-linked genes (McQueen et al. 2001). This is important in distinguishing between a dosage-based mechanism (two copies of genes in males vs. one copy in females) and one in which the Z genes in females are upregulated (Ellegren 2002).

In chickens, embryonic gonads appear as a thickening of the mesoderm on the ventromedial surface of the mesonephric kidneys at day 3.5 (stages 21 and 22). The bipotential gonads comprise an outer epithelial layer (cortex) and the underlying medulla. In the medulla, cords of epithelium-like cells (medullary cords) are interspersed with mesenchyme. In both sexes, the germ cells are scattered throughout the gonad but are more concentrated in the cortex. The gonads remain undifferentiated until day 6.5 (stage 30), when the first histological signs of sex differences are evident. By day 10.5 (stage 36), differentiation is advanced. In ZW females, the left gonad becomes a large ovary, characterized by a thickened surface epithelium (cortex) with proliferating germ cells and somatic cells. The medullary cords become thin and elongated, comprised of flattened epithelial cells separated by luminal spaces (lacunae). The right gonad of ZW females is much smaller than the left gonad and fails to develop. In ZZ male embryos, bilateral testes develop and are evident by day 10.5. Sertoli cells have differentiated within the medullary cords, giving rise to seminiferous cords and a reduced surface epithelium. Interstital cells are well developed.

In the chicken, PGCs in the early embryos are distinguished by their morphological characteristics, by their high glycogen content, as evidenced by staining with periodic acid-Schiff (PAS) reaction, and by immunocytochemical identification of specific cell surface antigens (Kingston and Bumstead 1995). Previous studies have shown that chicken PGCs are recognized at the early somite stage as a population of about 200 dispersed cells located in an extraembryonic region anterior to the head fold, referred to as the germinal crescent (Rogulska et al. 1971). From the germinal crescent region, PGCs migrate into newly formed vascular veins and are passively transported by the bloodstream to the vicinity of the embryonic gonads, where they invade the gonads. The chicken germ line is likely determined by maternally inherited factors in the germ plasma and involves the homolog to the *Vasa* gene (Tsunekawa et al. 2000). Moreover, again in contrast to mammals, both ovaries and testes will develop in the absence of PGCs (McCarrey and Abbott 1978, 1982).

No *Sry* gene has been detected in birds, although several *Sox* genes have been identified (Coriat et al. 1994). *Sox9* is expressed male specifically (Smith et al. 1999c), as in mammals, whereas *Amh* is expressed in both sexes (Oreal et al. 1998; Smith et al. 1999c) and, in contrast to mammals, expression of *Amh* precedes that of *Sox9*. In chickens *Wt1* is highly expressed in both male and female gonads from stage 28 to stage 40 (Smith et al. 1999c); however, this study did not distinguish between the various isoforms. In all vertebrates examined to date, *Sf1* expression in the bipotential gonad is observed prior to sexual commitment. In chickens, *Sf1* expression is maintained in males but is upregulated in females during sex determination (Smith et al. 1999b,c). *Dax1* expression increases between stage 28 and stage 30 in females and continues to increase at least to stage 30. In the male embryo, *Dax1* expression also increases after stage 30, but never approaches the levels observed in the female (Smith et al. 2000).

Estrogens and antiestrogens are involved in avian gonadal sex differentiation. In males, estrogens induce inhibition of the right gonad and transient feminization of the left gonad; conversely, in females antiestrogens induce masculinization of gonads (Scheib 1983). Sex reversal of genetic females has also been obtained by *in ovo* injection of aromatase inhibitors prior to the first histological signs of gonadal sex differentiation. CYP19/aromatase is the P450 enzyme responsible for converting androgens into estrogens. Inhibition of its activity results in the differentiation of a right testes and a left testis or ovotestis in ZW female chickens (Abinawanto et al. 1996; Elbrecht and Smith 1992; Vaillant et al. 2001). Hence, estrogen is critical for female gonad development. While estrogen receptor-α (ER-α) (Smith et al. 1997) and all the steroidogenic enzymes necessary to synthesize estrogens (Nomura et al. 1999) are expressed in the gonads of both sexes prior to sexual differentiation (See Figure 11.2), *aromatase* is only expressed in female gonads (Smith et al. 1997). *Aromatase* expression is first detected at stage 30 (day 6.5), the onset of differentiation. Expression is in the medullary cords of the gonad,

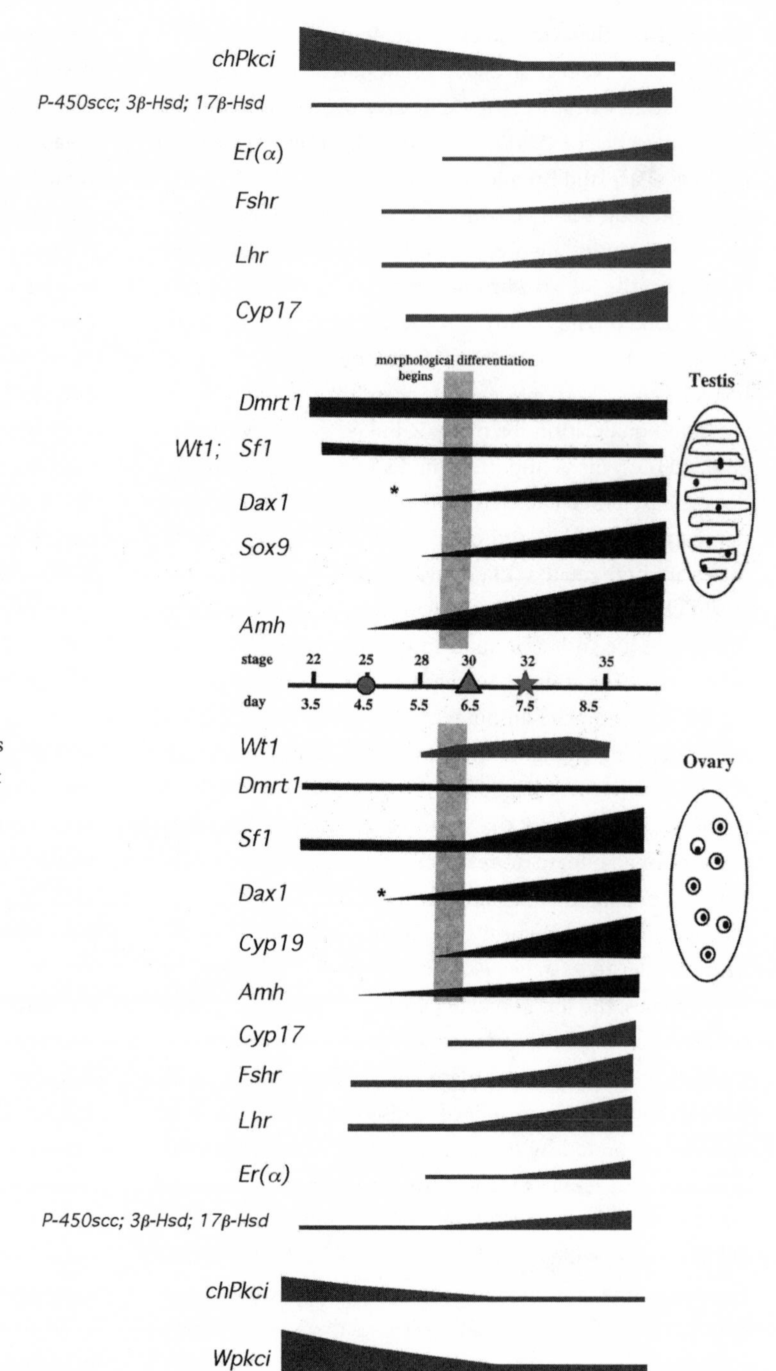

Figure 11.2 Sexually dimorphic gene expression during gonadal development in the chicken embryo. For each gene, line thickness reflects relative level of expression compared to that in the opposite sex (but not between genes within a sex). The shaded region (vertical bar) indicates the onset of morphological differentiation. Symbols: ● Bipotential gonad; ▲ Testis cord formation begins; ★ Testis cords complete; *Exact time of onset not determined. [Modified from Smith and Sinclair 2001 and Morrish and Sinclair 2002.)

while *ERα* is expressed primarily in the cortex. *ERα* is downregulated in the right gonad, consistent with its lack of development (Andrews et al. 1997). Moreover, a female-specific 17-β hydroxysteroid dehydrogenase (17-β-HSD), responsible for interconversion of testosterone and androstenedione as well as estradiol and estrone, is also expressed in the gonadal medulla of female gonads (Nakabayashi et al. 1998). Interestingly, the receptors for gonadotropins (follicle stimulating hormone receptor, *Fshr*, and luteinizing hormone receptor, *Lhr*) are expressed before morphological differentiation occurs in gonads of both sexes in birds (Figure 11.2).

A summary of the gene action at early steps in avian gonadal development is presented in Figure 11.2. Overall, many of the same genes shown to be important in the mammalian sex determination network are involved in avian sex determination, but the timing and interplay of the genes differ (Smith and Sinclair 2001).

Some Like It Hot, Some Not

In contrast to mammals and birds, many reptiles lack heteromorphic sex chromosomes, and it is the egg incubation temperature that determines the sex of the embryo (e.g., Lance 1997; Pieau et al. 1999a; reviewed in Chapters 3–6). In many turtles, including all sea turtles, temperatures around 27°C produce males, while females are produced at temperatures around 31°C. In crocodilians, females are also produced around 30°C, while males hatch at an incubation temperature of 33°C. Temperatures intermediate to these two "set points" provide mixed sex ratios or sometimes neonatal intersexes that are later resolved to one or the other sex. A period of development (thermosensitive period, TSP), typically between 18 and 30% of the incubation time, is the window for commitment. This can be as few as four to six days in turtles and as many as 10 days in alligators. In all species this TSP corresponds to the first stages of gonadal differentiation, as revealed by classical histology (Pieau 1996).

Gonadal differentiation in alligators and turtles is described in Pieau et al. (1999a). Briefly, both testicular seminiferous cords and ovarian medullary cords/lacunae derive from the proliferation of germinal epithelium. Follicle cells (pregranulosa) also appear to be derived from the germinal epithelium, whereas interstitial cells appear to be derived from the initial mesonephric mesenchyme underlying the germinal epithelium.

PGCs in snakes and in some lizards are found in the anterior germinal crescent and later migrate toward the gonads by a vascular transfer pathway, as in the case of the chicken. In contrast, in turtles and in other species of lizards such as some *Lacerta*, PGCs locate at the posterior part of the primitive streak, and their interstitial migration is similar to that of mammals (Fujimoto 1979; Hubert 1969). Interestingly, PGCs in *Sphenodon* locate and migrate in both manners (Tribe and Brambell 1932). It seems possible that the difference in localization and subsequent migration in PGCs may be due to the time of PGC allocation. Thus, further comprehensive studies of *Vasa*-expressing cells in any temperature-dependent species will provide new insights into the evolutionary changes in germ cell development.

Differences in gene expression for sex-determining genes among mammals, chickens, and alligators are described in Morrish and Sinclair (2002). The sequence, sex specificity, and temporal expression of these genes during the transition from a bipotential gonad to a testes or ovary differ among the groups (see summary in Figure 11.3, which contains added information on *aromatase* expression, and notice expression dissimilarities of the same genes in chicken by comparing with Figure 11.2). In contrast with *Amh* expression in female chickens, *Amh* is not expressed in female gonads at any alligator stage. In chickens, *Dax1* expression is higher in female embryos than in male embryos, while no apparent differences are seen in alligator embryos. One set of common attributes is the lack of *Sox9* expression in the presumptive ovaries, expression of *Amh* preceding expression of *Sox9*, and upregulation of *Dmrt1* in male embryos.

If we now compare what is known about the developmental expression of these genes in turtles (Figure 11.4), the "many means to an end" problem becomes even more evident. Not only is *Sox9* expressed in both the male and female developmental profiles (Spotila et al. 1998; Torres Maldonado et al. 2002; Valleley et al. 2001; Western et al. 1999) (although it is shut off later in ovarian development), but steroidogenic enzymes (including aromatase) are expressed early and more intensely in ovarian development than in the alligator at the same incubation temperature (Desvages et al. 1993; Willingham et al. 2000; Desvages and Pieau 1992; Jeyasuria and Place 1997, 1998; Jeyasuria et al. 1994; Place et al. 2001). *Dmrt1* and *Dax1* are upregulated in the testes pathway (Torres Maldonado et al. 2002). *Wt1* is upregulated in the ovarian condition (Spotila and Hall 1998; Spotila et al. 1998), while in the red-eared slider turtle, *Sf1* expression is maintained in males and downregulated in females (Crews et al. 2001; Fleming and Crews 2001). We currently have no information on *Mis* expression in turtles.

None of these transcription factors work in isolation, as discussed for mammals. Not only is a combinatorial repertoire of factors needed for transcriptional activity of the *Sf1* promoter, but this suite of characters changes among cell types (Scherrer et al. 2002). Figure 11.5 contains the alignment for all known reptilian SOX9 proteins, illustrating that there is 100% amino acid sequence homology in the HMG domain among the reptiles and birds, including the nuclear localization signals necessary for import and export to the nucleus. However, in the domain responsible for interaction with other transcription factors (transactivation domain), a far greater degree of divergence is found, implying that the transcriptional machinery may be more sensitive to coevolution and changes in the interacting partners.

It is expected that in the next few years, each of the genes presented in Table 11.1 and Figure 11.1 will be found to be operative in the combinatorial network of TSD. Moreover, utilization of a large diversity of the mechanisms, from

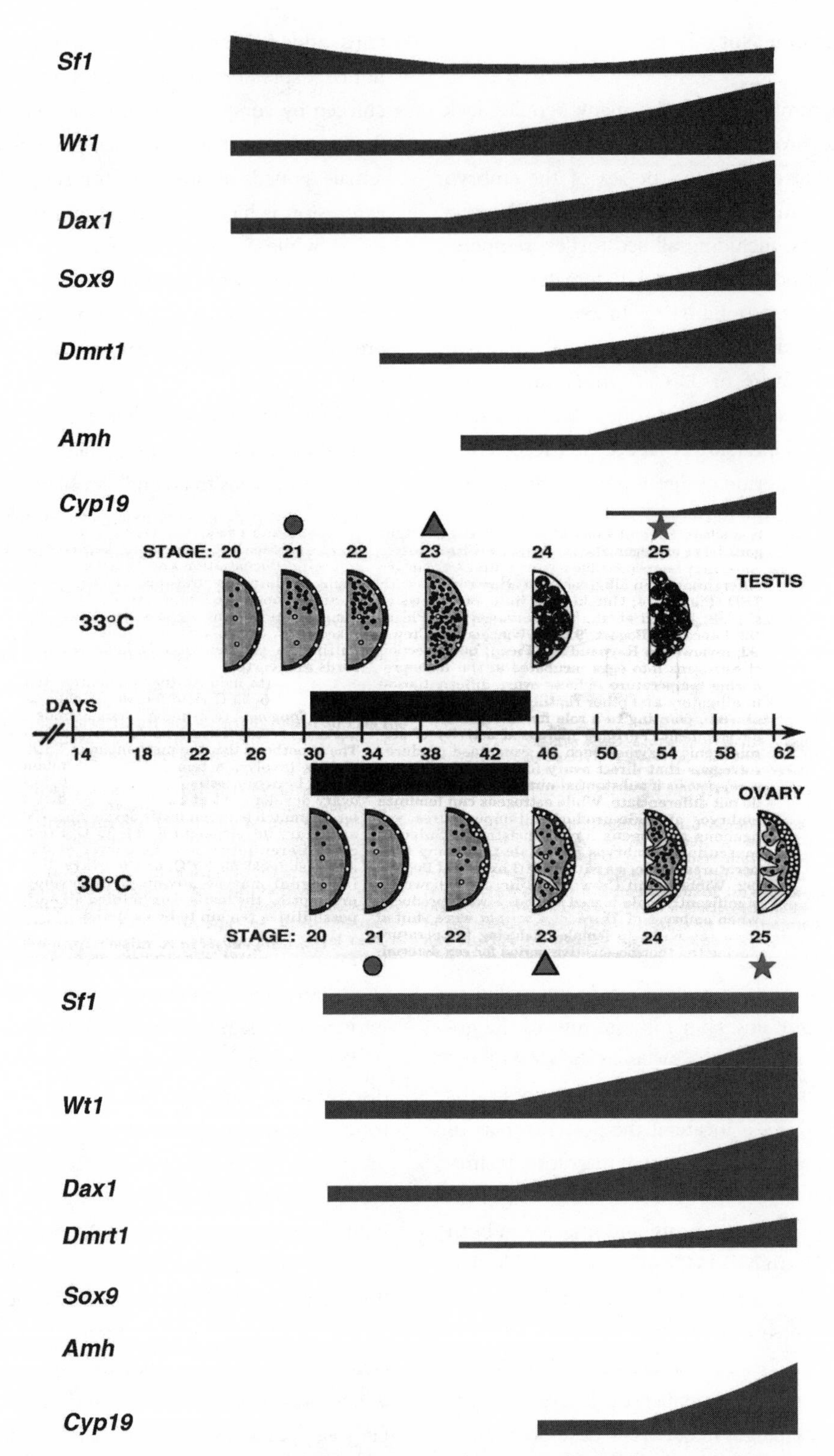

Figure 11.3 Sexually dimorphic gene expression during gonadal development in the alligator embyro. For each gene, line thickness reflects relative level of expression compared to that in the opposite sex (but not between genes within a sex). The shaded region (black horizontal bar) indicates the thermosensitive period. Symbols: : ● Bipotential gonad; ▲Testis cord formation begins; ★ Testis cords complete. (Modified from Morrish and Sinclair 2002.)

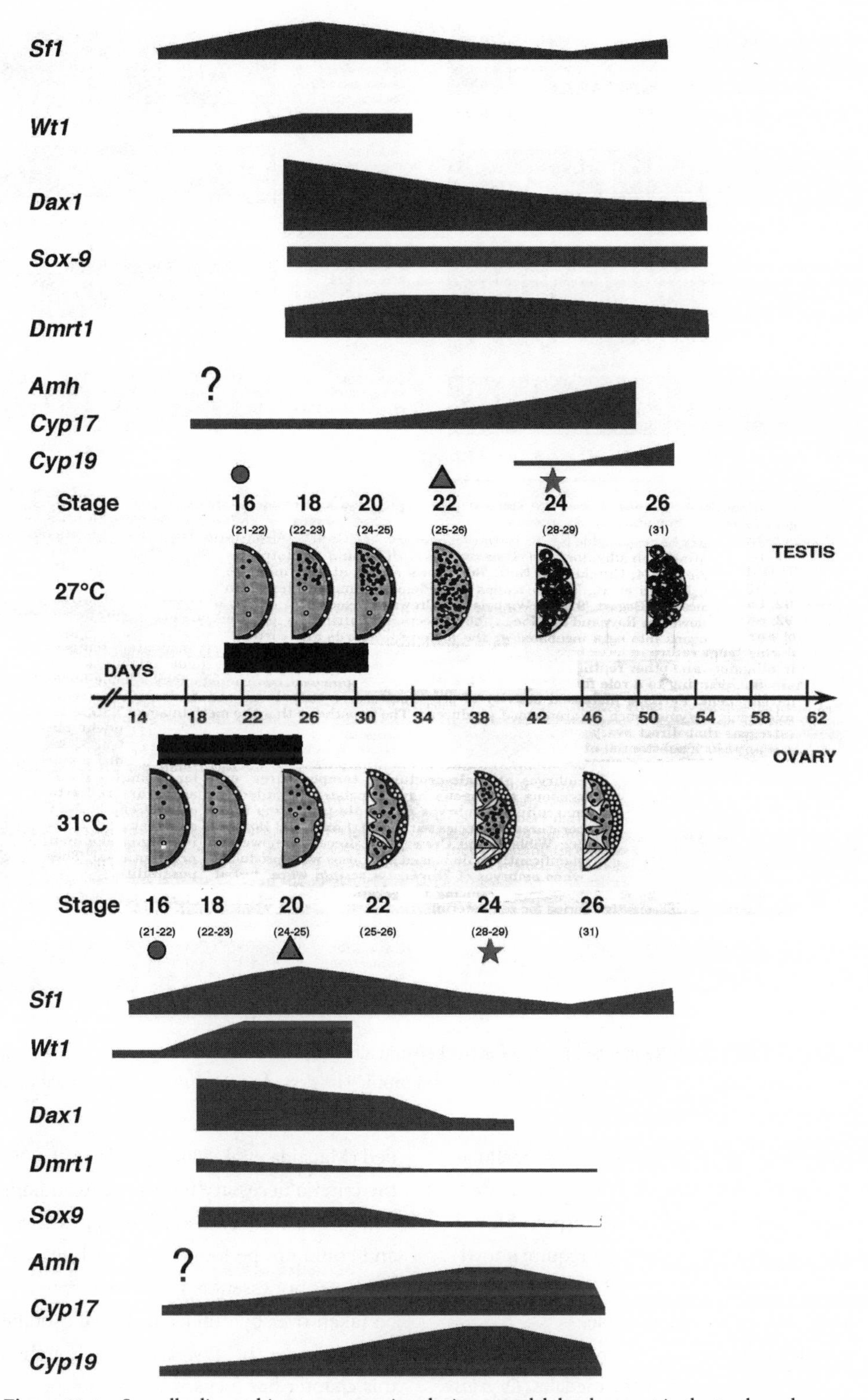

Figure 11.4 Sexually dimorphic gene expression during gonadal development in the turtle embryo. For each gene, line thickness reflects relative level of expression compared to that in the opposite sex (but not between genes within a sex). The shaded region (black horizontal bar) indicates the thermosensitive period. Symbols: : ● Bipotential gonad; ▲Testis cord formation begins; ★ Testis cords complete. (Modified from Morrish and Sinclair 2002.)

Figure 11.5 Alignment of the protein sequence of SOX9 for chicken and all published reptiles. The solid lines mark the HMG boundaries. The enclosed rectangle delineates the nuclear localization motif. The dotted lines mark the transactivation domain.

alternative splicing (e.g., *Wt1*) to control of subcellular localization of transcription factors (SOX9) is anticipated. Which gene or gene-gene interactions are responsible for tipping the scale toward ovary or testes may require knowledge of the entire gene network, which in this day of "whole genome sequencing" is possible and probable.

One final example of the diversity for the vertebrate sex determination cascade is illustrated by the medaka *(Oryzias latipes)*. A putative Y chromosome determinant (DMY), a member of DM-domain proteins (i.e., DMRT) was identified (Matsuda et al. 2002; Nanda et al. 2002), and it met all the criteria necessary for a sex-determining gene. Unfortunately, this Y chromosome switch was specific to medaka and could not be found in other bony fish. Moreover, the gene was not essential for sex determination; its role could be taken over by other autosomal modifiers (Nanda et al. 2003). Shades of *Sry*! Lastly, although all discussion in this chapter has focused on gonad-specific events, gender-specific changes that occur in the developing brain also deserve attention (Place et al. 2001; Wolfner 2003).

12

PAMELA K. ELF

Yolk Steroid Hormones and Their Possible Roles in TSD Species

Hormones and Development

The hormonal environment during embryonic development has a significant influence on brain differentiation and postnatal behavior in several mammalian species (Goy and McEwen 1980; Arnold and Gorski 1984; Frank et al. 1991). In oviparous vertebrates, maternal contributions in the form of steroid hormones that establish the embryonic environment are deposited in the yolk during vitellogenesis. Vitellogenin, the primary yolk protein, is produced by the liver and transported to the developing oocyte. This process is estrogen dependent and is the consequence of increased estradiol (E_2) production by the reproductively active female ovary.

In avian systems, maternally derived yolk steroid hormones are reported to have numerous developmental, growth, and behavioral impacts on offspring. Differences in maternal hormone contributions to yolk, specifically testosterone (T), in sequentially laid canary and zebra finch eggs correlate with social rank in the offspring (Schwabl 1993). Variations in yolk T levels correlate with differences in subsequent growth and behaviors in canaries and are independent of offspring sex. Further, these hormone effects can be enhanced by the addition of exogenous T (Schwabl 1996a). These differences in maternal yolk T are also sensitive to photoperiod, indicative of environmental regulation (Schwabl 1996b). In cattle egrets, greater androgen levels in the yolk material of the first egg laid appear to contribute to siblicide behaviors in this species (Schwabl et al. 1997). Gil et al. (1999) reported differential maternal deposition of T and dihydrotestosterone (DHT) in yolks of zebra finch eggs, with higher levels associated with matings to males of greater attractiveness, that is, better fitness qualities. Nager et al. (1999) documented a bias in sex ratio skewed toward females in the lesser black-backed gull in response to starvation conditions. In this system, female offspring have a better survival rate under adverse conditions, and so would fare better than male offspring. Further, Petrie et al. (2001) described sex differences in yolk hormone levels during embryonic development of the peafowl. They hypothesize that the difference in hormone profiles is responsible for chromosomal segregation events that determine the sex of the hatchlings. Exogenous estradiol (E_2), when administered to laying female Japanese quail, can be transferred to offspring via the yolk and subsequently can influence sexual differentiation in the offspring and alter adult phenotype (Adkins-Regan et al. 1995). Sockman et al. (2001) described a relationship between female plasma prolactin levels and yolk T concentrations that can be modified by a reduction in food availability in the American kestrel. This correlation implies that circulating female hormone levels correlate directly with quantities found in the yolk, and also can be modulated by environmental conditions. All of these reports demonstrate that through differential maternal steroid contribution to yolk material, avian females can strongly impact the development and subsequent behaviors and phenotypes of their offspring.

Much less is known about the effects of maternally contributed steroid hormones in the yolks of oviparous reptiles. Besides possible growth and behavioral influences, do yolk steroid hormones play a role in the sex determination process in reptiles with temperature-dependent sex determination (TSD)? This chapter will review what is currently known about the dynamics of yolk steroid hormones in TSD reptiles.

As in avian systems, yolk steroid hormones are deposited during vitellogenesis, and similarly, there is some evidence that yolk hormone levels are reflections of circulating plasma levels in reproductively active females. Janzen et al. (2002) treated gravid female red-eared slider turtles, *Trachemys scripta*, with DHT, E_2, and T via silastic capsule implants. They found that implanted females had increases in systemic circulation of the respective hormones and subsequent increases in yolk hormone content in developing follicles compared with controls. Callard et al. (1978) reported a direct relationship between circulating plasma E_2 levels and vitellogenin in the painted turtle *(Chrysemys picta)*, with the largest E_2 peak occurring in the spring at the time of preovulation. They also document peaks in T and progesterone (P) just slightly later, during ovulation. Concurrent with these hormonal changes, yolk is being added to developing follicles of different sizes within the ovary. These different-sized follicles are thought to represent the reproductive output of the female for the next three years, at least in this population of turtles that lays one clutch per year.

Possible Roles of Yolk Steroid Hormones in TSD

It has been hypothesized that steroid hormones or hormone precursors present in the adrenals, the gonads, or the yolk play a role in the process of sex determination in TSD species (Crews et al. 1989). According to another hypothesis proposed by Bogart (1987), the E_2/T ratio at the site of gonadal differentiation is the factor responsible for determining the sex of the developing embryo. In support of these hypotheses, hormones and hormone antagonists, specifically estrogens and P450 aromatase inhibitors, when applied to the eggshell of some TSD species, can alter the sex of the developing embryo (Crews et al. 1989, 1991, 1995, 1996a; Rhen and Lang 1994; Rhen et al. 1996). Also, hormone disrupters, including estrogen mimics, have been reported to have developmental impacts on crocodilians and turtles (Guillette et al. 1994, 1996; Willingham and Crews 1999).

Subsequent research demonstrated that the developing gonad of the red-eared slider turtle, *T. scripta*, does not produce either E_2 or T and that the adrenal-kidney complex produces only very low levels of T (White and Thomas 1992a). Further studies in the American alligator *(Alligator mississippiensis)* determined that E_2 was not synthesized in significant quantities by the gonad-adrenal-mesonephric kidney (GAM) until after the thermosensitive period (TSP) of development, during which time the sex of the hatchling is determined (Smith et al. 1995; Gabriel et al. 2001). The finding that the GAMs are quiescent early in development leads to the hypothesis that the yolk is the initial source of hormones necessary for successful development of the embryo and that the relative levels of hormones in the yolk may exert a significant influence on the developing gonads.

It has since been documented that yolk is a plentiful source of both E_2 and T in TSD turtles (Bowden et al. 2000; Elf et al. 2002a), alligators (Conley et al. 1997; Elf et al. 2001), and leopard geckos *(Eublepharis macularius)* (Table 12.1), and androstenedione (A_4) has also been quantified in yolks from alligator eggs (Conley et al. 1997). White and Thomas (1992b) measured whole-body and circulating plasma steroid concentrations of E_2 and T in the developing embryo of the red-eared slider turtle. They reported greater levels of E_2 than T in both male and female embryos at all stages measured, and similar, if not greater levels of E_2 than T, in circulating plasma. If the yolk is the source of these endogenous steroids prior to any significant production by the embryo, it would be logical for embryonic levels to mirror those found in the yolk. The quantities of E_2 and T in turtle yolks reported to date (Gerum 1999; Bowden et al. 2000; Elf et al. 2002a) are consistent with that conclusion.

So, how do yolk steroid hormones, specifically E_2, act at the molecular level to influence sex determination in these TSD species? The underlying gene expression patterns responsible for sex differentiation have only been investigated in a few TSD species (see Place and Lance, Chapter 11). In the alligator, expression patterns of several of the genes involved in sex differentiation have been described. One of these, *Sf1* (steroidogenic factor one), a transcription factor, regulates the expression of steroidogenic enzymes. *Sf1* is necessary for activation of AMH (anti-Müllerian hormone) in mammalian species (reviewed by Koopman et al. 2001), which is necessary for male differentiation. *Sf1* is expressed very early in sex differentiation and is thought to be necessary for germinal ridge formation (Ingraham et al. 1994).

Table 12.1 Initial Yolk Hormone Levels in TSD Reptiles

Species	Yolk Estradiol 17-β (ng/g)	Yolk Testosterone (ng/g)
Alligator mississippiensis[a]	19.0 ± 1.3 (1994)	2.3 ± 0.18 (1994)
	5.0 ± 0.4 (1995)	1.0 ± 0.1 (1995)
Chelydra serpentina[b]	2.78 ± 0.095	2.56 ± 0.098
Chrysemys picta[b]	0.89 ± 0.064	0.68 ± 0.045
Emydoidea blandingii[c]	0.99 ± 0.126	1.15 ± 0.093
Eublepharis macularius[d]	0.55 ± 0.06	0.15 ± 0.03
Graptemys ouachitensis[c]	2.66 ± 0.181	0.15 ± 0.012
Graptemys pseudogeographica[c]	1.91 ± 0.156	0.19 ± 0.024
Trachemys scripta[e]	2.815 ± 0.703	0.674 ± 0.178

[a]Data from Conley et al. (1997).
[b]Data from Elf et al. (2002a).
[c]Data from Gerum (1999).
[d]P. Elf, unpublished data.
[e]Data from Bowden et al. (2001).

Sf1 expression patterns in developing gonads have been investigated in two TSD reptiles, the alligator (Western et al. 2000) and the red-eared slider turtle (Fleming et al. 1999). Results of these two reports both demonstrate sexually dimorphic expression of *Sf1* early in the TSP, during gonadal differentiation, but they document opposite patterns of expression. *Sf1* appears to be upregulated in the developing male gonad in the red-eared slider turtle, and downregulated in the female, similar to the pattern seen in mammals. Contrarily, *Sf1* expression is increased in the female gonad of the alligator and decreased in the male gonad, as in the chicken. Further, in the red-eared slider turtle, addition of exogenous E_2 has been reported to modulate *Sf1* expression (Fleming and Crews 2001). Applications of exogenous E_2 to the shell resulted in a downregulation of *Sf1* expression and the histological appearance of female gonads at male-producing temperatures. Thus, exogenous E_2 treatment resulted in an artificial hormonal environment similar to that reported to occur naturally in the yolks of snapping turtle eggs incubated at the female-producing temperature, that is, greater E_2 concentration, compared with those incubated at the male temperature (Elf et al. 2002b). This would indicate a possible interaction between yolk E_2 and SF1 in the sex determination pathway. Murdock and Wibbels (2002) recently presented opposite results with respect to *Sf1* expression in the red-eared slider turtle. They performed quantitative competitive reverse transcriptase polymerase chain reaction (RT-PCR) for *Sf1* on the total RNA isolated from GAMs taken at male- and female-producing temperatures during development. They documented greater levels of *Sf1* expression in GAMs from male-producing temperatures than in those from female-producing temperatures, but since the embryonic adrenal gland is extremely active, these results do not accurately reflect the activity of the gonad alone (T. Wibbels, pers. comm.). These conflicting pictures emphasize the importance of further work to confirm the pattern of *Sf1* expression in turtle species.

Yolk Dynamics during Development in Turtles and Alligators

The changes in yolk steroid hormones during the course of development in two TSD species, the alligator (Conley et al. 1997; Elf et al. 2001) and the snapping turtle (Elf et al. 2002b) have been investigated. The results provide evidence for similarities not only in yolk hormone levels, but also in yolk dynamics for the snapping turtle and the alligator. In both systems, E_2 and T decrease significantly by hatching. The androstenedione (A_4) levels in the alligator have also been quantified, and although initial levels are several times greater than those of all other hormones, A_4 declines to levels comparable to those of E_2 and T by stage 25 of development (Conley et al. 1997).

In the snapping turtle a significant difference in the rate of decline of E_2 levels at different incubation temperatures was documented. Greater levels of E_2 were maintained at the female-producing temperature (29.5°C) than at the male-producing temperature (26°C), and intermediate E_2 levels were measured at the pivotal temperature (28°C) (Figure 12.1). T levels did not follow this pattern, suggesting that incubation temperature influences the hormonal environment of the developing embryo by regulating the amount of E_2 in the yolk.

The pattern of E_2 decline reported in the alligator (Conley et al. 1997) varied slightly from that reported in the snapping turtle. Although not significant, yolk levels at stage 21 (during the sex-determining period) were greatest at 34°C (high-temperature female producing), least at 33°C (male-producing), and intermediate at 31°C (low-temperature female producing). Temperature appears to affect E_2 levels differently in this system, but elevated E_2 levels are associated with yolks from eggs that will produce females, regardless of whether they are produced at higher or lower temperatures. Analysis of residual yolk material obtained from hatchling alligators showed a similar pattern (Elf et al. 2001). Yolks from hatchlings incubated at two female-producing temperatures (29 and 31°C) had the greatest E_2, and yolks from hatchlings incubated at the male-producing

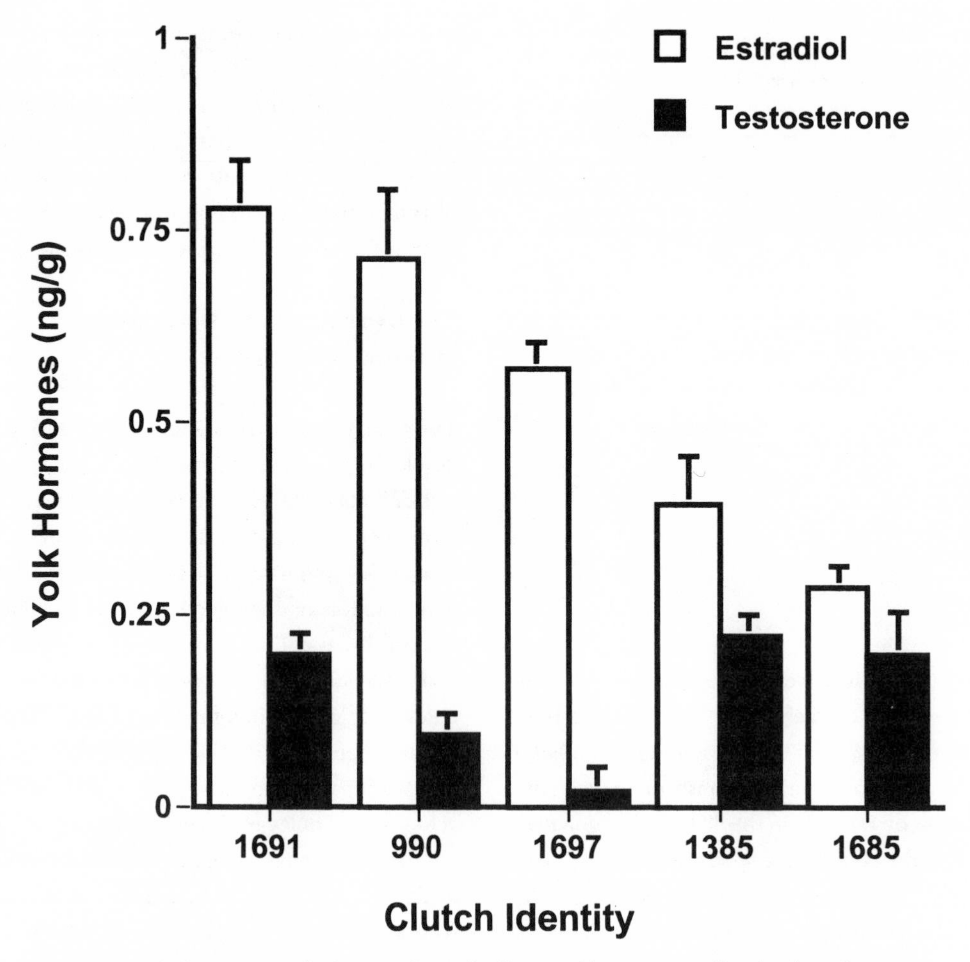

Figure 12.1 Preliminary results from analysis of yolk steroid hormones in five clutches of two eggs each from five female leopard geckos, *Eublepharis macularius*. Values are means ± SEM. Data were analyzed for clutch differences using ANOVA ($p = 0.0078$, $F = 7.62$, $df = 4, 10$ for E_2; $p = 0.0144$, $F = 6.97$, $df = 4, 10$ for T).

temperature (33°C) had the least E_2. Yolks from hatchlings incubated at an intermediate temperature (32°C), which produced a mixed sex ratio, had an intermediate level of E_2. These E_2 differences were significant among temperatures. The most surprising result was that no difference existed between yolk E_2 levels from male and female hatchlings produced at the intermediate temperature. This would indicate that incubation temperature, not hatchling sex, has the greatest influence on yolk E_2. As in the snapping turtle system, the alligator system shows no similar temperature effects on yolk T levels.

The influence of incubation temperature on yolk E_2 levels is consistent with the reported synergism between temperature and E_2 effects seen in the red-eared slider turtle (Wibbels et al. 1991a). Wibbels et al. (1991a) reported exogenous E_2 treatment produced more females at a pivotal temperature than would be expected from the effects of E_2 treatment or temperature alone. Wibbels and Crews (1995) also reported that in this species lesser concentrations of exogenous E_2 were required to produce 100% females as incubation temperature was increased toward the female-producing temperature. Based on these results, they hypothesized that E_2 and temperature are both components of the same sex-determining pathway. The author's data indicates that eggs incubated at higher temperatures retain more endogenous yolk E_2; therefore, less exogenous E_2 would be required to increase the female to male ratio. This interrelationship is consistent with the report by Bowden et al. (2000), in which elevated E_2 in yolks of eggs from the painted turtle produced later in the season was correlated with a female sex bias in the offspring. Together these data suggest that a similar mechanism is operating in the alligator and in all three of these TSD turtle species.

The author also documented significant differences

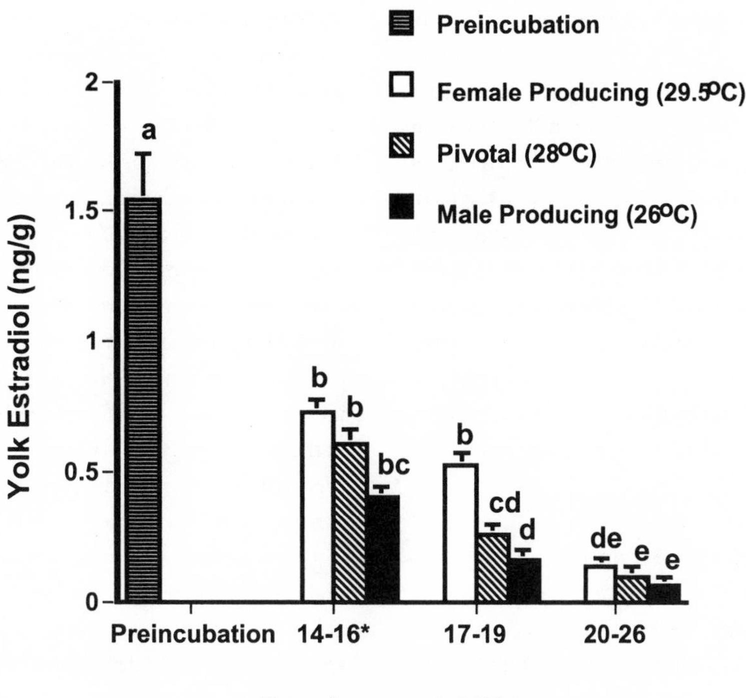

Figure 12.2 Incubation temperature effects on E_2 in yolks of snapping turtle eggs through development. Data were log transformed and analyzed using a mixed model two-way ANOVA for temperature, time and temperature-time interactions. $p < 0.001$, $F = 31.534$, $df = 2, 306$ for temperature; $p < 0.001$, $F = 178.643$, $df = 2, 306$ for time; $p = 0.600$, $F = 0.689$, $df = 4, 306$ for time-temperature interactions. Values represent means ± SEM. Letters indicate significant differences between groups having different letters. For stages 14–16, N = 30 for 29.5°C, N = 35 for 28°C, N = 41 for 26°C; for stages 17–19, N = 39 for 29.5°C, N = 34 for 28°C, N = 34 for 26°C; for stages 20–26, N = 36 for 29.5°C, N = 39 for 28°C, N = 28 for 26°C. Embryos staged according to Yntema 1968. Symbol: * Designates the thermosensitive period for this population (adapted from Elf et al. 2002b).

among clutches in E_2 levels throughout development in both the snapping turtle and alligator systems (Conley et al. 1997; Elf et al. 2002a). The fact that the clutch effects of E_2 persist and those of T do not supports the hypothesis that E_2 is the more influential of the hormones and probably plays a more important role during sex differentiation of the embryo.

The importance of E_2 in the sexual differentiation process in turtles is well documented. Studies addressing sex reversal in the red-eared slider turtle show estrogens to be potent agents for the production of a greater number of female offspring at male-producing and pivotal temperatures (Crews et al. 1995, 1996a), a phenomenon demonstrated to be dose dependent (Crews et al. 1991). Willingham and Crews (1999) reported that environmental estrogen mimics can cause sex reversal in the red-eared slider turtle. In the snapping turtle, exogenously applied E_2 can increase the number of females produced at male temperatures, and exogenously applied aromatase inhibitors are capable of producing greater percentages of males at temperatures that typically produce a majority of females (Rhen and Lang 1994). In contrast, testosterone treatment does not increase the number of males produced at female temperatures (Rhen and Lang 1994). Further, ^3H-E_2, when applied to snapping turtle eggs incubated at male- and female-producing temperatures (26°C and 30°C, respectively), shows different patterns of absorption into the yolk as well as differential uptake by developing tissues (Elf et al. 2002c, unpubl. data). At the female temperature there appears to be an equilibrium established between ^3H-E_2 in the albumin and that in the yolk material, whereas, at the male temperature, ^{3}H-E_2 appears to be sequestered in the albumin, and levels remain low in the yolk through development (Figure 12.2). Other differences include significantly greater uptake by gonads at the female-producing temperature and by brain tissue at the male-producing temperature during the later stages of development (P. K. Elf, unpubl. data). These findings not only provide further support for an essential role of estrogens in the sex determination and differentiation processes in these species, but also accentuate the impact that environmental estrogens could have on these developing systems. Because they would be absorbed through the eggshell in the same way, they could also be incorporated similarly into developing tissues.

Initial Levels of Hormones in TSD Species

Initial levels of E_2 and T (measured at or shortly after oviposition) have been quantified for several TSD reptiles, including the alligator (Conley et al. 1997), several turtle species (Janzen et al. 1998; Gerum 1999; Bowden et al. 2000, 2001, 2002; Elf et al. 2002a), and the leopard gecko (see Table 12.1). All species measured to date contain yolk steroids in the nanogram per gram range, with alligators having the greatest levels and leopard geckos having the least (see Table 12.1). A strong correlation between initial E_2 and T levels was also reported for painted and snapping turtles (Elf et al. 2002a), indicating that female yolks with greater E_2 concentrations usually have greater T as well. In the painted turtle, Callard et al. (1978) reported large increases in plasma E_2 and T during the preovulatory period, but the T levels were approximately four times greater than E_2 levels. They hypothesized that T was important as a precursor for E_2 in tissues expressing aromatase activity. If T functions primarily as a substrate for E_2 production, it would be logical to expect the high correlations between the levels of the two hormones.

There appear to be significant differences among hormone levels from different TSD species (Gerum 1999; Elf et al. 2002a; Table 12.1) and, perhaps of even more relevance, significant differences among steroid levels in eggs from different females of the same species that would contribute to "clutch effects." These clutch effects are present in the alligator (Conley et al. 1997), all turtle species studied to date (Janzen et al. 1998; Gerum 1999; Bowden et al. 2000; Elf et al. 2002a), and the leopard gecko (Figure 12.3) (see also Rhen and Lang, Chapter 10; Valenzuela, Chapter 14). This variation might be the consequence of varying levels of circulating hormones present in different females of some TSD species (Guillette et al. 1997) and could be the reason for differences noted in several parameters of sex determination and growth. Family, or female, differences have been reported in the sex ratios produced at pivotal temperatures of both snapping turtles and painted turtles (Rhen and Lang 1998). Clutch identity has also been reported to influence residual yolk mass, fat body mass, and total mass of hatchling snapping turtles (Rhen and Lang 1999b). Moreover, studies of the growth of snapping turtles showed significant clutch effects in growth rates that were independent of egg mass (Rhen and Lang 1995). Differences in individual female yolk hormone levels might play a role in these previously reported variations in growth, similar to the variations attributed to yolk T differences in finches and canaries (Schwabl 1996a).

Despite the fact that there is wide variation among females in steroid levels found in their eggs, there is remarkably little difference among eggs within a clutch for all the species examined to date, including the alligator (Conley et al. 1997), five turtle species (Gerum 1999; Elf et al. 2002a), and the leopard gecko (Figure 12.3).

Several studies describe layering of steroid hormones in the yolks of eggs from the painted turtle. Elf et al. (2002a) described the hormone content of shelled eggs and follicles of four different sizes obtained from a gravid female from a Northern population of *C. picta* that produces only one clutch of eggs per year. Smaller follicles had the greatest E_2 concentrations per gram of yolk, with levels decreasing as follicle size increased and eggs became shelled. This decline in E_2 is possibly due to diminished incorporation of E_2 and/or dilution of E_2 already present in the initial layers. Testosterone levels, on the other hand, are greatest in middle-sized follicles, with lesser amounts in small and large follicles and shelled eggs. Bowden et al. (2001, 2002) reported similar E_2 dynamics for developing follicles and whole eggs from *T. scripta* and a population of *C. picta* that lays two clutches per season. They also measured yolk P in both turtle populations and found the greatest amounts in the largest follicles and/or outer layers of follicles. There is evidence that a relationship between circulating maternal hormones and yolk hormone levels also exists for other TSD reptiles. Guillette et al. (1997) reported elevated circulating levels of E_2 and vitellogenin and lower levels of T in reproductive alligators, concurrent with vitellogenesis. Levels of T were greatest concurrent with courtship and mating behaviors, and the levels of both E_2 and T varied among females. Differential layering of hormones in the yolk of alligator eggs and differences between levels in eggs from different females (Conley et al. 1997) could be the result of

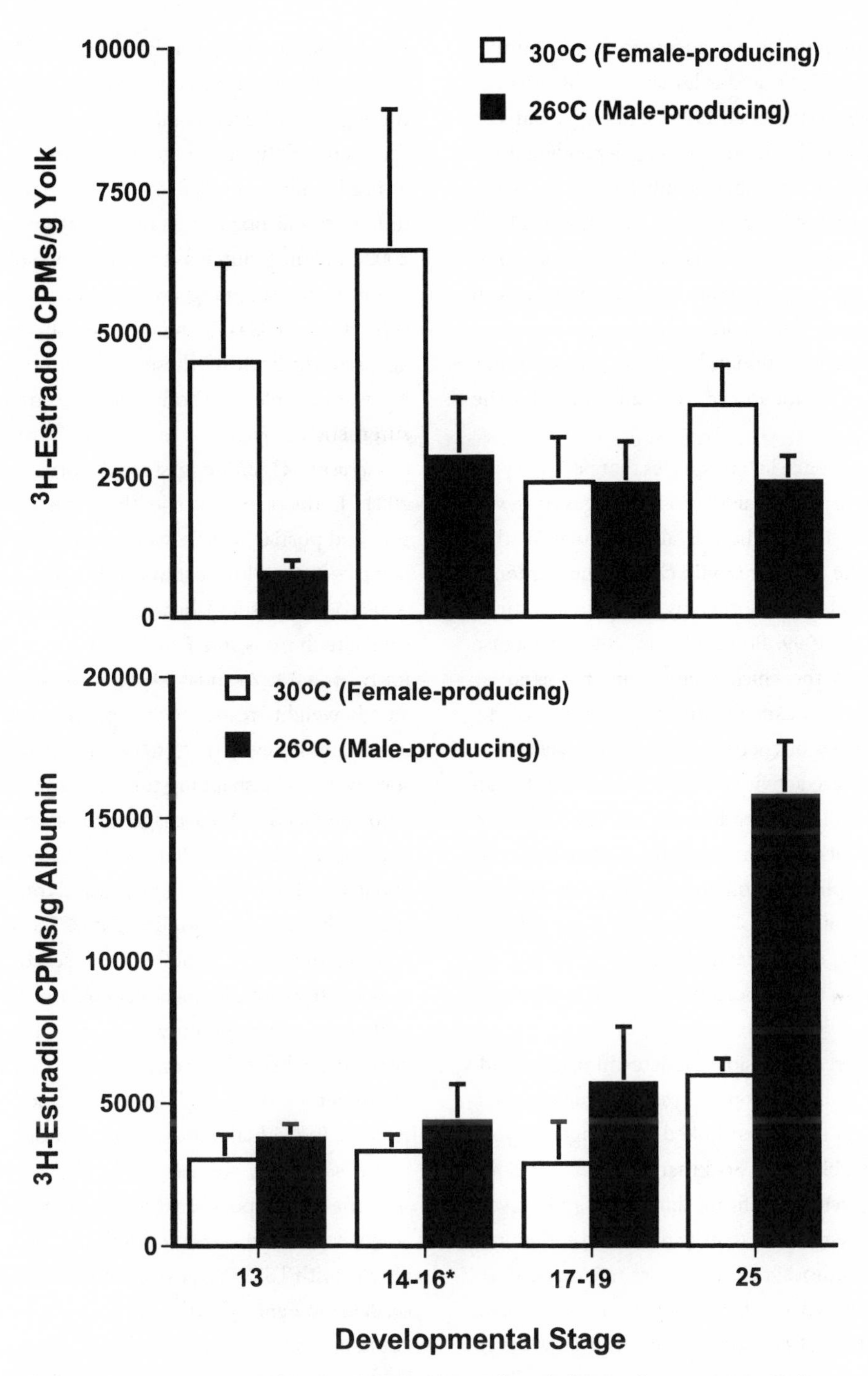

Figure 12.3 Dynamics of ^{3}H-estradiol uptake in yolk and albumin from snapping turtle eggs incubated at male- and female-producing temperatures. Values represent means ± SEM. *Group A* = Yntema stage 13, N =12 for 30°C, N = 2 for 26°C; *Group B* = Yntema stages 14–16 (TSP), N = 9 for 30°C, N = 6 for 26°C; *Group C* = Yntema stages 17–19, N = 5 for 30°C, N = 17 for 26°C; *Group D* = Yntema stage 25, N = 4 for 30°C, N = 6 for 26°C. Data was analyzed for incubation temperature effects using ANOVA. For yolk, $p = 0.0422$, $F = 4.3$, $df = 1, 62$; for albumin, $p = 0.012$, $F = 6.71$, $df = 1, 61$. Symbol: * Designates the thermosensitive period for this population.

those varying circulating female hormone levels. Similar studies have reported E_2, T, and P layering in the yolks of dark-eyed junco eggs (Lipar et al. 1999). They speculate on possible repercussions of hormone layering, depending upon the manner in which avian embryos utilize the yolk. If uptake patterns correlate with deposition, varying levels of different hormones would be available to the embryo at specific times during development. The layering pattern evident in the yolks of TSD reptile eggs may indicate the possibility of a similar variation in hormone exposure during development, as a function of yolk utilization by the embryos.

There is some evidence in bird species that perhaps yolk material is mixed during the middle to late stages of development (Lipar et al. 1999). There is also speculation that vascularization of the yolk sac would change the sequence in which yolk materials are absorbed by the avian and turtle embryos (Lipar et al. 1999; Bowden et al. 2001). Protease and lipase activity in the chick embryo are reported by Rol'nik (1970) to occur maximally on days 10 and 16 of development, respectively, but between embryonic days 5 and 6, differential aromatase activity can be measured in the developing gonads of chicks (Smith et al. 1997), so perhaps embryonically produced hormones are of more importance than those supplied by yolk in sexually precocial species. Further studies of how yolk is absorbed during development and what types of metabolites are produced are needed to determine the importance of yolk layering in TSD reptile eggs.

So what factors are responsible for determining individual female levels of yolk hormones? In some avian systems, diet and photoperiod have been linked to changes in yolk steroid levels (Schwabl 1996b; Sockman et al. 2001). Bowden et al. (2000) reported significant differences in E_2 levels between clutches of eggs from *C. picta* produced earlier and later in the season, although females were not marked, so direct correlations could not be made (R. Bowden, pers. comm.). Conley et al. (1997) also reported differences in yolk hormones between two mating seasons in the alligator, but again, females were not marked, so direct correlations were not possible. Since there is variation in maternal contributions of steroid hormones to yolk material (i.e., clutch effects), these yearly or seasonal fluctuations cannot necessarily be attributed to environmental factors alone. In the leghorn chicken *(Gallus domesticus)*, it has been found that hens of the same age, mated to the same rooster and housed under identical conditions, still have significantly different yolk hormone concentrations (Elf and Fivizzani 2002). Further, clutch averages of the yolk hormone values reported for snapping and painted turtles were analyzed (Elf et al. 2002a) in relation to the individual female's age/size. In both of these species, no correlation existed between female age/body size, determined by direct measurement *(C. picta)* or estimated from total clutch mass *(C. serpentina)*, and egg steroid hormone levels (Elf et al. 2003). In the painted turtle, there is no correlation between pre- and postlay female weight and T ($p = 0.8181$ for prelay; $p = 0.6663$ for postlay) or E_2 ($p = 0.6873$ for prelay; $p = 0.5653$ for postlay), between either average egg weight or total clutch mass and T or E_2 (for T, $p = 0.2059$ and 0.1589; for E_2, $p = 0.1276$ and 0.4902 for average egg mass and total clutch weight, respectively), or between carapace length and T and E_2 levels ($p = 0.7086$ and 0.5849 for T and E_2, respectively). The snapping turtle analyses indicated no correlation between either average egg mass or total clutch weight and T or E_2 (for T, $p = 0.9728$ and 0.5595; for E_2, $p = 0.7851$ and 0.3315 for average egg mass and total clutch weight, respectively) (P. K. Elf, unpubl. data). The author is currently investigating within- and between-season variations in yolk steroids from the leopard gecko in order to elucidate the underlying source(s) of variation, but since species differences probably exist, more work is needed to understand the dynamics of steroid deposition and the factors that determine individual female yolk hormone levels.

In summary, we are just beginning to gain an understanding of the possible impacts of maternally contributed yolk steroid hormones on developing embryos in oviparous reptiles with TSD. Besides the multiple effects they could have on development, growth, and behavior in these species, they could be impacting the sex determination process itself. In light of current environmental conditions and the prevalence of estrogen-like contaminants, it is essential that we gain a better understanding of their dynamics and of the role that these hormones play in normal embryonic development.

Part 3

Evolutionary Considerations

13

FREDRIC J. JANZEN AND JAMES G. KRENZ

Phylogenetics

Which was First, TSD or GSD?

The basic challenge of evolutionary biology is to explain variation or the lack thereof, be it phenotypic, genetic, phylogenetic, spatial, temporal, and so on. To illustrate, one gross generalization is that phenotypic traits we think of as being very important to organisms tend to be highly conserved (e.g., binocular vision in vertebrates), probably because the genomic and developmental underpinnings are essentially fixed. Thus, one striking feature about sex-determining mechanisms (SDMs), a fundamental aspect of sexual organisms, is the enormous variety (Bull 1983).

This great diversity of SDMs (in vertebrates in particular) has long puzzled biologists. Given the existence of sexual reproduction and its demonstrable adaptive significance (e.g., West et al. 1999), one might assume that the means by which it is accomplished might be highly conserved. Moreover, SDMs and the primary sex ratio are inextricably linked (Bull 1983), and selection for a 1:1 primary sex ratio is strong (Fisher 1930); thus, one might also assume that any SDM producing a skewed primary sex ratio would be nonexistent. But these two assumptions are, remarkably, incorrect. In fact, sex is determined in various vertebrate taxa in an extraordinary variety of ways (Bull 1983; Janzen and Paukstis 1991a).

SDMs in vertebrates nonetheless can be collected roughly into two major categories. One is environmental sex determination (ESD), wherein sex is fixed permanently by environmental cues (primarily temperature in vertebrates) during a discrete period after fertilization (Bull 1983). Note, however, that this temperature-dependent sex determination (TSD) does not imply a lack of genetic involvement in sex determination, just an absence of sex chromosomes and temperature-insensitive sex-determining genes (Valenzuela et al. 2003). TSD can be contrasted with the more familiar genotypic sex determination (GSD), sometimes referred to as chromosomal sex determination (Deeming and Ferguson 1988), wherein sex is fixed permanently by genetic factors at conception. Exceptions exist (e.g., sex-changing fish), but these two categories serve to organize, identify, and elucidate the key evolutionary issues involved (Valenzuela et al. 2003).

In the course of cataloguing this diversity, a controversy of sorts has ensued over the origins of various sex-determining systems in vertebrates. How many times has each mechanism evolved and, as exemplified in the chapter title, which mechanism is ancestral? The answers to these and related questions strike to the heart of at least two critical and controversial issues involving SDMs: (1) adaptive significance and (2) molecular genetic and physiological underpinnings.

Traits as fundamental as SDMs are typically not amenable to experimental evolution, necessarily leaving informed speculation and theoretical exploration to fill the void. Indeed, such has been the case regarding evolutionary transitions among SDMs in vertebrates (Ohno 1979; Bull 1983; Karlin and Lessard 1986; Ewert and Nelson 1991; Janzen and Paukstis 1991a; Solari 1994; but see Hillis and Green 1990; Janzen and Paukstis 1991b; Kraak and Pen 2002).

However, the development of rigorous, phylogenetically based comparative methods over the past two decades provides another valuable approach: the retrodiction of the evolutionary history of SDMs. This method is not perfect of course (e.g., it is correlative in nature), but it does provide a strong framework for much modern evolutionary research on the nature of variation (e.g., Harvey et al. 1996; Martins 1996; Avise 2000; Page and Holmes 1998).

In this chapter, the evolutionary history of SDMs in vertebrates will be explored. First previous thinking on this topic will be discussed to illustrate the issues involved. Then comparative methods will be applied to rigorous (mainly molecular) phylogenetic hypotheses in vertebrates and particular vertebrate lineages, with special emphasis on lepidosaurs (lizards, snakes, and sphenodontians) and turtles, to evaluate evolutionary transitions between SDMs. These comparative analyses will provide explicit tests of hypotheses concerning the evolutionary diversity of SDMs in vertebrates. After the hypothesized ancestral SDM in vertebrates is identified and the approximate phylogenetic loci of evolutionary transitions between SDMs noted, the implications of these results for important biological issues involving TSD will be discussed. In particular, this discussion will focus on the ramifications of these findings for research into the adaptive significance and mechanistic underpinnings of TSD in vertebrates.

Taxonomic Distribution

SDMs are diverse and nonrandomly distributed in vertebrates. Few fish (see Conover, Chapter 2) and no amphibians (Hayes 1998; Chardard et al., Chapter 7), snakes (Janzen and Paukstis 1991a), birds, or mammals (Bull 1983) are known to have TSD. Instead, male and female heterogamety are widely distributed in fish and amphibians, whereas snakes and birds only have female heterogamety, and mammals have only male heterogamety. Crocodilians and sphenodontians have exclusively TSD (Deeming, Chapter 4; Nelson et al., Chapter 6), but lizards and turtles exhibit a variety of genotypic and environmental SDMs (Bull 1983; Janzen and Paukstis 1991a; Harlow, Chapter 5; Ewert et al., Chapter 3). GSD is much more common in lizards than in turtles; the converse is true for TSD (Janzen and Paukstis 1991a). Moreover, TSD is not monotypic; in fact, three types have been recognized, based on laboratory incubation of eggs (Bull 1983; Conover 1984; Ewert and Nelson 1991; Ewert et al. 1994; Viets et al. 1994; Deeming, Chapter 4; Nelson et al., Chapter 6; Conover, Chapter 2). The distribution of these types of TSD, like that of SDMs in general, is nonrandom: fish primarily have TSD Ib, crocodilians TSD II, sphenodontians TSD Ib (or possibly TSD II), lizards TSD Ib and TSD II, and turtles TSD Ia and TSD II.

Hypotheses on Evolutionary Transitions

Scenarios regarding evolutionary transitions between SDMs in vertebrates have historically been qualitative. For example, Witschi (1959) stated that fish had morphologically indistinguishable (i.e., homomorphic) sex chromosomes, if any at all, representing "a primitive and ancient condition" (see also Ohno 1979), with genetically based sex determination thus arising in tetrapods in the Jurassic around 150 million years ago. Similarly, Ohno (1967) proposed independent evolutionary transitions from ancestral homomorphic to derived heteromorphic sex chromosomes in lizards, snakes, and birds and a comparable more recent event in mammals. These kinds of scenarios are understandable in the absence of (1) information on the true diversity of SDMs in vertebrates, (2) rigorous phylogenetic hypotheses, and (3) acceptable formal comparative or molecular methods. Furthermore, these propositions have no doubt persisted and become established dogma owing to support from some population genetic models of the evolution of sex chromosomes (e.g., Charlesworth 1978; reviewed in Bull 1983; Charlesworth 1991) and a few indirect observations of sex chromosome degradation after autosomal conversion (e.g., reviewed in White 1973). To the authors' knowledge, however, these verbal scenarios remain untested explicitly using modern comparative methods.

The qualitative nature of hypotheses concerning the evolution of SDMs in vertebrates has characterized most scenarios involving TSD as well. Most authors have weighed in cautiously on this subject, which is summed up nicely as, "There is no clear empirical evidence suggesting the evolutionary order of [TSD] compared to GSD ..." (Karlin and Lessard 1986; see also Bull 1983; Janzen and Paukstis 1991a). Kraak and Pen (2002) qualitatively interpreted their phylogenetic representation of SDMs in vertebrates to indicate that both GSD and TSD have evolved multiple times (see also Janzen and Paukstis 1991a) but staked no position regarding their polarity. Even so, an underlying sentiment regarding the ancestral nature of TSD (at least in reptiles) is evident (Karlin and Lessard 1986; Webb and Cooper-Preston 1989; Janzen and Paukstis 1991a; Solari 1994). On the other hand, some workers have cautioned against too quickly rejecting an ancestral GSD scenario for reptiles (Bull 1983; Janzen and Paukstis 1991a).

To the authors' knowledge, only two studies have attempted quantitative tests of these verbal hypotheses, and at that only for turtles. Ewert and Nelson (1991), citing Gaffney's (1984) morphology-based turtle phylogeny, inferred that TSD must have been lost independently four to six times in this group. Janzen and Paukstis (1991a), using phylogenetic information accumulated for another study (Janzen and Paukstis 1991b) and employing a parsimony-based comparative analysis, found that TSD was the ancestral condition for turtles. The authors now explore these issues further in a larger context, using phylogenetic hypotheses for vertebrates, with a particular focus on the SDM-diverse lepidosaurs and turtles.

Comparative Analyses

The authors employ a parsimony-based statistical framework implemented in MacClade 4.0 (Maddison and Maddison 2000) to test hypotheses concerning the evolutionary history of SDMs in vertebrates and thereby provide guidance for future work. Information on SDMs was gathered from the literature (Table 13.1). General relationships among vertebrate groups were based on the traditional view (e.g., Benton 2000). Modern phylogenetic hypotheses for major vertebrate lineages (mainly family level and above) constructed using DNA sequence data from the nuclear gene *Rag1* were extracted primarily from Krenz et al. (unpubl. data) and T. Townsend (unpubl. data). Relationships within emydine, deirochelyine, and batagurid turtles and within lacertid and gekkotan (eublepharids, diplodactylids, pygopodids, and gekkonids) lizards were derived from additional sources (respectively: Feldman and Parham 2002; Bickham et al. 1996; Gaffney and Meylan 1988; Fu 2000; Kluge 1987 and Grismer 1988). The numbers of lepidosaur and turtle species have been targeted to 35 each, with broad within-group sampling to minimize the complexity of the analyses yet adequately capture the trends in SDM evolution. This decision also limits controversies regarding supportable phylogenetic relationships among the focal taxa. In all comparative analyses in MacClade, the authors further adopted the conservative tactic of ascribing no ordering between SDMs: gain or loss of any particular SDM is thus assumed to be equally likely. This parsimony approach seems prudent in the absence of information on the molecular underpinnings of SDMs in most vertebrates.

The authors first examined evolutionary transitions between the two major categories of SDMs without distinguishing among the different types of SDMs within each category (Figure 13.1). A parsimony analysis in MacClade under these conditions cannot pinpoint the likely SDM at several major nodes, particularly where sauropsids (reptiles and birds) split from mammals. Even with this poor resolution at deep branches in the tree, several firm conclusions can be drawn. First and foremost, GSD is an ancient condition in vertebrates, and there is at least one clearly documented origin of TSD in fishes (Conover, Chapter 2). However, how early TSD arose within sauropsids cannot be determined unambiguously by this particular analysis. Nonetheless, TSD is nearly ubiquitous in turtles, crocodilians, and sphenodontians and occurs in a number of squamates (i.e., the clade containing lizards and snakes) as well (Figure 13.1). Within squamates, this comparative analysis indicates a minimum of five independent origins of TSD: (1) eublepharids, (2) gekkonids, (3) lacertids, (4) agamids, and (5) varanids. Alternatively, within turtles, this analysis identifies at least three independent origins of GSD: (1) staurotypids, (2) emydids, and (3) batagurids. Overall then, the criterion of parsimony applied to this phylogenetic hypothesis for vertebrates confirms multiple independent origins of both GSD and TSD.

The previous analysis treats GSD as a homogeneous category. A more informative approach may be to explore evolutionary transitions among SDMs after breaking out the different types of GSD. Labeling a species as having GSD without indicating the heterogametic sex (i.e., homomorphic) may simply reveal a lack of sufficiently detailed genetic information. However, male and female heterogamety are quite dissimilar and likely not subject to rapid, direct intertransitions due to the time required for the complex evolutionary and genetic mechanisms (e.g., Muller's ratchet) to operate (Bull 1983; Charlesworth 1991; but see, e.g., Schmid and Steinlein 2001). Thus, equating all such SDMs under the umbrella term of GSD may inadvertently obscure legitimate patterns of SDM evolution. In contrast, distinguishing between the different patterns of TSD, which probably share closely homologous genetic pathways (e.g., Deeming and Ferguson 1988), is less justifiable in the comparative analyses.

Analyzing SDMs with respect to different types of GSD versus TSD provides considerable insight into the evolutionary history of vertebrates (Figure 13.2). Indeed, parsimony reconstruction of ancestral states in MacClade in this context resolves several major issues revealed in the initial analysis. In addition to the fish case, TSD is now identified as having originated early in sauropsid history, apparently from ancestors possessing male heterogamety. Independent origins of TSD in eublepharids and gekkonids are less certain in this analysis, but the other three instances within squamates (lacertids, varanids, and agamids) are solid. In

Table 13.1 Information on Sex-Determining Mechanisms (SDMs) in Vertebrate Taxa Analyzed in This Chapter

Taxon	SDM	Source
FISH	H, XY, ZW, TSD	Conover, Chapter 2
AMPHIBIA	H, XY, ZW	Hayes 1998
MAMMALIA	XY	Bull 1983
Podocnemis (Pelomedusidae)	TSD	Valenzuela 2001b
Pelomedusa (Pelomedusidae)	TSD	Ewert et al. 1994
Pelusios (Pelomedusidae)	TSD	Ewert and Nelson 1991
Emydura (Chelidae)	H	Janzen and Paukstis 1991a
Chelodina (Chelidae)	H	Janzen and Paukstis 1991a
Acanthochelys (Chelidae)	XY	Janzen and Paukstis 1991a
Carettochelys (Carrettochelyidae)	TSD	Janzen and Paukstis 1991a
Pelodiscus (Trionychidae)	H	Choo and Chou 1992
Apalone (Trionychidae)	H	Janzen and Paukstis 1991a
Staurotypus (Staurotypidae)	XY	Janzen and Paukstis 1991a
Claudius (Staurotypidae)	H	Vogt and Flores-Villela 1992
Sternotherus (Kinosternidae)	TSD	Ewert and Nelson 1991
Kinosternon (Kinosternidae)	TSD	Ewert and Nelson 1991
Dermatemys (Dermatemydidae)	TSD	Vogt and Flores-Villela 1992
Chelydra (Chelydridae)	TSD	Ewert and Nelson 1991
Dermochelys (Dermochelyidae)	TSD	Ewert and Nelson 1991
Chelonia (Cheloniidae)	TSD	Ewert and Nelson 1991
Caretta (Cheloniidae)	TSD	Ewert and Nelson 1991
Calemys (Emydinae)	H	Janzen and Paukstis 1991a
Terrapene (Emydinae)	TSD	Ewert and Nelson 1991
Clemmys (Emydinae)	TSD	Ewert and Nelson 1991
Emys (Emydinae)	TSD	Ewert and Nelson 1991
Deirochelys (Deirochelyinae)	TSD	Ewert and Nelson 1991
Trachemys (Deirochelyinae)	TSD	Ewert and Nelson 1991
Malaclemys (Deirochelyinae)	TSD	Ewert and Nelson 1991
Chrysemys (Deirochelyinae)	TSD	Ewert and Nelson 1991
Pseudemys (Deirochelyinae)	TSD	Ewert and Nelson 1991
Siebenrockiella (Bataguridae)	XY	Janzen and Paukstis 1991a
Chinemys (Bataguridae)	TSD	Janzen and Paukstis 1991a
Kachuga (Bataguridae)	ZW	Janzen and Paukstis 1991a
Mauremys (Bataguridae)	TSD	Ewert and Nelson 1991
Melanochelys (Bataguridae)	TSD	Ewert and Nelson 1991
Rhinoclemmys (Bataguridae)	TSD	Ewert and Nelson 1991
Gopherus (Testudinidae)	TSD	Spotila et al. 1994
Testudo (Testudinidae)	TSD	Janzen and Paukstis 1991a
SPHENODONTIA	TSD	Nelson et al., Chapter 6
Coleonyx (Eublepharidae)	H	Viets et al. 1994
Hemitheconyx (Eublepharidae)	TSD	Viets et al. 1994
Eublepharis (Eublepharidae)	TSD	Viets et al. 1994
Lialis (Pygopodidae)	XY	Janzen and Paukstis 1991b
Gekko gecko (Gekkonidae)	XY	Ewert and Nelson 1991
Gekko japonicus (Gekkonidae)	TSD	Viets et al. 1994
Gehyra (Gekkonidae)	ZW	Janzen and Paukstis 1991b
Phelsuma (Gekkonidae)	TSD	Viets et al. 1994
Phyllodactylus (Gekkonidae)	ZW	Olmo 1986
Tarentola (Gekkonidae)	TSD	Viets et al. 1994
Gonatodes (Gekkonidae)	XY	McBee et al. 1987
Scincella (Scincidae)	XY	Janzen and Paukstis 1991b

Taxon	SDM	Source
Eumeces (Scincidae)	H	Viets et al. 1994
Eremias (Lacertidae)	ZW	Gorman 1973
Podarcis pityusensis (Lacertidae)	TSD	Viets et al. 1994
Podarcis erhardii (Lacertidae)	ZW	Olmo et al. 1990
Lacerta vivipara (Lacertidae)	ZW	Janzen and Paukstis 1991b
Gallotia (Lacertidae)	ZW	Olmo 1986
Bipes (Amphisbaenidae)	ZW	Cole and Gans 1987
Gymnophthalmus (Gymnophthalmidae)	XY	Cole et al. 1990
Cnemidophorus (Teiidae)	XY	Janzen and Paukstis 1991b
Ameiva (Teiidae)	XY	Peccinini-Seale and de Ameida 1986
Chamaeleo (Chamaeleonidae)	H	Viets et al. 1994
Agama (Agamidae)	TSD	Viets et al. 1994
Pogona (Agamidae)	H	Viets et al. 1994
Basiliscus (Iguanidae)	H	Viets et al. 1994
Sceloporus (Iguanidae)	XY	Janzen and Paukstis 1991b
Anolis (Iguanidae)	XY	Janzen and Paukstis 1991b
Crotaphytus (Iguanidae)	H	Viets et al. 1994
Dipsosaurus (Iguanidae)	H	Janzen and Paukstis 1991b
Tropidurus (Iguanidae)	XY	Janzen and Paukstis 1991b
Varanus salvator (Varanidae)	TSD	Viets et al. 1994
Varanus acanthurus (Varanidae)	ZW	Janzen and Paukstis 1991b
SERPENTES	ZW	Bull 1983, Olmo 1986
AVES	ZW	Bull 1983
CROCODILIA	TSD	Deeming, Chapter 4

Note: Reviews that contain the relevant information are cited in many cases in the table to minimize the length of the literature cited section. Please consult those reviews for citations to the original sources of the research. H = homomorphic sex chromosomes, XY = male heterogamety, ZW = female heterogamety, and TSD = temperature-dependent sex determination. For the purpose of this analysis heterogamety implies heteromorphic sex chromosomes as well.

the former two instances, TSD appears to have been derived from ancestors possessing female heterogamety; no firm conclusion can be drawn for agamids. Different types of GSD have also arisen independently within squamates, although the polarity and specific nodes of the transitions are often unclear (e.g., Is male heterogamety, female heterogamety, or TSD the basal SDM in gekkonids?). These transitions are greatly clarified within turtles, however. Different types of GSD have originated in turtles a minimum of six times: (1) trionychids, (2) chelids, (3) staurotypids, (4) emydids, and (5 and 6) twice in batagurids. Once again, both major categories of SDMs are identified as having originated many times within vertebrates.

Implications

Biologists have long struggled with intriguing and important questions surrounding the origins of different SDMs. The majority of scenarios regarding evolutionary transitions between SDMs, including within the vertebrate lineage, have been qualitative, although clearly based on an appreciation of the historical origins of different taxonomic groups. Widespread adoption of such "tree thinking," accompanied by dramatic improvements in phylogenetic tree reconstruction and comparative analysis, has aided rigorous quantitative evaluation of many analogous evolutionary questions (e.g., Geffeney et al. 2002). The formal modern comparative analyses of SDM evolution in vertebrates herein, based on current understanding of phylogenetic relationships and SDM distribution, are instructive about the phylogenetic loci where SDM transitions probably occurred and identify crucial taxa to target for further empirical work. These analyses also provide a launching point for further discussion of the implications of this striking phenotypic diversity for the adaptive significance of TSD and its molecular genetic and physiological underpinnings in vertebrates.

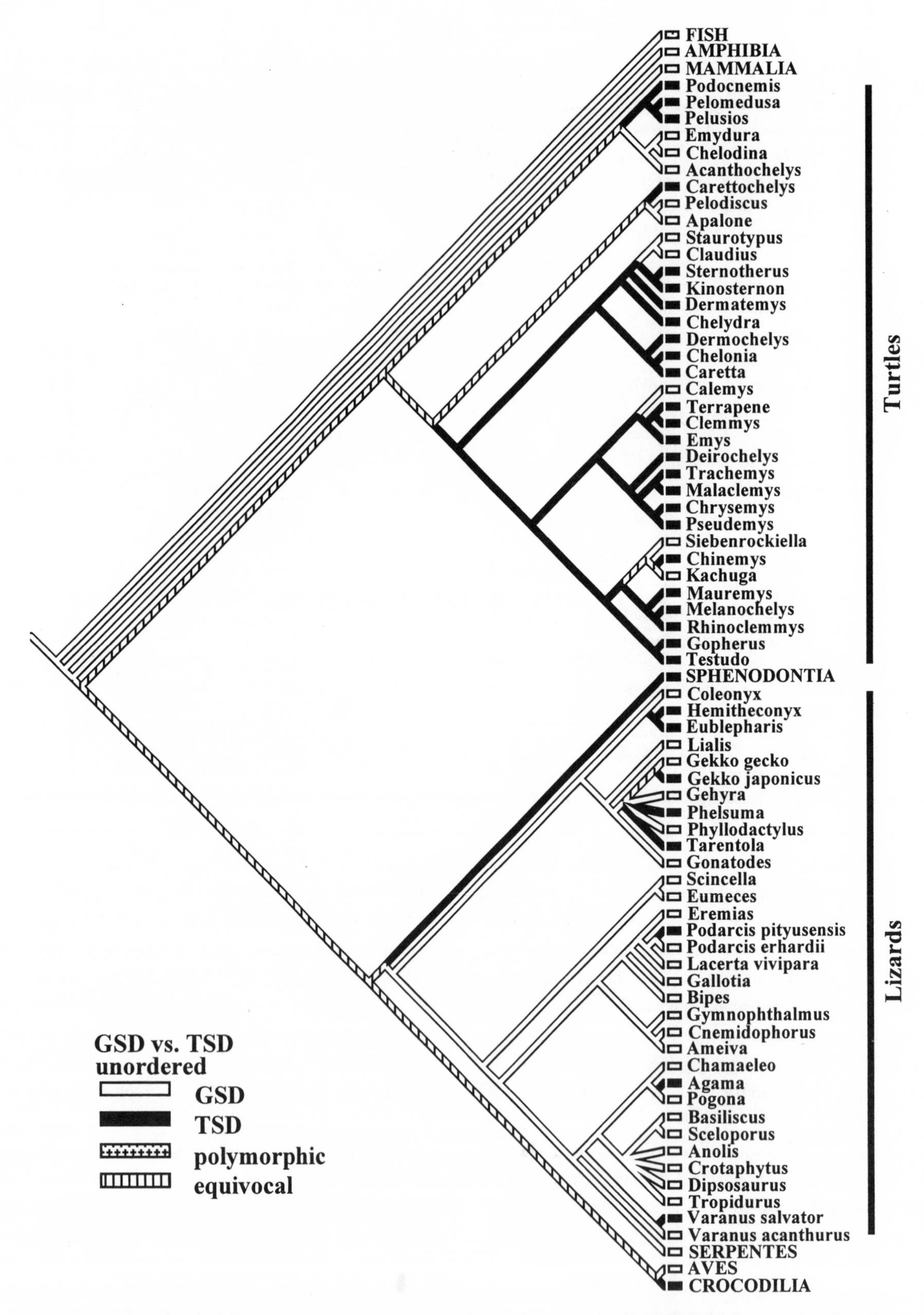

Figure 13.1 Parsimony analysis of SDM evolution on a vertebrate phylogeny, emphasizing lepidosaurs and turtles. This comparative analysis of GSD vs. TSD was conducted in MacClade 4.0 (Maddison and Maddison 2000) using broadly recognized relationships among major amniote lineages. Information on systematics and SDMs (black = TSD, white = GSD) is provided in Table 13.1.

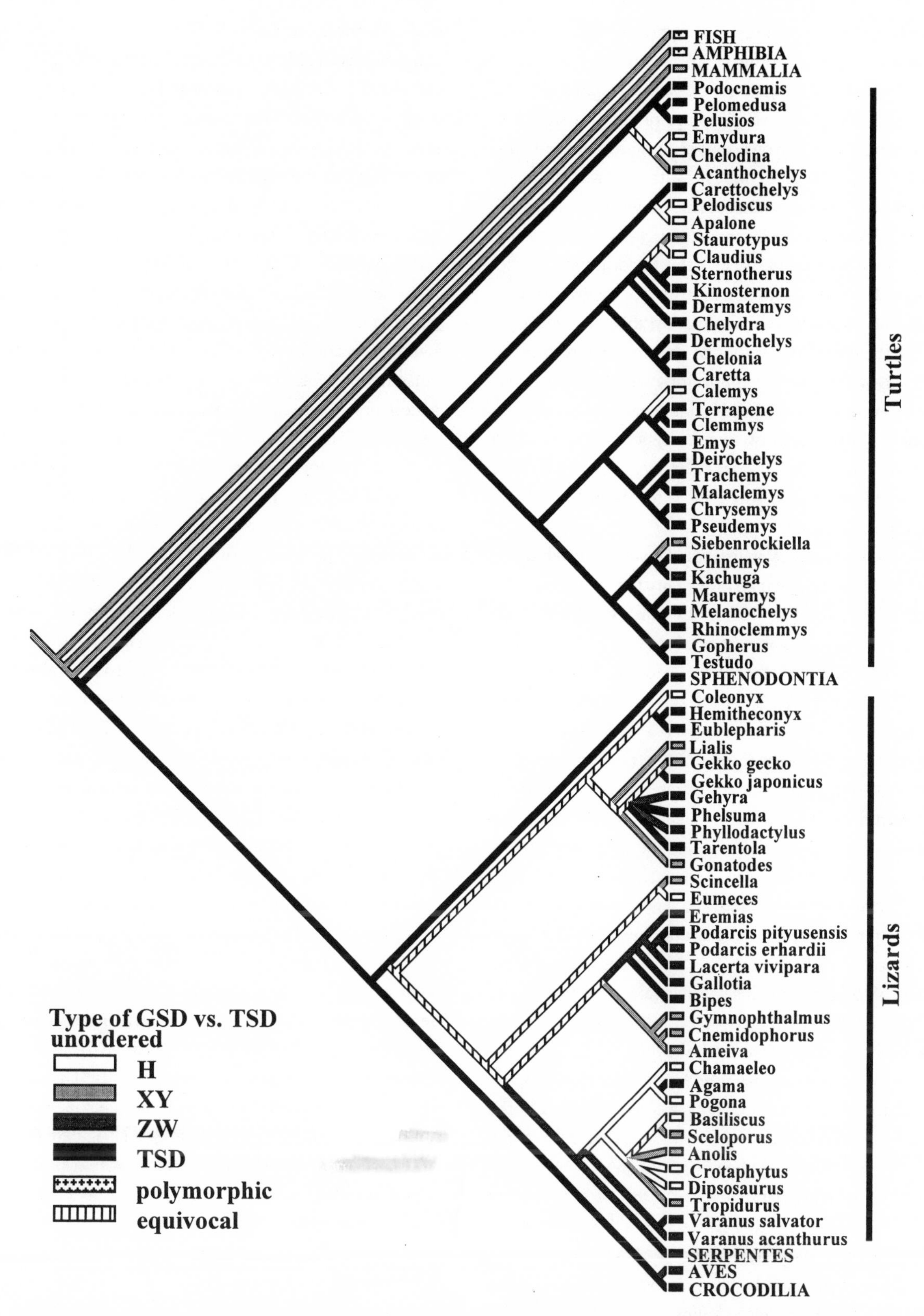

Figure 13.2 Parsimony analysis of SDM evolution on a vertebrate phylogeny, emphasizing lepidosaurs and turtles. This comparative analysis of types of GSD vs. TSD was conducted in MacClade 4.0 (Maddison and Maddison 2000) using broadly recognized relationships among major amniote lineages. Information on systematics and SDMs [black = TSD, white = homomorphic sex chromosomes (H), light gray = male heterogamety (XY), dark gray = female heterogamety (ZW)] is provided in Table 13.1.

But first the authors present a cautionary note concerning some of the limitations of comparative phylogenetic approaches. Like any analysis, garbage in equals garbage out. If the phylogenetic relationships or SDMs assigned to particular taxa are incorrect, the resulting conclusions may fail as well (e.g., Losos 1994). The authors have attempted to minimize this potential problem by working at higher taxonomic levels with well-supported phylogenetic hypotheses. Even so, variation exists in the hypothesized phylogenetic relationships among major vertebrate groups (e.g., Hedges and Poling 1999). Fortunately, this variation does not alter the main conclusions of this chapter (results not shown). However, more importantly, lepidosaurian relationships are not resolved with confidence for all branches on the tree. A well-supported phylogenetic hypothesis for gekkotan lizard genera, in particular, would likely lead to a dramatic improvement in resolving the polarity of SDM evolution in sauropsids.

The authors are less assured of the validity of all designations of SDMs, particularly for many fish species and for the less common and/or Afro-Asian sauropsids (Valenzuela et al. 2003; Harlow, Chapter 5; Conover, Chapter 2). Information on SDMs for taxa exposed to more detailed, replicated work is simply more robust. In the authors' opinion, follow-up investigation of SDMs in *Acanthochelys*, *Kachuga*, *Tarentola*, *Podarcis*, *Bipes*, *Chamaeleo*, and *Varanus* would strengthen our understanding greatly (see Figures 13.1 and 13.2). Moreover, a number of key lineages, especially (1) numerous fish species, (2) trionychid, platysternid, and bataguridturtles, and (3) eublepharid, gekkonid, scincid, lacertid, amphisbaenid, agamid, varanid, xantusid, xenosaurid, helodermatid, anniellid, dibamid, cordylid, and anguid lizards, have not been sufficiently explored for SDMs. Basic research on SDMs in all of these groups would again considerably improve our view of SDM evolution in vertebrates.

Another important issue involves the statistical assumptions underlying the comparative method employed. Some or all of the actual evolutionary processes involved in the transitions between SDMs in vertebrates may have followed different "rules" (Valenzuela, Chapter 14). Although the authors were explicit in describing the assumptions behind their analyses, it nonetheless remains the case that, for example, evolutionary transitions of character states between lineages may not be best or always reflected by the criterion of parsimony (e.g., Collins et al. 1994). Alternative methods such as maximum likelihood Markov-transition approaches (Pagel 1999) may be better suited in such circumstances, particularly where branch lengths or internode distances are short (P. Lewis, pers. comm.). The authors also assumed that different patterns of TSD are variations on a homologous theme and thus could be treated as a single trait (sensu Deeming and Ferguson 1988). To assume otherwise would alter some of the conclusions (results not shown). In the end, the reader must thus recognize the potential limitations of the comparative analyses and accept the authors' interpretations with appropriate caution.

Regardless of any methodological concerns, the authors believe that the comparative phylogenetic approaches adopted in this chapter have provided clear tests of several hypotheses and have thereby generated a few robust conclusions about SDM evolution in vertebrates. Above all, the ancestral state of sex determination in vertebrates is almost certainly GSD (Figure 13.2) (contra Ohno 1979). TSD has been clearly documented in fishes, but most cases occur within a single genus; there are many more recognized instances of GSD in those basal vertebrate lineages (Conover, Chapter 2). Moreover, no amphibians (Hayes 1998; Chardard et al., Chapter 7) or mammals (Bull 1983) can be said to have TSD, properly defined (Valenzuela et al. 2003). Even so, there is some circumstantial evidence for TSD (or more likely a thermally modified GSD system) in lampreys (Beamish 1993), which is the sister taxon to the rest of vertebrates, and next to nothing is known about SDMs in the phylogenetically important chondrichthyan, coelacanth, and lungfish lineages (Maddock and Schwartz 1996; Devlin and Nagahama 2002; Conover, Chapter 2). Although much crucial work on SDMs in basal vertebrates is necessary, the authors nonetheless conclude from their comparative examination that some form of GSD exemplifies the ancestral condition of sex determination in vertebrates: GSD was first!

The authors' parsimony-based comparative analyses (esp. Figure 13.2) also lend credence to other key hypotheses regarding the evolution of TSD in vertebrates. In particular, the authors found that TSD appears to be the ancestral condition of sex determination in sauropsids. The results further support contentions (Ewert and Nelson 1991; Janzen and Paukstis 1991a) that TSD has been lost at least six times in turtles and has originated at least three times in lizards. Thus, SDMs exhibit remarkable phylogenetic lability: both major categories of SDMs have been "lost" and "found" numerous times in vertebrates. It is curious, however, that temperature is the cue in all known independent origins of ESD in vertebrates (although Conover, Chapter 2, mentions the possibility of pH-dependent sex determination in fish), when a variety of factors are involved in other groups (reviewed in Korpelainen 1990, 1998). Why this should be so is unclear, but it might reflect the widespread importance of temperature in affecting sex-specific fitness attributes of vertebrates (reviewed by Deeming and Ferguson 1991b; Janzen and Paukstis 1991a; Shine 1999; but see Rhen and Lang, Chapter 10; Valenzuela, Chapter 14) or perhaps insufficient exploration by biologists of alternative environmental signals that might influence sex determination in this group.

What does the pattern of transitions between SDMs reveal about the evolutionary forces and molecular mechanisms involved? Most researchers have proceeded on the expectation that TSD in vertebrates is adaptive. However, despite considerable effort devoted to detailed study of a variety of sauropsid taxa (especially turtles, crocodilians, and eublepharids), the adaptive significance of TSD has only been clearly demonstrated for one species of fish (Conover 1984 and Chapter 2; see alsoValenzuela, Chapter 14). This outcome is all the more surprising because one elegant theoretical framework (Charnov and Bull 1977) has robustly explained the adaptive significance of other forms of ESD in numerous taxonomically diverse invertebrates (reviewed in Bull 1983; Korpelainen 1990, 1998; Shine 1999). The lack of success of the Charnov-Bull model in explaining the existence of TSD in sauropsids has led some researchers to question the current adaptive significance of this unusual SDM (e.g., Janzen and Paukstis 1988; Girondot and Pieau 1999; Valenzuela, Chapter 14).

Our comparative analyses provide a fresh perspective on this overarching issue. With the exception of the origin of TSD at the sauropsid/mammal node, the remaining clearly independent cases of TSD (fish, lacertids, agamids, and varanids; Figure 13.2) have most likely arisen relatively recently. If so, and if these latter four derived instances of TSD were in fact driven by the forces of natural selection, then the signature of adaptation may be more likely to be detected therein than in the ancient origin of TSD at the sauropsid/mammal split. In the authors' view then, effort expended on evaluating the adaptive significance of TSD in turtles, crocodilians, sphenodontians, and perhaps gekkotans is potentially misdirected. TSD in these taxa may instead be quasi neutral (sensu Janzen and Paukstis 1988; Girondot and Pieau 1999; Valenzuela, Chapter 14), and the signature of adaptation, if it ever existed in this case, has simply attenuated over the 300+ million years since the ancient transition event occurred. Indeed, is it any wonder that none of the considerable experimental effort devoted to turtles, crocodilians, and eublepharids has produced a *broadly* convincing explanation for the adaptive significance of TSD across these taxa? If the authors are correct, then microevolutionary analyses directed toward the derived cases of TSD in lacertids, agamids, and varanids (and of GSD in turtles; Janzen and Paukstis 1988) will prove to be the most fruitful and enlightening research projects. Alternatively, an explanation for the adaptive significance of TSD in turtles, crocodilians, sphenodontians, and perhaps gekkotans may lie outside the framework of the Charnov-Bull model (e.g., in sex ratio evolution).

The authors' phylogenetic historical analyses also shed light on, and provide guidance for research into, the mechanistic underpinnings of TSD in vertebrates, a hot (no pun intended) topic these days. One general prediction from the parsimony-based comparative assessments is that the basic molecular and physiological pathways involved in TSD in turtles, crocodilians, sphenodontians, and perhaps gekkotans should be broadly conserved. Different patterns of TSD, for example, occur in various species in these lineages (Ewert et al. 1994; Viets et al. 1994; Deeming, Chapter 4; Nelson et al., Chapter 6), yet the underlying major developmental pathways should be similar (e.g., Smith et al. 1999a) if they share a common ancestry. The authors eagerly await empirical tests of this conjecture.

On the other hand, the mechanics involved in the newer independent origins of TSD in vertebrates might very well be different in each case. Still, it is puzzling to note that the type of TSD appears to be identical in all four instances (probably TSD Ib)! Why this should be so, other than by chance, remains to be explained. The pattern (in addition to the mechanism itself) may be adaptive in the same way in each case, but the phenotypic similarity might instead reflect constraints on the molecular and physiological pathways involved in evolving TSD from GSD. Empirical work on fish, lacertids, agamids, and varanids is greatly needed to distinguish between these competing explanations.

Fascinating to many, the remarkable diversity of SDMs in vertebrates holds an elevated position in the pantheon of longstanding evolutionary enigmas. In particular, how the unusual mechanism of TSD arose and withstood untold environmental upheavals over geological eons has long captivated the attention of scientists and interested laypersons alike (e.g., Deeming and Ferguson 1989c). How taxa with TSD will respond to current rapid climatic changes and swift habitat modifications are especially timely and unsettling questions, particularly since many of these species are already imperiled. Whether the prediction is more dire (e.g., Janzen 1994a; Morjan 2002) or more hopeful (e.g., Rhen and Lang 1998; Girondot and Pieau 1999), research on TSD in vertebrates promises to continue to challenge us practically and scientifically. The authors hope that their comparative analyses lend guidance to researchers intending to tackle these and related evolutionary challenges.

Acknowledgments—Many thanks to Nicole Valenzuela for inviting us to write this chapter; Ted Townsend (Washington University) for his generosity in sharing his unpublished *Rag1* lepidosaur phylogeny; Aaron Bauer, Lee Grismer, and Arnold Kluge for helping clarify gekkotan systematics; and Mike Ewert, John Wiens, and an anonymous reviewer for their constructive criticisms on the manuscript. Funding of Janzen's most recent research in this area has been provided by U.S. National Science Foundation grants DEB-9629529 and DEB-0089680.

14

NICOLE VALENZUELA

Evolution and Maintenance of Temperature-Dependent Sex Determination

Temperature-dependent sex determination (TSD) is an ancient sex-determining mechanism among vertebrates, perhaps ancestral in reptiles. How and why transitions between TSD and genotypic sex determination (GSD) occur are two important questions regarding TSD origin and maintenance. Most formulated hypotheses for TSD evolution are adaptive, but neutral or quasi-neutral alternatives exist as well (Table 14.1). Here possible explanations for the origin and persistence of TSD are reviewed, and these evolutionary events are treated separately because each may result from different forces. First, the postulated adaptive and neutral hypotheses are described, the existing evidence (or lack thereof) for TSD taxa is explored, and a series of testable predictions based on theory is presented. The third section briefly discusses TSD absence in various vertebrate groups. In general, current hypotheses deal with genetic, maternal, and environmental effects on fitness components, either survival or fertility, sometimes overlapping several effect types (Figure 14.1).

Theory and Evidence for TSD Taxa

This section considers transitions from GSD to TSD (TSD gains) and from TSD to GSD (TSD loses), TSD persistence and differentiation, and whether each of these events can be explained by chance or selection.

TSD Origin: From GSD to TSD

Adaptive Hypotheses

Theoretically, selection favors equal parental investment in male and female production, and if the costs of producing one male and one female are identical, the sex ratio tends to equality (Fisher 1930). But when these costs differ, a biased sex ratio can equalize total parental investment in male and female production, yield a higher reproductive success than the average under balanced sex ratios, and thus be selected (Shaw and Mohler 1953). Consequently, a mechanism allowing such biases is favored.

In TSD, males and females are produced polyphenically—identical genomes can produce discrete phenotypes depending on the environmental conditions (e.g., incubation temperature). Polyphenisms are a form of phenotypic plasticity that may help coping with heterogeneous environments (Kawecki and Stearns 1993 in Downes and Shine 1999). At first glance the life history and ecology of some TSD and GSD taxa appear very similar, yet their sex-determining mechanisms differ. Seasonal sex ratio shifts may be selected when life histories vary seasonally (but differently) for males and females, and if generations overlap causing individuals from different cohorts to compete for reproductive success (Werren and Charnov 1978). Under those circumstances, parental ability to bias sex ratios accordingly is also favored.

Table 14.1 Evidence Related to the Adaptive and Neutral Hypotheses Described in the Text

Theoretical Models	Theoretical Models	Evidence in Pro	Evidence Against	Equivocal Evidence
TSD Origin				
Adaptive				
Differential fitness	Charnov and Bull 1977			
Maternal Effects				
Nest-site choice				
Natal homing	Reinhold 1998	Reinhold 1998	Valenzuela and Janzen 2001	
Differential dispersal	Julliard 2000		Godfrey et al. 1996 Mrosovsky 1994 Valenzuela and Janzen 2001	
Survival				
Differential mortality	Burger and Zappalorti 1988	Janzen 1995	Rhen and Lang 1995	Bobyn and Brooks 1994 Elphick and Shine 1999 Gutzke and Packard 1987a Joanen et al. 1987
Fecundity				
Sexual size dimorphism	Head et al. 1987 Webb et al. 1987	Allsteadt and Lang 1995a Crews et al. 1998 Ferguson and Joanen 1982 Packard and Packard 2001 Rhen and Lang 1995 Saillant et al. 2002 Tousignant and Crews 1995	Allsteadt and Lang 1995a O'Steen and Janzen 1999 St. Clair 1998 Steyernark and Spotila 2001	Bobyn and Brooks 1994 Braña and Ji 2000 Campos 1993 Demuth 2001 Elphick and Shine 1999 Ewert et al. 1994 Ferguson and Joanen 1983 Gutzke and Packard 1987a Gutzke et al. 1987 Harlow and Shine 1999 Hutton 1987 Joanen et al. 1987 McKnight and Gutzke 1993 O'Steen 1998 Packard et al. 1987 Webb and Cooper-Preston 1989
Seasonal hatching time	Conover and Kynard 1981	Conover and Kynard 1981 Harlow and Taylor 2000		

Phenotypic effects		Allsteadt and Lang 1995a Crews et al. 1998 Gutzke and Crews 1988 Rhen and Lang 1999b	Arnold et al. 1995 Bull and Charnov 1989 Packard and Packard 2001	Braña and Ji 2000 Demuth 2001 Elphick and Shine 1999 Gutzke et al. 1987 Janzen 1995 O'Steen 1998 O'Steen and Janzen 1999 Packard et al. 1987 Tousignant and Crews 1995 Webb et al. 2001
Biased sex ratios				
Group structure adaptation	Bull and Charnov 1988		Burke 1993 Thorbjarnanson 1997	
Neutral				
Preexisting T°sensitivity	Bull 1981			Chardard et al., Chapter 7
Sex-ratio distorter	Morjan 2002	–	–	–
TSD Maintenance				
Adaptive				
Maternal Effects				
Nest-site choice by egg size	Roosenburg 1996	Bull 1983 Roosenburg 1996	Bulmer and Bull 1982 Ewert and Nelson 1991 Valenzuela 2001a,b	
Biased sex ratios				
Group selection of sex ratio	Woodward and Murray 1993		Girondot and Pieau 1996 Thorbjarnanson 1997	Woodward and Murray 1993
Sib-avoidance	Ewert and Nelson 1991		Burke 1993	
Cultural inheritance of natal homing	Freedberg and Wade 2001		Godfrey et al. 1996 Valenzuela and Janzen 2001 Bull 1980 Bull and Charnov 1988	Gibbons 1990 Mrosovsky 1994 Thorbjarnarson 1997
Neutral				
TSD equivalent to GSD	Bull 1980 Mrosovsky 1980			
Phylogenetic inertia	Bull 1980	Ewert and Nelson 1991	Burke 1993 Janzen and Paustkis 1991b	
Life-history dependent				
Longevity	Bull and Bulmer 1989 Girondot and Pieau 1996	Bull and Bulmer 1989 Girondot and Pieau 1996 Girondot and Pieau 1999		

(*continues*)

Table 14.1 (Continued)

Theoretical Models	Theoretical Models	Evidence in Pro	Evidence Against	Equivocal Evidence
Overlapping generations	Bull and Bulmer 1989	Bull and Bulmer 1989		
	Girondot and Pieau 1996	Girondot and Pieau 1996		
		Girondot and Pieau 1999		
TSD Loss				
Adaptive				
GSD individuals more fit	Bull 1981	de Lisle 1996	Robert and Thomson 2001	
	Bull 1983			
Sex ratio fluctuations	Bull 1980	Conover and Heins 1987a	Post et al. 1999	
		Godfrey et al. 1996		
Intersexes	Bull 1981	Pieau 1982	Bull 1983	
		Pieau et al. 1999a,b	Crews et al. 1998	
			Girondot et al. 1998	
Allee effect	Berec et al. 2001			Berec et al. 2001
Low/late dimorphism	Bull 1983	Greenbaum and Carr 2001		
		Gutiérrez et al. 2000		
Antagonistic pleiotropy	Moran 1992			
Imperfect phenotype-environment matching	Moran 1992			
Parental sex ratio control	Roosenburg 1996		Bull 1980	
	Reinhold 1998		Julliard 2000	
Neutral	–	–	–	–

Note : Theoretical models list the original source(s) of each particular hypothesis. *Evidence in Pro* refers to empirical data supporting any given hypothesis. *Evidence Against* refers to empirical data or theoretical considerations against each hypothesis. *Equivocal Evidence* refers to empirical data that can be interpreted either way because (a) no directional effect was detected, (b) data are in the expected direction but come from GSD taxa, or (c) experimental design or statistical difficulties preclude a clear conclusion. The evidence included in this table is not exhaustive.

GSD species can bias sex ratios adaptively (Bulmer and Bull 1982; Bull and Charnov 1988; Krakow 1992; Perret 1996; Komdeur et al. 1997; Nager et al. 1999), but sometimes at a relatively high cost (e.g., waste of resources invested in the wrong-sex offspring by differential offspring mortality or discarding of gametes). Higher benefits are needed to offset these costs; otherwise, sex ratio biases in GSD species would be constrained. When benefits are not large enough, a low-cost system allowing sex biases will be advantageous. This may be the case of TSD for some taxa. A key corollary of this scenario, however, is that when the benefits of biasing sex ratios are small, formidable efforts may be required to detect them.

1. DIFFERENTIAL FITNESS HYPOTHESIS

Charnov and Bull (1977) proposed that environmental sex determination (ESD) is selected over GSD if three conditions are met (the Charnov-Bull model): (1) the environment consists of patches (spatial or temporal) that grant sex-specific fitness to the offspring, (2) patches cannot be chosen by the offspring nor by their parents, and (3) random mating occurs among patches. When applied to TSD, thermal conditions early in development (when sex is determined) are assumed to differ unpredictably among patches and to confer differential lifetime fitness directly or through a correlated variable. This makes TSD beneficial, since offspring develop into the sex with highest fitness in each patch. Primary population sex ratios can deviate from ½ towards the lower-fitness sex through frequency-dependent selection plus the fitness differential of the patches (Bull 1983).

Environmental effects on development and fitness can be important to maintaining polyphenisms (Moran 1992). First, an environmental factor (cue) affects development, resulting in plasticity (sex-by-temperature production in TSD), while a factor (selective agent) affects fitness differently for alternative phenotypes. The cue can differ from the selective agent but must act first. The cue accuracy in predicting the selective-agent state (along with spatial vs. temporal environmental variation, fitness differentials among environments, relative frequency of the environments, and the cost of plasticity) determines the adaptive maintenance of phenotypic plasticity (Moran 1992). Consistently, under the Charnov-Bull model, sex ratio adjustment is selected first, and consequently, a mechanism allowing such adjustment is favored (Bull 1983).

The Charnov-Bull model is the most theoretically robust of the adaptive hypotheses, and has empirical support for some ESD invertebrates(e.g., Blackmore and Charnov 1989) and some vertebrate species (Table 14.2), although reptilian evidence is not conclusive, as will be shown below. Several related (sub)hypotheses (see TSD Origin, Differential Fitness Subhypotheses A–F, below) link incubation temperature with a fitness correlate to explain TSD origin or maintenance (see TSD Maintenance, Adaptive Hypotheses, Maternal effects on egg allocation).

Differential Fitness Subhypotheses Related to Maternal Effects by Female Nest-Site Choice

Natal homing. Natal homing could trigger ESD evolution if daughters are produced in high quality (i.e., high survival) sites to which they return to nest, thus deriving higher fitness than sons whose reproductive output is unaffected by natal patch quality (assuming they survive to maturity) (Reinhold 1998). Long-term fluctuations in nest-site quality and imperfect natal homing prevent runaway selection of nesting in high quality sites (Reinhold 1998). *Eretmochelys imbricata* exhibit natal homing, and offspring survival is positively correlated with percent females within nests (Swingland et al. 1990; Horrocks and Scott 1991), thus supporting this hypothesis (Reinhold 1998). This model was not applicable to *Chrysemys picta*, as hatching success and sex ratio were not repeatable, females did not nest preferentially in female-producing sites, and mortality was lower at male-producing sites (Valenzuela and Janzen 2001). This model is inapplicable to species whose lifespan is shorter than the scale of fluctuations in nest-site quality, otherwise runaway selection restricts nesting to high quality sites, leading to high female biases and thus to selection for GSD (Bull 1980).

Differential dispersal. Differential dispersal of males and females, and varying environmental quality (reproductive success) were proposed to explain the evolution of habitat-dependent biased sex ratios (Julliard 2000). The evolutionary stable strategy (ESS) overproduces the dispersing sex in poor habitats and the philopatric sex in good habitats, and biases increase in (1) rare or poor habitats, (2) under large habitat quality differentials, (3) under almost random dispersal of one sex regarding habitat availability, (4) with high (but different) dispersal rates for both sexes, (5) without individual control of reproductive habitat (Julliard 2000). The model purportedly supports the Charnov-Bull model by advocating differential fitness by environmental heterogeneity, and requires temporal predictability of habitat quality. However, no data exist on repeatability of nesting conditions in sea turtles (Mrosovsky 1994; contra Julliard 2000), and annual sex ratio can vary substantially in several turtle species (e.g., Godfrey et al. 1996; Valenzuela and Janzen 2001).

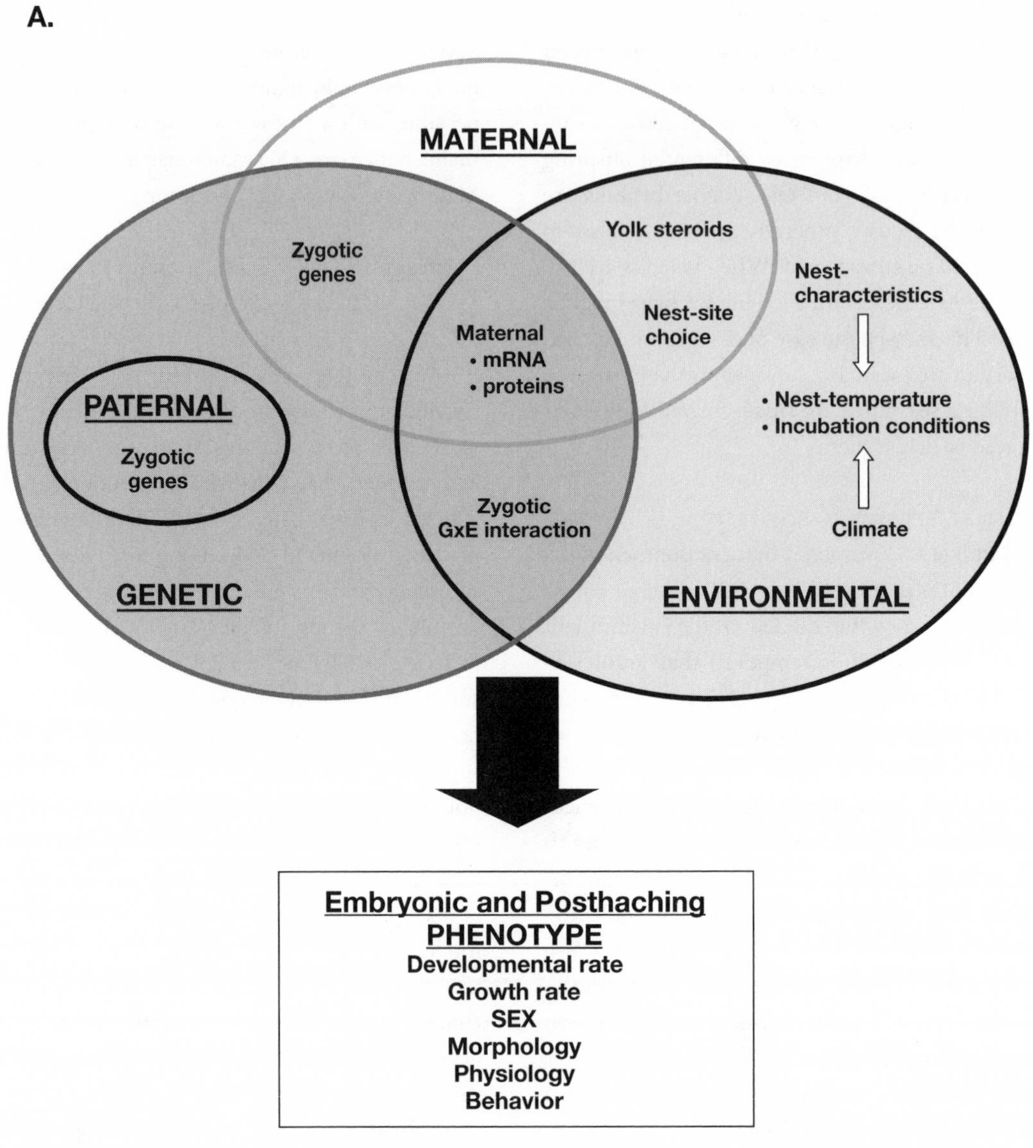

Figure 14.1 Diagram of the categories of factors affecting hatchling phenotype and classification diagram of TSD evolutionary hypotheses by the main fitness effect they address. (**A**) Factors with phenotypic effects on TSD individuals include genetic and environmental variables. Maternal effects encompass both types, whereas known paternal effects are only genetic. (**B**) Effects on lifetime reproductive success can be exerted through effects in survival (individuals must survive to maturity in order to reproduce) and/or fecundity (direct offspring production). Most postulated hypotheses for TSD evolution can be categorized by the fitness component they address, while a few other hypotheses focus primarily on biased sex ratios.

Nonetheless, this hypothesis requires direct testing at the relevant spatial scale.

Differential Fitness Subhypothesis Related to Survivorship

Differential mortality. Sex and incubation temperature in TSD could interact through a mechanism to enhance embryo survivorship (Webb and Smith 1984). If males and females suffer differential mortality during development that covaries with incubation temperature, TSD would be beneficial by allowing the production of the best-fit sex at the temperature extremes (Burger and Zappalorti 1988). Differential mortality may favor TSD origin, but once established, producing a single sex at extreme temperatures erases the traces of this pressure. The same applies to any differential fitness effect at single-sex temperatures. Therefore, evidence supporting this hypothesis requires artificially producing rare-sex individuals at the temperature ex-

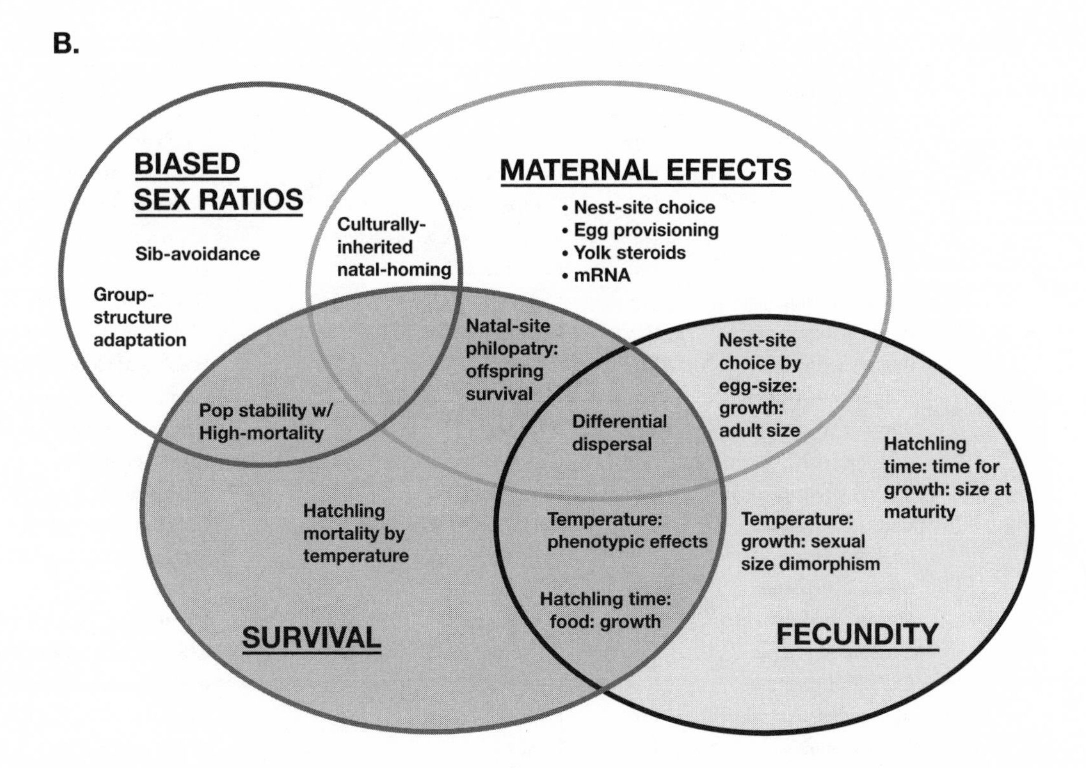

tremes (e.g., Rhen and Lang 1995), which are predicted to suffer higher mortality than common-sex individuals at a given temperature. Interestingly, support for this hypothesis is equivocal as it comes from GSD species (Tables 14.1 and 14.2). No sex or temperature effect on hatchling mortality was found in snapping turtles using hormonal manipulations (Rhen and Lang 1995; contra Janzen 1995; but see Elphick and Shine 1999). Thus, this hypothesis does not explain TSD maintenance generally. Furthermore, data from intermediate temperatures (e.g., Janzen 1995) need not reflect relative fitness at single-sex temperatures (nor reveal whether producing one sex at such values is adaptive), because for each sex fitness can be higher, equal, or lower at mixed-sex-ratio temperatures than at extreme temperatures while sex ratio remains identical (Figure 14.2).

Differential Fitness Subhypotheses Related to Fecundity

Sexual size dimorphism. The sexual size dimorphism hypothesis considers differential temperature effects on growth, with the larger sex being produced at the optimal temperature (Head et al. 1987; Web et al. 1987; Deeming and Ferguson 1988; Ewert and Nelson 1991). Some data are consistent with this hypothesis (Ewert and Nelson 1991; Ewert et al. 1994). However, this evidence is not conclusive, partly because data amenable to analysis are reduced due to missing information from many taxa, and to confounding phylogenetic effects (Janzen and Paukstis 1991b; Mrosovsky 1994). As more species are studied and stronger phylogenies constructed, a reassessment of this hypothesis will prove valuable.

Seasonal hatching time. In some taxa temperature during development predicts seasonal hatching time and is associated with available time for growth before breeding and adult body size, which affects reproductive success differently for males and females (Conover and Kynard 1981; Conover 1984). In this case, TSD allows matching offspring sex with the hatching time during a breeding season that best affects fitness (Conover and Kynard 1981). This occurs in the fish *Menidia menidia* and *M. peninsulae* (Conover and Kynard 1981; Middaugh and Hemmer 1987; Conover, Chapter 2). GSD replaces TSD in northern populations of *M. menidia* where annual sex ratio fluctuations induced by climate cancel any fitness gain attained by TSD (Conover and Heins 1987b). Similarly, GSD prevails in subtropical semiannual populations of *M. peninsulae* (where temperature is a poor predictor of time for growth and size at maturity), and TSD increases with latitude as life cycle becomes annual (Yamahira and Conover 2003).

Comparable systems may be present in reptiles (Shine 1999; Harlow and Taylor 2000) as predictable seasonal temperature and sex ratio changes exist in some species (Mro-

Table 14.2 Examples of Existing Evidence Related to the Potential Adaptive Temperature Differential Fitness Effects

Temperature Effects	Species	Evid	Effect	Reference
Mortality	*Crocodylus porosus*	Eqv	T° or sex? MT	Webb and Cooper-Preston 1989
	Alligator mississippiensis	Eqv	T° or sex? ST	Joanen et al. 1987
	Chelydra serpentina	Eqv	T° or sex? MT	Bobyn and Brooks 1994
	Chelydra serpentina	Con	No T°, no sex, C, T°×Sex	Rhen and Lang 1995
	Chelydra serpentina	Pro	T°, sex, T°×Sex, ST	Janzen 1995
	Emydoidea blandingii	Eqv	T° or sex?	Gutzke and Packard 1987a
	Bassiana duperreyi^{GSD}	Eqv	T°, sex, T°×Sex, MT	Elphick and Shine 1999
Morphology (shape or size)	*Crocodylus porosus*	Eqv	T°, sex n.e., ST	Webb and Cooper-Preston 1989
	Alligator mississippiensis	Eqv	T°, sex, T°×Sex, ST	Ferguson and Joanen 1982
	Alligator mississippiensis	Eqv	T° or sex? MT	Ferguson and Joanen 1983
	Alligator mississippiensis	Eqv	T°, sex, T°×Sex, ST	Joanen et al. 1987
	Chlamydosaurus kingii	Eqv	T°, no sex, ST	Harlow and Shine 1999
	Chelydra serpentina	Con	C×T°, no T°, C, C×H, ST	Rhen and Lang 1999b
	Chelydra serpentina	Eqv	T° or sex? C, ST	McKnight and Gutzke 1993
	Chelydra serpentina	Eqv	T°, no sex, no T°×Sex, C, ST	O'Steen 1998
	Emydoidea blandingii	Eqv	T° or sex?	Gutzke and Packard 1987a
	Chelydra serpentina	Con	T°, no sex, no T°×Sex, ST	O'Steen and Janzen 1999
	Chelydra serpentina	Eqv	T° or sex?	Packard et al. 1987
	Alligator mississipiensis	Pro	T°, sex, T°×Sex, C	Allsteadt and Lang 1995a
	Podarcis muralis^{GSD}	Eqv	T°, sex	Braña and Ji 2000
	Terrapene carolina	Con	Growth unexpected by TSD mode	St. Clair 1998
	Terrapene ornata	Con	Growth unexpected by TSD mode	St. Clair 1998
	Chrysemys picta	Pro	No T°, Sex n.e., C, N	Packard and Packard 2001
	Chrysemys picta	Eqv	T° or sex?	Gutzke et al. 1987
	Eublepharis macularius	Pro	T°, sex, LT	Tousignant and Crews 1995
	Eublepharis macularius	Pro	T°, sex, LT	Crews et al 1998
	Bassiana duperreyi^{GSD}	Eqv	T°, sex, T°×Sex, MT	Elphick and Shine 1999
	Crocodylus niloticus	Eqv	T°, T°×C, no sex, MT	Hutton 1987
	Caiman crocodilus yacare	Eqv	T°, sex? n.e. directly	Campos 1993
	Gopherus polyphemus	Eqv	T°, no sex, ST	Demuth 2001
Growth	*Crocodylus porosus*	Eqv	T°, sex, T°×Sex, MT	Webb and Cooper-Preston 1989
	Alligator mississippiensis	Eqv	T°, sex, T°×Sex, ST	Ferguson and Joanen 1982
	Alligator mississippiensis	Eqv	T° or sex? MT	Ferguson and Joanen 1983
	Alligator mississippiensis	Eqv	T°, sex, T°×Sex, MT	Joanen et al. 1987
	Emydoidea blandingii	Eqv	T° or sex? ST	Gutzke and Packard 1987a
	Terrapene carolina	Con	Growth unexpected by TSD mode	St. Clair 1998
	Terrapene ornata	Con	Growth unexpected by TSD mode	St. Clair 1998
	Chelydra serpentina	Con	No T° or sex, C, ST	Steyermark and Spotila 2001
	Chelydra serpentina	Eqv	T° or sex? MT	Bobyn and Brooks 1994
	Chelydra serpentina	Pro	T°, no sex, C, T°×C, ST	Rhen and Lang 1995
	Chelydra serpentina	Eqv	T° or sex? C, Soc, ST	McKnight and Gutzke 1993
	Chelydra serpentina	Eqv	T°, no sex, no T°×Sex, ST	O'Steen 1998
	Eublepharis macularius	Pro	T°, sex, C, LT	Crews et al. 1998
	Dicentrarchus labrax	Pro	T°, Sire, Dam, S×T°,D×T°	Saillant et al. 2002
	Bassiana duperreyi^{GSD}	Eqv	T°, sex, T°×Sex, MT	Elphick and Shine 1999
	Gopherus polyphemus	Eqv	T°, no sex. ST	Demuth 2001
Energy reserves				
Res. yolk mass	*Chelydra serpentina*	Pro	T°, no sex, C, C×T°, ST	Rhen and Lang 1999b
	Chrysemys picta	Eqv	T° or sex?	Gutzke et al. 1987
Fat body / yolk mass	*Chelydra serpentina*	Eqv	T° or sex?	Packard et al. 1987
	Chrysemys picta	Con	No T°, sex n.e., C, N	Packard and Packard 2001
	Alligator mississipiensis	Pro	T°, sex, C	Allsteadt and Lang 1995a

Temperature Effects	Species	Evid	Effect	Reference
	Chelydra serpentina	Pro	T°, sex, C, ST	Rhen and Lang 1999b
	Chrysemys picta	Con	No T°, sex n.e., C, N	Packard and Packard 2001
	Alligator mississipiensis	Con	C×Sex, T°×C	Allsteadt and Lang 1995a
Behavior and performance	*Eublepharis macularius*	Pro	T°, sex, LT	Gutzke and Crews 1988
	Tropidonophis mairii^{GSD}	Pro	T°, sex, ST	Webb et al. 2001
	Eublepharis macularius	Pro	T°, sex, LT	Crews et al. 1998
	Chelydra serpentina	Eqv	T°, sex, T°×Sex, ST	Janzen 1995
	Chelydra serpentina	Eqv	T° , no sex, no T°×Sex, ST	O'Steen 1998
	Thamnophis elegans^{GSD}	Con	No T°, no sex, litter, no dam	Arnold et al. 1995
	Podarcis muralis^{GSD}	Eqv	T°, no sex	Braña and Ji 2000
	Bassiana duperreyi^{GSD}	Eqv	T°, sex, T°×Sex, MT	Elphick and Shine 1999
	Gopherus polyphemus	Eqv	No T°, sex, ST	Demuth 2001
Physiology	*Eublepharis macularius*	Pro	T°, sex, LT	Gutzke and Crews 1988
	Chelydra serpentina	Pro	T°, sex, C, ST	Rhen and Lang 1999b
	Chelydra serpentina	Eqv	T°, no sex, no T°×Sex, ST	O'Steen and Janzen 1999
	Eublepharis macularius	Pro	T°, sex, LT	Crews et al 1998
Reproductive success	*Eublepharis macularius*	Eqv	♀: no T°	Tousignant and Crews 1995
	Eublepharis macularius	Pro	T°, sex, LT	Gutzke and Crews 1988
Not reported	*Trachemys scripta*	Con	No effect	Bull and Charnov 1989

Note: Evid = evidence for the differential fitness hypothesis: Pro = evidence in support, Con = evidence against, Eqv = equivocal evidence (due to e.g., confounding factors, ultimate effect on fitness not known, or if species has GSD). Significant effects detected: T° = temperature, T° or sex? = temperature and sex effects confounded, T°×Sex = temperature by sex interaction, N = nest environment, C = clutch identity, T°×C = temperature by clutch interaction, C×Sex = clutch by sex interaction, Dam = maternal identity, Sire = paternal identity, S×T° = sire by temperature interaction, D×T° = dam by temperature interaction, Soc = social environment. No T°, no sex, or no T°×Sex = factors analyzed and found to have no significant effect. n.e = factor not explored in study. ST = short-term study (< 1 year for long-lived species), MT = medium-term study (at least 1.5–2 years for long lived species), LT = long-term study (> 2 years or up to maturity of short-lived species). The evidence presented in this table is not an exhaustive list.

sovsky et al. 1984b; Vogt and Bull 1984; Mrosovsky 1994; Godfrey et al. 1996; Bowden et al. 2000; Harlow and Taylor 2000; Girondot et al. 2002). For instance, increasing temperature during the nesting season correlates with increasing female offspring in the fast-growing, short-lived jacky dragon (Harlow and Taylor 2000). Early hatchlings are male, and grow for a longer time, thus attaining larger sizes than females. This is probably favored by sexual selection (Harlow and Taylor 2000). The possibility of temporal rather than spatial patches (sensu Charnov and Bull 1977) remains unexamined for long-lived species, perhaps because among-year temperature variation is generally viewed as causing undesirable sex ratio fluctuations, though longevity and overlapping generations buffer sex ratio variation among cohorts.

Phenotypic thermal effects. The remaining hypothesis predicts phenotypic thermal effects (e.g., morphology, physiology, or behavior; Table 14.2) independently of offspring sex or differentially in sons and daughters (e.g., Rhen and Lang 1995; Shine et al. 1995; Tousignant and Crews 1995; Shine 1999). Exploring these alternatives requires decoupling temperature from sex effects (Rhen and Lang 1995, Chapter 10), preventing individual thermoregulation and social interactions that can obscure thermal effects (Steyermark and Spotila 2001). Unfortunately, such decoupling is rarely done and long-term thermal effects on lifetime reproductive success are understudied (Table 14.2). Alternatively, phenotypes within sex across temperatures are compared assuming that fitness for each sex should be maximal at temperatures that only produce that sex and lower at values that produce both sexes, which may be incorrect (see Adaptive Hypotheses, Differential Mortality, above, and Figure 14.2). Support for this hypothesis is consistent but not necessarily conclusive (Table 14.2; Rhen and Lang, Chapter 10), particularly if no null hypothesis is falsified. Support from GSD species (Table 14.2) is also equivocal since they lack TSD despite displaying conditions purportedly favoring its evolution. Although heteromorphic sex chromosomes and TSD may not coexist in the same individual (Bull 1983; Valenzuela et al. 2003), evolving TSD from heteromorphic sex chromosomes is not precluded (Bull 1981; see TSD Origin, Adaptive Hypotheses; contra Elphick and Shine 1999). Further, contrary to Shine et al. (1995), theoretical models do not predict differential phenotypic effects of temperature only

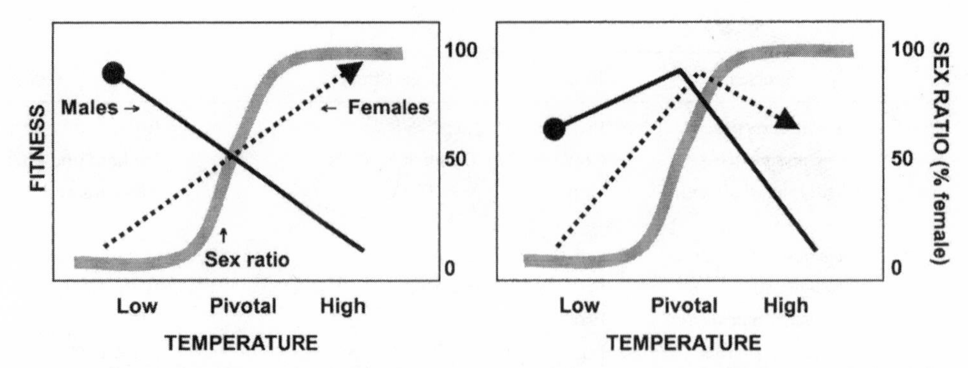

Figure 14.2 Hypothetical relationships between fitness and temperature for both sexes (black solid and dashed lines), and sex ratios (gray curve) that may be observed among or within species. As illustrated, identical sex ratios patterns could be expected from very different fitness within each sex. Solid circles and triangles denote the fitness value of the sexes that are produced at the extreme temperatures, while the opposite ends indicate the fitness of the sex that is no longer produced at the extreme temperatures. The pattern depicted here can be interpreted as TSD Ia (current illustration) or Ib (if solid line were to represent females and dashed line males, and the sex ratio axis were percent males). These simple examples could be extended to many other, more complex patterns.

in TSD species. Rather, theory proposes that differential thermal effects (sensu Charnov and Bull 1977) were either present in a GSD ancestor of TSD and subsequently favored TSD evolution, or are currently present and explain TSD persistence adaptively. Therefore, differential effects in GSD species constitute an opportunity to identify (1) potential effects for their TSD relatives, and (2) constraints preventing TSD evolution.

2. ADAPTIVE HYPOTHESES RELATED TO SELECTION FOR BIASED SEX RATIOS

Alternatives to the Charnov-Bull model's differential fitness hypothesis have also been proposed, some of which are also adaptive.

Group-structured adaptation. The group-structured adaptation hypothesis states that if populations are geographically structured into small breeding kin groups with minimal gene flow and refounded periodically, TSD allows the production of female-biased sex ratios, which would favor families or larger groups (sensu Hamilton 1967; Ferguson and Joanen 1983; Bull and Charnov 1988; Ewert and Nelson 1991). This hypothesis assumes that breeding groups in TSD species are smaller and more isolated than in GSD species (Burke 1993). However, no data support the derived expectations of higher inbreeding and lower heterozygosity in TSD species than in GSD taxa, or negative correlation of heterozygosity with female-biased sex ratio (Burke 1993). Further, *Alligator mississipiensis*, which might exhibit such a correlation (Burke 1993), may actually not have general female biases (Thorbjarnarson 1997).

Neutral Hypotheses

1. EXPRESSION OF A PREEXISTING THERMAL SENSITIVITY

Bull (1981) described the neutral transition from a heterogametic XX/XY GSD system to a TSD system, when an environmental change, or a shift in the species-realized niche, exposes individuals to conditions permitting the expression of a ubiquitously preexisting thermosensitivity. In a species possessing an XX-XY GSD system (with homomorphic or even heteromorphic sex chromosomes), some XX females develop into males in certain thermal environments (Bull 1983, 1989). Because population sex ratio evolves towards ½, the introduction of XX males induces a compensatory reduction in XY-male frequency according to the level of environmental influence (k), such that if $k \geq ½$ of the XX individuals develop as males, XY genotypes disappear from the population (Bull 1983). However, if $k < ½$ of the XX individuals become males, there is a continuum of attainable neutral equilibria from pure GSD (XX = XY = ½ when $k = 0$) to pure ESD (XX = 1 when $k = ½$) if all genotypes within a sex are equally fit—that is, if sex chromosomes are not highly heteromorphic (Bull 1981, 1983, 1989). The same will happen if female heterogamety (ZZ/ZW) is the starting condition. Because thermosensitivity was ubiquitous as the initial condition, this scenario leaves unexplained how this sensitivity first originated and spread.

2. INVASION BY A SEX RATIO DISTORTER

However, TSD can arise as a relatively neutral change or as sex ratio distorter and can invade a GSD population with differentiated sex chromosomes (male or female heterogamety with lethal YY or WW) (Morjan 2002). Based on *Chrysemys picta*'s life history, Morjan (2002) modeled nest temperatures as a function of female nest-site choice (heritable, nonheritable, or natal philopatry) and climatic conditions. TSD originated as a dominant or recessive mutation in the mitochondria, either sex chromosome, or an autosome. TSD eliminated the W or Y chromosome when TSD (1) was inherited mitochondrially, (2) biased sex ratios towards the heterogametic sex, (3) was dominant to Y or W, or (4) in two isolated simulations of natal philopatry (Morjan 2002). The analytical model showed that when the environment produces $k \geq ½$ females, TSD always invades if transmitted perfectly as a mitochondrial mutation, and if $k = ½$, TSD fully replaces a ZW system. In many other cases, TSD reached intermediate frequencies (Morjan 2002).

The models by Bull (1981, 1983, 1989) and Morjan (2002) demonstrate that positive adaptive selection is not essential to explain TSD origin in vertebrates, although selection (see Adaptive Hypotheses, above) increases the likelihood and rate of its spread and fixation. If Y (or W) chromosomes degenerate (e.g., through heteromorphism) and disappear during GSD evolution (Graves 2001, 2002), the moment of loss constitutes an opportunity for TSD to be quickly established, particularly if TSD works just as well as GSD. Testing of this prediction is required.

TSD Maintenance

Adaptive Hypotheses

The same adaptive hypotheses proposed for the origin of TSD (see TSD Origin, Adaptive Hypotheses, above) could also explain its maintenance, but other adaptive hypotheses only apply to TSD persistence.

1. DIFFERENTIAL FITNESS SUBHYPOTHESIS RELATED TO MATERNAL EFFECTS BY FEMALE NEST-SITE CHOICE

Maternal effects on egg allocation. Purportedly, TSD persists adaptively in some systems where maternal effects on egg allocation prevent the evolution of GSD. Based on data from *Malaclemys terrapin*, Roosenburg (1996) proposed that females choose nesting sites according to the size of their eggs, because egg size is positively correlated with offspring size and translates into differential fitness for sons and daughters (sensu Charnov and Bull 1977). Female hatchlings are postulated to benefit more from a larger initial size through posthatching growth (thus related to the sexual size dimorphism hypothesis). Incubation temperature would allow females to bias offspring sex ratio accordingly (Roosenburg 1996). The model is inapplicable to some turtles. In *Chrysemys picta,* egg mass is positively correlated with hatchling mass (Janzen and Morjan 2002), and larger females attain higher fecundity while smaller males attain higher mating success (Janzen and Morjan 2002 and references therein), yet egg size is uncorrelated with offspring sex (Janzen and Morjan, unpubl. data). Likewise, *Podocnemis expansa* has larger adult females than males. Contrary to expectation, however, larger females likely produce more males than small females because they lay more and larger eggs in deeper (colder) nests (Valenzuela 2001a,b). Further, larger hatchlings grow less than smaller ones, and thermal clines within nests yield mixed sex ratios (Valenzuela 2001a,b), reducing the required maternal sex ratio control. Alternatively to matching sex to egg size via nest-site selection, such matching could occur via allocation of yolk hormones (Bowden et al. 2000), though this requires direct testing. *Macroclemmys temiinckii* merits further investigation in this regard due to a positive correlation between intraclutch egg mass and sex ratio (Ewert et al. 1994). Therefore, maternal allocation of yolk hormones may interact with egg size (Bowden et al. 2000), maternal age (Ewert et al. 1994), and nest-site choice synergistically or antagonistically, such that the effects postulated by Roosenburg (1996) hypothesis may be masked by or be the indirect result of hormones or other factors.

2. ADAPTIVE HYPOTHESES RELATED TO SELECTION FOR BIASED SEX RATIOS

Population stability under high mortality schedules. Based on *Alligator mississippiensis*, Woodward and Murray (1993) proposed that TSD is favored over GSD in crocodilians by allowing adult female biases, which provide population stability under high mortality schedules. The model assumed that (1) female alligators nest preferentially in wet marshes, but if sites are unavailable they nest in dry marshes, and lastly in dry leeves; (2) incubation temperatures in these sites are low, intermediate, and high, and (3) these incubation sites/temperatures produce 100% females, mixed sexes, and 100% males, respectively. However, sex ratios vary annually within habitats due to climatic variation (Rhodes and Lang 1995, 1996), and widespread adult female-biased sex ratios in *A. mississippiensis* and other crocodilians have been contested (Thorbjarnarson 1997; Lance et al. 2000). Additionally, introducing modifier alleles of the individual temperature sensitivity or female nesting behavior triggers the invasion by masculinizing alleles and population extinction, such that the model may not explain the occurrence of TSD in crocodilians (Girondot and Pieau 1996). Finally, use of cohort or population sex ratios as the selected values in sex ratio evolution models for species with overlapping generations may be inappropriate (Girondot and Pieau 1996).

Sib avoidance. The sib-avoidance hypothesis suggests that TSD could be favored if it reduces the likelihood of inbreeding by producing unisexual clutches (Ewert and Nelson 1991; Burke 1993). Inbreeding avoidance would be important in species with likely sib mating due to geographic fidelity in reproduction or whole clutch mortality from abiotic factors (Ewert and Nelson 1991). Sib avoidance does not explain TSD origin because the advantages arise subsequent to strong within-clutch sex ratio biases (Ewert and Nelson 1991). Further, this hypothesis relies on the unsupported implication that inbreeding should be generally more common in GSD than in TSD species (Burke 1993). It also seems an unlikely explanation for TSD persistence for several reasons: (1) Thermoclines within nests reduce unisex clutch production (Wilhoft et al 1983; Georges 1992; Hanson et al. 1998; Kaska et al 1998; Valenzuela 2001b), (2) TSD cannot prevent cohorts from interbreeding in long-lived/iteroparous taxa (many TSD turtles and crocodilians) (Burke 1993), and (3) multiple paternity reduces inbreeding equally in TSD and GSD taxa (e.g., Harry and Briscoe 1988; Galbraith et al. 1993; Olsson et al. 1994; Hoggren and Tegelstrom 1995; Kichler et al. 1999; Abell 1997; Gullberg et al. 1997; Valenzuela 2000; Davis et al. 2002; Morrison et al. 2002). Therefore, TSD is not a crucial or unique inbreeding avoidance mechanism in many reptiles. Note that the sib-avoidance and group-structured adaptation hypotheses (see TSD Origin, Adaptive Hypotheses, Group-Structured Adaptation) attempt to explain sex ratio biases rather than being an argument for TSD specifically (Bull and Charnov 1988; Burke 1993).

Cultural inheritance. Natal homing through cultural inheritance was suggested to favor skewed primary sex ratios as reported for TSD reptiles (Freedberg and Wade 2001). As defined, natal homing cannot be *culturally* transmitted from mother to daughters in taxa showing no contact after oviposition; it is better explained by imprinting (Cury 1994), without impacting the model. Although some sea turtle data support this hypothesis (Freedberg and Wade 2001), its applicability is restricted for several reasons: (1) Reported sex ratio biases are not general (Gibbons 1990; Mrosovsky 1994; Thorbjarnarson 1997; Girondot et al., Chapter 15), and some correspond to secondary or adult sex ratio skews unrelated to TSD (Gibbons 1990; Mrosovsky 1994). (2) The assumption of temporal stability in sex ratios (as by Reinhold 1998) is inconsistent with observed climatically induced annual sex ratio variation (e.g., Godfrey et al. 1996; Valenzuela and Janzen 2001). (3) Vegetation cover was regarded as the ultimate determinant of turtle sex ratios, but nesting sites for many species lack vegetation. (5) Finally, Freedberg and Wade (2001) concluded that the runaway female bias via natal homing is countered only by genes reducing natal homing (still maintaining TSD), but it actually favors GSD evolution (Bull 1980; Bull and Charnov 1988). Alternatively, persistent sex ratio biases in TSD taxa could be explained by low effective heritabilities of both threshold temperatures and maternal nest-site choice (Bull et al. 1982a, 1988; Rhen and Lang 1995, 1998; Janzen 1992; Olsson et al. 1996).

Neutral Hypotheses

1. TSD-GSD EQUIVALENCE

The simplest explanation for TSD persistence is that TSD is neutral (Bull 1980; Mrosovsky 1980)—that is, if TSD is as good as GSD, there is no intrinsic selective pressure to evolve GSD. Such is the case under TSD if nest temperatures are uncorrelated with male and female fitness, and under a relatively constant environment inducing a sex ratio equilibrium around ½, (Bull 1980).

2. PHYLOGENETIC INERTIA

The phylogenetic inertia hypothesis states that TSD is vestigial, maintained because lack of genetic variation prevents GSD from evolving (Bull 1980). This hypothesis does not explain TSD origin because it leaves unexplained how TSD invaded the ancestral species, and it is not entirely neutral because such genetic constraints allow TSD persistence even if maladaptive. The distribution of sex-determining mechanisms among taxa supports multiple independent origins of TSD and GSD in vertebrates, invalidating phylogenetic inertia as a general cause for TSD maintenance (Burke 1993; Janzen and Paukstis 1991b; Chapters 2–7 and 13). However, if all crocodilians do possess TSD, perhaps it is maintained by phylogenetic inertia in this group.

3. NEUTRALIZING EFFECT OF LIFE HISTORIES AND GENETIC VARIATION

Certain life histories can transform TSD from maladaptive into a neutral or nearly neutral trait. For example, *overlapping generations* counter annual sex ratio fluctuations because multiple cohorts interbreed (Mrosovsky 1994; Girondot and Pieau 1996), an effect enhanced by *longevity* as more cohorts interbreed in a long-lived than in a short-lived species. Additionally, *genetic variation for individual thermal responsiveness* may counter biased sex ratios if variants ensure male and female production under climatic change. It may also permit TSD adjustments through the evolution of allelic frequencies according to environmental changes. However, such evolution may be precluded if climate changes faster than the rate of genetic evolution.

Interestingly, TSD appeared neutral when GSD was rarely known in turtles and unknown in crocodilians (Mrosovsky et al. 1984b), but as reports of GSD turtles increased, adaptive TSD evolution became more plausible (Ewert and Nelson 1991). Currently, support for adaptive explanations of TSD persistence among reptiles remains inconclusive, and neutrality has regained momentum (Girondot and Pieau 1999). Neutrality should be used as the null hypothesis, and falsified directly using adaptiveness as the alternative hypothesis. Perhaps it will be concluded that neutrality often explains TSD persistence satisfactorily, particularly in many long-lived species. Rather, we should ask how large any advantage or disadvantage must be to overcome the neutralizing effect of longevity and overlapping generations.

Only long-term research of incubation temperature effects on lifetime reproductive success will provide conclusive evidence against the neutral hypothesis and in favor of an adaptive explanation. Temperature effects for more than one year remain understudied (Elphick and Shine 1999), and reports of delayed thermal effects are restricted to short-lived, fast-maturing species (e.g., Crews et al 1998). Given the potential prevalence of substantial maternal effects, at least two generations must be reared in common environments to obtain a reasonable assessment of the genetic basis of threshold traits such as TSD (Roff 1996)—difficult for long-lived taxa.

Evolution of TSD Modes: TSD Ia, Ib, or II

Once a TSD system (as any other polyphenic developmental system) is established, the developmental switch regulation must adjust under environmental changes to preserve maximal accuracy of the phenotypic-environment matching. For instance, temperature thresholds and the timing of the thermosensitive period (TSP) could evolve, otherwise cues become unreliable, and if alternative cues are lacking, GSD evolution is favored.

In TSD II, low and high temperatures produce females, while intermediate temperatures produce males. In TSD Ia, low temperatures produce males and high values produce females, while the opposite is true for TSD Ib. TSD II may be the ancestral condition from which TSD Ia and TSD Ib evolved, by species shifting their responses along the temperature range and by some extreme temperature becoming lethal (Deeming and Ferguson 1988; Pieau 1996; Valenzuela 2001b). Alternatively, TSD II may be the transitional stage in the evolution between TSD Ib and TSD Ia. Unfortunately, data on TSD modes is incomplete or inconclusive, partly because some taxa remain unexamined, and partly because some TSD II cases may be experimental artifacts. For example, some TSD II reports derive from laboratory experiments using low temperatures only during the TSP because extended exposure is lethal (Ewert and Nelson 1991; Ewert et al. 1994). Such species with precluded production of low-temperature females are functionally TSD Ia (Valenzuela 2001b), as might occur in *Podocnemis expansa* (Valenzuela 2001b), *P. erythrocephala* (Vogt, unpubl. data), and *Kinosternon leucostomum* (Ewert et al. 1994). Perhaps this inherent TSD II mode reflects (not mutually exclusively) (1) an evolutionary vestigial and ecologically irrelevant ability (since it is unrealized under typical field conditions), or (2) a coaptable faculty for TSD evolution under climate change. Testing the validity of 1 requires sound phylogenetic hypotheses and reliable data on TSD modes. Additionally, it is uncertain whether among-species variation in TSD modes reflects drift or local adaptation, or could derive from the evolution of correlated traits.

TSD and GSD are frequently treated as two single traits, but this is not always appropriate. For instance, since gonadal differentiation can become thermosensitive by modifications at various steps, the independent origins of TSD (Janzen and Pausktis 1991b; Janzen and Krenz, Chapter 13) could generate differing mechanisms molecularly that constitute distinct traits. Similarly, differentiation within taxa sharing a common TSD ancestor could produce distinct TSD mechanisms, although related by the ancestral state. This is important because to understand TSD and GSD evolution we must define and identify the traits unambiguously (Valenzuela et al. 2003).

The Issue of Maternal Effects

A female contributes zygotic genes to her offspring and additional clutch effects through energy and yolk steroids allocation to the egg (Bowden et al. 2000; Elf et al. 2002), maternal-effect genes (maternal mRNA) (Werren et al. 2002), nest-site choice, and perhaps mate choice, since paternal genetic contributions can affect sex ratios in TSD species (Conover and Heins 1987a; Saillant et al. 2002) (Figure 14.1). Nest-site choice directly influences the incubation conditions of the offspring and their subsequent phenotype and fitness, but need not imply active female manipulation of offspring sex as is sometimes assumed explicitly or implicitly (e.g., Roosenburg 1996; Roosenburg and Niewiarowski 1998; Janzen and Morjan 2001, 2002). Similarly, yolk steroid concentrations can affect sex ratios substantially, not necessarily reflecting active sex ratio manipulation via hormonal allocation (Roosenburg and Niewiarowski 1998), as they could be [passive] by-products of the female's reproductive cycle (Bowden et al. 2000). Caution is necessary because extrinsic factors can skew sex ratios in seeming agreement with deterministically adaptive maternal manipulation if not tested properly (Post et al. 1999).

Nonetheless, parental patch choice theoretically elicits TSD evolution (sensu Charnov and Bull 1977) faster than does increasing the embryonic response to temperature (Bull 1983), if females choose the patch type that confers the highest fitness for each sex in response to environmental influences on their own condition, thus affecting offspring fitness differentially (sensu Trivers and Willard 1973; Roosenburg 1996). Consequently, population sex ratio approaches ½ under TSD, whereas half the offspring of GSD females encounter a disadvantageous patch (Bull 1983). Conversely, the rate of sex ratio evolution can be half as fast via maternal control than by evolution of embryonic responsiveness to temperature, even if the associated heritability is low (Bulmer and Bull 1982; Morjan 2004), such that the impact of maternal choice on thermal incubation conditions for sex ratio evolution (Bull 1983; Janzen and Morjan 2002) might not be as significant. For other scenarios (e.g., Reinhold 1998), under free patch choice, individuals will select the patch conferring the highest fitness, thus eliminating any TSD advantage.

Whether variation in nest-site choice and hormonal allocation (among females, seasonally within females, and among years) are subject to selection, particularly for sex ratio evolution, are entirely separate questions worthy of study. One caveat is that ideally, studies concerned with nest-site choice in TSD evolution should encompass undisturbed populations, since choices in natural versus disturbed habitats can differ dramatically (Hanson et al. 1998; Kolbe and Janzen 2002). Consequently, interpreting data from disturbed habitats (e.g., Woodward and Murray 1993; Janzen and Morjan 2001) requires caution to avoid misleading conclusions about the significance (adaptive or not) of individual behavior. Nevertheless, comparing disturbed and undisturbed populations can reveal factors currently affecting female nesting or population sex ratios, which could impact future population survival.

The Issue of Heritability

Adaptive TSD and sex ratio evolution require the existence of genetic variability, genetic effects, and genotype-by-environment interactions (Rhen and Lang 1998 and references therein). Few estimates of heritability *sensu stricto* of individual thermal responsiveness in reptiles exist (Bull et al. 1982a; Rhen and Lang 1995, 1998; Janzen 1992), and they are most likely inflated by confounding clutch effect components (Bull and Charnov 1988) including dominant and epistatic genetic variance (Conover and Heins 1987b; Olsson et al. 1996; Saillant et al. 2002), maternal effects (Bowden et al. 2000), and environmental (nest) effects (Shine et al. 1997a). Moreover, effective heritabilities are much lower under fluctuating (field) temperature conditions (Bull et al. 1982a), and probably become negligible after correcting for the effect of yolk steroids (see Bowden et al. 2000). Thus, additive genetic variance may be of reduced magnitude, overriden by temperature except around the pivotal temperature, and by epigenetic maternal effects around the pivotal temperature. Consequently, TSD evolution may be constrained in different taxa (but see Rhen and Lang 1998) by lack of genotype-by-temperature interaction (e.g., Janzen 1992), by nest temperature variance (Bull et al. 1982a), or by maternal effects (Bowden et al. 2000). Nonetheless, clutch identity affects reptilian phenotypes (Bull et al. 1982a; Brooks et al. 1991; Van Damme et al 1992; Janzen 1993; Allsteadt and Lang 1995a; Shine and Harlow 1996; Shine et al. 1997a,b), and components other than additive genetic variance can substantially affect sex determination.

TSD Loss: From TSD to GSD

Adaptive Hypotheses

1. INVASION OF TSD BY GSD

TSD can also be invaded by GSD even in the absence of fitness differences within each sex (Bull 1981, 1983). In a TSD species lacking sex chromosomes, a dominant factor *G* may appear, such that *G* carriers are always males while *gg* individuals exhibit TSD. Factor *G* can invade if the proportion of TSD males differs from ½ between generations, a likely occurrence due to environmental variation. Male frequency increases with *G*'s appearance, favoring *gg* individuals to become females, such that male heterogamety (GSD) is established. The same happens if *G* carriers are females. GSD also spreads under spatial rather than temporal environmental variation, with limited gene flow among patches (Bull 1983). Likewise, if the environment is constant but *Gg* males are more fit than *gg* males, GSD is favored (Bull 1981).

2. CONDITIONS THAT RENDER TSD DISADVANTAGEOUS OVER GSD

Sex ratio fluctuations. TSD species can suffer sex ratio fluctuations induced by climatic variation within and among years. Short-lived species, and taxa with nonoverlapping generations are more susceptible to this effect (see TSD Maintenance, Neutral Hypotheses, Neutralizing Effect of Life Histories, above). Such fluctuations partially explain the adaptive presence of GSD in northernmost populations of *Menidia menidia* (Conover and Heins 1987b).

Production of intersexes. Production of intersexes is another potential cost of TSD (Bull 1981). However, intersexes in TSD vertebrates are rare in nature (Bull 1983; Crews et al. 1994; but see Pieau et al. 1999b), perhaps because canalization of vertebrate sex differentiation prevents their development, at least naturally (but see Crews and Bergeron 1994; Wibbels and Crews 1995; Chardard and Dournon 1999). Further, early-life intersexuality occurs at similar levels and timing in TSD and GSD reptiles (Ewert and Nelson 1991). In species with potentially true intesexes (Pieau et al. 1999a; Pieau 1982), intersexuality is also transient, and adult reproduction is unaffected (Girondot et al. 1998; Pieau et al 1998).

Allee effect. Berec et al. (2001) postulated that at low population sizes or densities, the reduction in individual fitness (Allee effect) narrows the viable temperature ranges for population survival, but the "spatially homogeneous model," as applied to TSD turtles, depends on initial conditions that do not resemble turtle life histories.

Delayed onset or limited sexual dimorphism early in life. Compared to GSD, TSD embryos may suffer delayed onset or limited sexual dimorphism early in life (Bull 1983). For example, embryonic mammals display sexually dimorphic gene expression, metabolism, and developmental rates even prior to the blastocyst stage (Gutiérrez et al 2000), conceivably because sex chromosomes provide sexual identity from the time of fertilization (Valenzuela et al. 2003). Although TSD embryos could potentially express such early dimorphism at constant single-sex temperatures, predicting their sex at early stages in nature is virtually impossible because temperatures fluctuate and thermal effects are cumulative and are exerted relatively late (middle third of incubation in reptiles, and soon after hatching in fishes). Further, at constant temperatures, *Apalone spinifera* (GSD turtle) shows an accelerated chronology of sexual differentiation compared with TSD turtles (Greenbaum and Carr 2001). Whether these effects reduce fitness of TSD individuals remains untested.

Antagonistic pleiotropy. Antagonistic pleiotropy (i.e., selection for one phenotype imposes negative selection on the opposing phenotype due to genetic correlation) is a cost of phenotypic plasticity difficult to analyze because our understanding of TSD's molecular network is incomplete and because the two alternative phenotypes (males and females) must exist for the species to persist. This field of study deserves further research.

Imperfect phenotype-environment matching. Imperfect phenotype-environment matching caused by a low correlation between cues and selective factors or by constraints in developmental sensitivity hinders the adaptiveness of phenotypic plasticity (Moran 1992). The level of adaptiveness becomes dependent on the combination of the magnitude of the fitness differentials and the relative frequency of the selective environments (Moran 1992). Thus, TSD may not be favored if it costs more than having GSD, or if environments favoring both sexes equally are common. However, the level of phenotype-environment matching and related determinants of TSD adaptiveness remain unstudied in TSD taxa.

Parental control of offspring's sex ratio. Finally, parental control of offspring sex ratio by patch quality via nest-site selection (e.g., Reinhold 1998), favors a 1:1 sex ratio through frequency-dependent selection (Julliard 2000), which, if not attainable under TSD, selects for GSD (Bull 1980).

Neutral Hypotheses

Interestingly, no study has explored the invasion of GSD into a TSD system under neutrality. The model by Bull (1981, 1983) (see TSD Loss, Adaptive Hypotheses, Invasion of TSD by GSD) starts with no fitness differences within each sex, but the male bias induced by the introduction of factor *G* favors *gg* individuals to become females through frequency-dependent selection (Bull 1983), and thus this scenario is not neutral. Whether GSD can invade a TSD system under complete neutrality remains an open question.

Predictions from Theory

A series of testable predictions can be derived from the described theoretical background, which should provide null hypotheses for further empirical and theoretical research (Table 14.3).

Presence of TSD in Fish and Absence in Some Vertebrate Groups

Evidence for TSD reptiles and *Menidia* fish was already examined. In this section, the presence TSD in fish and its absence in other vertebrate groups is briefly discussed.

Fish

TSD is reported in numerous fish (Devlin and Nagahama 2002; see also Conover, Chapter 2), but some cases are GSD systems altered by environmental factors (e.g., thermal sex reversals in the presence of sex chromosomes) rather than TSD *sensu stricto*—that is, permanent sex determination by environmental temperature postfertilization without con-

Table 14.3 Testable Predictions about the Presence/Absence of TSD Among Taxa, Based on the Theoretical Models Described in the Text

Species should not have TSD if:	Species could have TSD if:
1. Variation in environmental temperatures skews population sex ratios drastically at the generation time scale (excessive thermal variability). 2. Sex ratios are highly biased because there is no variation in the environmental temperature experienced by the offspring during development (insufficient thermal variability). 3. Heteromorphic sex chromosomes are present (production of YY or WW lethals). 4. Sex determination by temperature is genetically linked to a trait that is selected against. 5. Parents or offspring can control the patch that offspring enter (environmental predictability and sex ratio control induce frequency-dependent selection). 6. Low matching of phenotype to environment (wrong sex produced at a given temperature) 7. Temperature and the differential-fitness factor are decoupled (low correlation between cue and selective agent). 8. Patches that confer differential fitness are uncommon (environments favoring both sexes are common).	1. Sex ratio biases are beneficial, highly correlated with temperature, and are constraint by GSD. 2. Temperature affects fitness directly or is highly correlated with a factor that affects fitness, differentially for males and females. 3. Fitness differentials between males and females are large enough, and conferred by environments frequently encountered in nature. 4. Longevity, overlapping generations, and/or genetic variation for temperature sensitivity exist (buffer against sex ratio fluctuations = TSD neutral).

sistent genetic differences among sexes (Valenzuela et al. 2003; see also Conover, Chapter 2). TSD adaptiveness in fish remains understudied because research in the wild is scarce (Conover, Chapter 2). TSD in some fish seems nonadaptive and may persist neutrally via longevity and/or phylogenetic inertia (Strüssmann et al. 1996b) while others cases require further study (Römer and Beisenherz 1996).

Amphibians

GSD is the only mechanism found so far in amphibians (reviewed in Solari 1994; Schmid and Steinlein 2001; Chardard et al., Chapter 7). Thermal sex reversal occurs in some species (see Chardard et al., Chapter 7) but at temperatures not typically encountered in the wild, thus reflecting developmental instability at extreme conditions rather than being ecologically relevant (Valenzuela et al. 2003; Schmidt and Steinlein 2001; Chardard et al., Chapter 7). Such lability however, could be coaptable for the evolution of TSD from GSD, which depends on a heritable susceptibility to develop either sex in response to temperature despite the genetic sex identity (Bull 1983).

Snakes

All snakes examined exhibit GSD (ZZ/ZW), with varying degrees of sex chromosome dimorphism (Solari 1994), but whether GSD is primitive or derived is uncertain because the ancestral state remains equivocal (see Janzen and Krenz, Chapter 13). Janzen and Paukstis (1991b) explained TSD absence in snakes by their shorter lifespan relative to turtles and crocodilians. However, lizards generally have shorter lives than snakes, yet some lizards have TSD. Temperature can distort secondary sex ratios as reported for *Pituophis melanoleucus* and *Nerodia fasciata* (Burger and Zappalorti 1988; Dunlap and Lang 1990; reviewed in Viets et al. 1994), but these were cases of documented or potential thermally induced differential mortality rather than TSD per se. Differential mortality also explains sex ratio biases induced by hydric conditions during incubation in elapid snakes (Reichling and Gutzke 1996). TSD absence in snakes is somewhat surprising since favorable conditions for TSD evolution exist in some cases (e.g., Table 14.2), and environmental factors, including temperature, can skew sex ratios in this group. GSD ubiquity in snakes merits further research.

Other Reptiles

Coexistence of TSD and viviparity appears incompatible because live-bearing parents thermally regulate their offspring's development (Bull 1980), such that embryos might experience insufficient thermal variability (Table 14.3). In fact, almost all reported TSD vertebrates are oviparous. *Eulamprus tympanum* could be an unexpected counterexample

(Robert and Thompson 2001), but its TSD classification requires confirmation (Valenzuela et al. 2003). Likewise, sex chromosomes and TSD *sensu stricto* cannot coexist within an individual, and thermal effects on sex ratios in species with sex chromosomes are explained by several alternative phenomena that do not constitute TSD per se (Valenzuela et al. 2003). Thus, the recently reported co-ocurrence of heteromorphic sex chromosomes and TSD in *Bassiana duperreyi* (Shine et al. 2002) is likely an example of thermal sex reversal such as that found in amphibians. Importantly, unlike in amphibians, sex ratio distortions in *B. duperreyi* occur within natural thermal ranges (Shine et al. 2002) and may be adaptive, sensu Charnov and Bull (1977) (Elphick and Shine 1999; Shine et al. 1995). Thermal sex ratio distortions like these are of great evolutionary significance, may be widespread, and warrant continued research.

Varanus lizards ovoposit inside termite nests where temperatures stay quite constant (de Lisle 1996) and should lack TSD (Table 14.3). Indeed, some species *(V. niloticus, V. varius,* and *V. albigularis)* possess identifiable sex chromosomes (GSD) (de Lisle 1996 and references therein), while others remain unexamined. *Paleosuchus trigonatus*, a TSD crocodilian, also uses termite mounds, but because nests are placed against or on top of the mounds (Magnusson et al. 1985), they experience wider temperature fluctuations than *Varanus* nests, allowing TSD to operate.

Birds and Mammals

Both birds and mammals are homeotherms. Since thermal invariance during development via parental thermoregulation would highly skew sex ratios, precluding TSD (Table 14.3), each group probably evolved or maintained their distinct GSD adaptively: female heterogamety (ZZ/ZW) in birds and male heterogamety (XX/XY) in mammals.

Conclusions

Theoretically, TSD can originate neutrally or adaptively, and once established it can evolve through either path independently of its origin. Conclusive evidence for adaptive TSD evolution is restricted to a few cases in vertebrates—for most TSD taxa the neutral null hypothesis remains unfalsified, particularly for long-lived species with overlapping generations. This is not to say that TSD is never adaptive, but rather that further long-term research is needed to reject either alternative conclusively. Most of the existing hypotheses for TSD evolution are not generalizable theoretically or empirically, perhaps because TSD evolved by different means in various vertebrate groups and thus requires case-specific explanations. Only by continuous observation, hypothesis building, and testing will we disentangle TSD evolution. Comparative approaches that explore TSD's physiological and molecular basis in an ecological context and examine the correlation of TSD/GSD's presence/absence with multiple life-history traits simultaneously are necessary to truly address the origin and persistence of TSD.

Acknowledgments—Many thanks to Claude Pieau, Mike Ewert, Carrie Morjan, Rachel Bowden, and Dean Adams for their insightful criticisms and suggestions on this manuscript.

15

MARC GIRONDOT, VIRGINIE DELMAS,
PHILIPPE RIVALAN, FRANCK COURCHAMP,
ANNE-CAROLINE PRÉVOT-JULLIARD,
AND MATTHEW H. GODFREY

Implications of Temperature-Dependent Sex Determination for Population Dynamics

The study of population dynamics involves linking the temporal change in numbers of individuals in a population with different life-history parameters and external environmental forces. External forces can include catastrophic events, but simple physical factors can also drive population dynamics. Temperature is an ecophysical factor that can have a major impact on some species, obviously including reptiles that exhibit temperature-dependent sexual determination (TSD). For turtles, crocodiles, and lizards with TSD, the primary sex ratio is directly dependent on environmental thermal conditions, and it is the direct factor distinguishing between GSD- (genotypic sex determination) and TSD-based population dynamics. Indeed, for many characteristics of population dynamics, the presence of TSD in a population will be neutral. Therefore, the authors integrate in this review only data about specific details of TSD. The majority of data comes from turtle species because TSD in this group has been the best studied, but comparisons with other vertebrate taxa are included whenever possible. The authors also provide the following definitions to simplify the discussion of sex ratio. *Sex ratio* is expressed as male or female frequency. The *primary sex ratio* is defined here as the frequency of male or female at the end of the thermosensitive period (TSP) for sex determination during development (Mrosovsky and Pieau 1991). The *secondary sex ratio* is the sex ratio at hatching. It is generally assumed that secondary and primary sex ratios are equal, although differential late-stage embryonic mortality can produce differences between the two. Throughout the text the term secondary sex ratio is used only when it differs from the primary sex ratio. *Population sex ratio* is the sex ratio of all individuals in a population that have reached sexual maturity, and *operational sex ratio* is the sex ratio of only those individuals that contribute to reproduction of a population at any given time.

Basics of Population Dynamics (Adapted from Charlesworth 1980)

Species with TSD are generally long-lived species (Janzen and Paukstis 1991a) and therefore are best described by age-structured population models. Descriptions of population structure are simpler when discrete age classes (or stages) are used (e.g., Heppell 1998). In this case, individuals may survive over many years, but reproduction is limited to one season of the year. Note that this assumption, while largely correct for many temperate species, may be too simplistic for tropical species in which reproduction could occur throughout the year. For example, the Amazonian tortoise *Geochelone denticulata,* can nest nearly year-round (Castaño-Mora and Lugo-Rugeles 1981; Métrailler and Le Gratiet 1996; Moreira 1991; Moskovits 1985). However, as sexual maturity is delayed in these species, age-structured population models can still be used by categorizing individuals into a year-based class.

The number of males and females at the beginning of the breeding season in a given year (t) can be described in terms of individuals falling into age classes 1, 2, 3, and so on. These correspond to individuals who were born 1, 2, 3, and so on years previously. The number of females aged x at the beginning of the breeding season in year t can be written as $n(x, t)$ and the number of males as $n^*(x, t)$. We describe $P(x, t)$ and $P^*(x, t)$ as the probability of survival over one age class (or year in most cases) for females and males, respectively, who are present in age class x at the start of the breeding season in year t. Using this notation yields, for age classes other than the first:

$$n(x, t) = n(x - 1, t - 1)\, P(x - 1, t - 1) \text{ and}$$
$$n^*(x, t) = n^*(x - 1, t - 1)\, P^*(x - 1, t - 1) \text{ with } (x > 1)$$

Age class 0 requires a specific treatment in order to distinguish between GSD- and TSD-based dynamics.

The authors describe $M(x, t)$ as the expected number of fertile eggs produced in year t by a female aged x at the start of the breeding season. The proportion of females among the offspring resulting from these eggs is $a(t)$ (note that primary sex ratio is supposed to be independent of the age of the mother), and the survival of these offspring is $P(0, t)$ for females or $P^*(0, t)$ for males (note that survival in the first year is also supposed to be independent of the age of the mother). For species with TSD, the parameters $P(0, t)$ and $P^*(0, t)$ can be calculated at three separate time periods: (1) before sex determination, when no sex-specific function is needed, survival is described as $Pu(0, t)$; (2) during and after sex determination but before hatching, survival of females is $Pd(0, t)$ and that of males is $Pd^*(0, t)$; and (3) during the first age class (or year) after hatching , survival of females is $Ph(0, t)$ and that of males is $Ph^*(0, t)$. From this,

$$P(0, t) = Pu(0, t)\, Pd(0, t)\, Ph(0, t)$$

for females and

$$P^*(0, t) = Pu(0, t)\, Pd^*(0, t)\, Ph^*(0, t)$$

for males.

Note that the secondary sex ratio is simply $a(t)/\{a(t) + [1 - a(t)]\, y\}$, with y being $Pd^*(0, t)/Pd(0, t)$. Thus secondary sex ratio depends on primary sex ratio [$a(t)$, female frequency] but also on differential survivorship of males and females during development (y parameter).

The net expected contribution from a female aged x in year t to the population aged 1 in year $t + 1$ is thus:

$$f(x, t) = M(x, t)\, a(t)\, P(0, t)$$

for the female offspring and

$$f^*(x, t) = M(x, t)\, [1 - a(t)]\, P^*(0, t)$$

for the male offspring.

The earliest possible age of reproduction of females (sexual maturity) is defined as b for females and b^* for males. Reproductive senescence has not been demonstrated in any reptile species with TSD and therefore no upper limit is set.

Then, from the above definitions, one obtains the following relations:

$$n(1, t) = \sum_{x=b}^{\infty} n(x, t - 1)\, f(x, t - 1)$$

for females and

$$n^*(1, t) = \sum_{x=b}^{\infty} n(x, t - 1)\, f^*(x, t - 1)$$

for males.

Note that the number of males $n^*(x, t)$ is relatively unimportant in this model, a phenomenon called "female dominance." An alternative birth function can be used to introduce some limitation due to male number (e.g., dependent on harmonic mean of the population sex ratio; see Caswell and Weeks 1986). From these equations, the primary sex ratio $a(t)$ must be defined independently each year, which introduces a temporal stochasticity.

A previous study has attempted to make a direct description of the process of population dynamics for a species with TSD (Woodward and Murray 1993), but in that study, $a(t)$, $P(x, t)$, and $P^*(x, t)$ were assumed to be constant across years, which is likely incorrect. In another study, a simplified model was used to check for the relative influence of sex-specific differences in survival rates or sex-specific differences in age at maturity to account for an overall bias in the population sex ratio (Girondot and Pieau 1993). In that particular case, neither time (t) nor age dependency (x) was taken into account in any of the parameters, making description of sex-specific frequencies quite simple: $f = M \cdot Pu \cdot a$ for numbers of females and $m = M \cdot Pu \cdot (1 - a)$ for number of males at the time of sex determination.

Then, the number of females of age 1 (second year) was assumed to be the fraction P of the f females that survived to age 1 (described as $f \cdot P$, and $m \cdot P^*$ for males). For the second year (note that no temporal variability was taken into account), f newborn females and m newborn males entered in the population. The same process was assumed to continue across all years, and therefore at any given year, the population was composed of $f + f \cdot P + f \cdot P^2 + f \cdot P^3 + \ldots + f \cdot P^\alpha$ females and similarly for males. Only sexually mature individuals account for the population sex ratio, and therefore the following equations were used to determine the number of sexually mature individuals:

$$\sum_{i=b}^{\infty} f\, P^i = \frac{f\, P^b}{1 - P}$$

for females and

$$\sum_{i=b^*}^{\infty} m\, P^{*i} = \frac{m\, P^{*b^*}}{1 - P^*}$$

for males.

The population sex ratio is consequently:

$$\frac{(f\, P^b)\,(1 - P^*)}{(f\, P^b)\,(1 - P^*) + (m\, P^{*b^*})\,(1 - P)}$$

In instances when P and P^* are large (near 1), a small difference between P and P^* will make the overall sex ratio value to tend toward 1 (all females) or 0 (all males). In other words, when survival is high, a small difference in the mean annual survival of males and females produces a strong bias in the population sex ratio. Differences in the values of b and b^* will alter the population sex ratio mostly for populations with low survival rates.

Primary and Secondary Sex Ratios in Natural Populations

Biases observed in the primary sex ratio in offspring of the squamate *Agama agama* facilitated the original discovery of TSD (Charnier 1966) in lizards and later in turtles (Pieau 1971, 1972). Sex ratio biases in clutches incubated in natural conditions have also been demonstrated in many reptile species, mainly turtles (Pieau 1974; Bull 1985; Mrosovsky et al. 1984b) and crocodilian species (Webb and Smith 1984; Lance et al. 2000). Recently, Freedberg and Wade (2001) reviewed the published literature and suggested that different populations of sea turtle species (all exhibit TSD) tend to produce female-biased primary sex ratios. Furthermore, these same authors suggested that there is an "ubiquity of biased sex ratios" in reptile populations (Freedberg and Wade 2001, 1053). Regardless of whether this accurately reflects all published accounts of reptilian sex ratios, this proposition assumes that published results represent a random subset of all TSD populations. This is probably not true (see also Valenzuela, Chapter 14). Indeed, it is likely that results of sex ratio studies will have a greater chance of being published if the results are different from equality (Festa-Bianchet 1996). Therefore, at the current time, any review of the published literature on primary sex ratios of TSD species should not be used as an indication of general trends in primary sex ratio production in nature.

Given that TSD species tend to be long-lived and that environmental conditions vary from year to year, the best estimates of mean primary sex ratio production in TSD populations must be based on multiyear data sets. But it is logistically difficult to directly estimate primary sex ratios, principally because for many reptile species, there is no sexual dimorphism before maturity. Distinguishing the sex of immature individuals usually requires invasive methods (e.g., histological preparation of the gonads or laparascopy in larger individuals). These methods are limited in that they can only give current time information; they cannot be used to go back and reconstruct primary sex ratios over a number of years. For these reasons, long-term data sets on primary sex ratio production in TSD species are often based on indirect estimates, using environmental correlates (temperatures, rainfall, vegetational cover) or clutch characteristics (incubation duration with some direct sexing of dead or laboratory-incubated hatchlings) to predict hatchling sex. Note that few studies using indirect predictors of sex ratio have subsequently validated the methods used to estimate the sex ratios, which makes it difficult to assess the validity of the results (cf. Mrosovsky et al. 1999).

It is beyond the purview of this chapter to provide an exhaustive review of sex ratio studies in reptilian TSD species. Rather, several studies with (1) large numbers of sampling localities or analysis of entire reproductive localities and (2) long-term data sets will be presented, and they will be used to highlight several consistent properties of sex ratios by TSD populations.

Extensive Spatial Analysis

For crocodilian species, several authors have suggested that female-biased sex ratios are the norm (Deeming and Ferguson 1989b; Woodward and Murray 1993). Thorbjarnarson

(1997) provided an extensive review of available information on primary, secondary, and operational sex ratios of different crocodile populations and reported that there are more male-biased sex ratios reported in the literature than female-biased sex ratios. More importantly, it was noted that a wide range of different sex ratios have been observed at the intraspecific level, and thus caution must be used when making statements about the significance of sex ratios at the species level. An extensive study on the sex ratios of the American alligator *(Alligator mississippiensis)* based on six years of data over 11 different sites in southern Louisiana reported that overall there was a slight male bias (58% male sex ratio) in juvenile sex ratios, but that local sex ratios varied in time and location (Lance et al. 2000). Thus, some sites produced a strong female bias (29% male sex ratio), others a strong male bias (83% male sex ratio), and some very close to equality (50% male sex ratio). Interestingly, these authors also report a five-year female-biased sex ratio for hatchlings produced by nests in Louisiana, although not in the same time period as the juvenile sex ratio results (Lance et al. 2000). Thus, there are cases where the juvenile and operational sex ratios in TSD populations are not the same as the primary sex ratios.

At this time, information on the sex ratios of hatchlings produced by the entire reproductive population of sea turtles is available for only two locations. Broderick et al. (2000) reported an extreme female bias (4–14% male) in green turtle hatchlings produced over a five-year period in Northern Cyprus, based on data from incubation durations. The larger breeding population of green turtles in the Mediterranean is largely restricted to nesting on beaches in Cyprus, Turkey, and more rarely, in Lebanon, Israel, and Egypt (Kasparek et al. 2001). Data from incubation durations in the other nesting areas also suggested that the majority of hatchlings produced over an extended time period were female (Broderick et al. 2000). Hence, it is likely that the overall estimate of a female-biased hatchling sex ratio for green turtles in the Mediterranean can be applied to the entire reproductive population. Godfrey et al. (1999) reported an extreme sex ratio bias (<10% male) of hatchlings produced by hawksbills in Bahia, Brazil, over a six-year period, with indications of similar extremes going back almost 20 years. Given that there are no other regular nesting areas for this species along the Atlantic coast of South America, and no reported nests on any islands off of Brazil, it is likely that these sex ratio estimates are representative of the entire reproductive population. Interestingly, both breeding populations are considered small (< 500 reproductively active females). Note, however, that *Eretmochelys imbricata* also nests in West Africa coast and that the relationships with the Brazilian population are unknown.

Long-Term Data Sets

One of the longest data sets available for primary sex ratio under TSD comes from Janzen (1994a), who reconstructed the annual hatchling sex ratio for a population of freshwater painted turtles *(Chrysemys picta)* over 49 years, using mean air temperatures. The overall mean for this period was 47.8% males, although individual years were highly variable, ranging from 0–100% males. This study highlights the necessity of integrating sex ratio data over time, to account for interannual variation in sex ratios that is inherently linked to variability in the environment. Similarly, using rainfall data, long-term data sets on hatchling sex ratios have been generated for leatherback *(Dermochelys coriacea)* and green turtle *(Chelonia mydas)* hatchlings produced on a major nesting beach in Suriname (Godfrey et al. 1996). Again, variation was observed in the annual mean sex ratio of hatchlings produced by the two species, ranging from extremely male-biased to extremely female-biased sex ratios, but the overall (based on all years) sex ratios was 54.6% male for leatherbacks and 31.7% male for green turtles. However, these estimates of hatchling sex ratio production represent only one nesting beach in the larger breeding range of these two populations. Female turtles from the larger reproductive populations use other nesting areas in Suriname and French Guiana (Schulz 1975), and information is lacking on the sex ratios of hatchlings produced over the entire range of beaches. Some studies of sea turtle hatchling sex ratios have reported extreme female biases, including loggerheads *(Caretta caretta)* nesting in the southeastern United States (Mrosovsky and Provancha 1992; Hanson et al. 1998), loggerheads nesting in Brazil (Marcovaldi et al. 1997), loggerheads nesting in the Mediterranean (Godley et al. 2001), green turtles nesting on Ascension Island (Godley et al. 2002), leatherbacks nesting on the Pacific coast of Costa Rica (Binckley et al. 1998), and hawksbills *(Eretmochelys imbricata)* nesting in Brazil (Godfrey et al. 1999).

In nearly all cases, the estimates represent only a portion of the hatchlings produced in the larger breeding range of each population. Thus, these observed extreme female biases (sometimes < 10% male) in hatchling sex ratio may be countered by different sex ratios of hatchlings produced on unmonitored nesting beaches elsewhere, which are nevertheless part of the larger population. For instance, the green turtles nesting at Ascension Island migrate to feeding areas off of northeastern Brazil (Carr 1964). These feeding grounds are also frequented by turtles from nesting areas in Suriname, French Guiana (Pritchard 1976), and perhaps other nesting areas in Brazil (e.g., Trindade Island, Atol das Rocas), which all likely contribute to the larger interbreeding green turtle population in the midwestern Atlantic (Karl et al. 1992). Without information on hatchling sex ratios from the entire range of the breeding population, it is not possible to estimate the primary sex ratio for the total reproductive population.

Studies conducted on immature loggerhead turtles found in the waters off the southeastern shore of the United States report that these individuals exhibit in general a female-biased (33% male) sex ratio (Wibbels et al. 1991d; Shoop et al. 1998). This is significantly different from the extreme female bias (< 20% male) reported for loggerhead hatchlings produced on nearby nesting beaches (Mrosovsky and Provancha 1992; Hanson et al. 1998). Potential explanations include sex-biased differences in the likelihood of capture and sampling, sex-biased differential mortality following hatching, and differential population origin of the individuals sampled. Some sampled turtles could have been produced at distant nesting beaches and have arrived in the study area during foraging migrations (e.g., Bolten et al. 1998). For these reasons, it is extremely difficult to interpret the results of sex ratio studies of immature sea turtles.

In reviewing the currently available data, there are certain generalities that can be made concerning the primary and secondary sex ratios of reptiles in natural populations. First, although some populations do produce extreme female biases, this is far from being the norm for all species. Indeed, in the case of the Crocodylia, more populations have been observed to be male biased than female biased. Second, when long-term data sets exist, there is often interannual variation in overall primary and secondary sex ratios produced, with extremes in both directions, which when averaged may or may not result in biased sex ratios. Third, many sex ratio studies have focused on only a portion of the entire nesting range of the breeding population. Given that there are often thermal differences across different nesting areas, it is nearly impossible to extrapolate observed sex ratios to the overall output of the larger population.

Primary Sex Ratio versus Operational Sex Ratio and Population Dynamics

For the study of population dynamics, the number of juveniles produced in a population is a key factor (see Basics of Population Dynamics, above). For many species with TSD, the number of juveniles produced is directly related to the number of sexually mature females (female dominance phenomenon; see Basics of Population Dynamics, above). Moreover, no conclusive data exist concerning the minimum number of males necessary to successfully fertilize all females in reproductive condition in a population, although it is thought that small numbers of males may be a limiting factor of population dynamics for reproduction, based on observations of green turtles made in artificial conditions where the operational sex ratio was strongly biased (Wood and Wood 1980).

One major difficulty has been to find a direct relationship between primary and operational sex ratio. The operational sex ratio is dependent on the population sex ratio, which in turn is influenced not only by primary sex ratio, but also by sex-specific differences in first age at maturity, survivorship, and emigration or immigration (Gibbons 1990). An example of the differences between observed primary and juvenile sex ratios concerns the loggerhead sea turtle (see previous section, Long-Term Data Sets).

For long-lived TSD species, the principal factor that is responsible for the decoupling of primary sex ratio and population sex ratio is thought to be differential survivorship between males and females (Girondot and Pieau 1993; Janzen and Paukstis 1991b). In this case, the population sex ratio is composed of the primary sex ratios of several cohorts of offspring. In general, the population sex ratio is by definition less variable statistically than the primary sex ratio. In addition, the population sex ratio itself can be different from the operational sex ratio due to either behavioral factors such as male-male competition for mates in *Alligator mississippiensis* (Woodward and Murray 1993) or environmental pressures such as the variabile availability of minimal nutritional resources that are required for entering into the reproductive state (Limpus and Nicholls 1988; Broderick et al. 2001).

Gibbons (1990) provided a review of population sex ratios for 52 species or subspecies of turtles. To this list, the authors add data from the European pond turtle, *Emys orbicularis* (Girondot et al. 1994), *Testudo hermanni* (Hailey and Willemsen 2000), and *Chelonoidis denticulata* (Josseaume 2002), but they exclude data from the only marine turtle species due to the difficulties in obtaining reliable data for this species. GSD or TSD patterns of sex determination are not known for all these species. The authors used phylogenetic analysis to infer the most probable pattern for the unknown species as follows (Girondot et al., unpubl. data). A composite phylogeny of turtles (Caccone et al. 1999; Dutton et al. 1996; Georges and Adams 1992; Georges et al. 1999; Krenz and Janzen 2000; Noonan 2000; Shaffer et al. 1997) has been used to infer the mode of sex determination (two states, TSD or GSD) in these 52 species or subspecies based on parsimony criteria. Among all 153 populations, nine originated from species with GSD, 143 from species with TSD, and one *(Clemmys marmorata)* was ambiguous based on parsimony criteria and has thus been removed from the analysis. For the GSD populations, three of nine (30%) exhibited significant biases of population sex ratio, whereas for TSD populations, 52 of 143 (37%) show significant biases in population sex ratio. No significant difference was detected in the proportion of population with or without significant bias between the groups with TSD or GSD (Fisher exact test, $p > 0.99$). Therefore, whereas biased primary sex ratios are often observed in TSD species and 0.5 is the rule for GSD species, mode of sex determination seems less important than other demographic factors, such as survivorship (Girondot and Pieau 1993) or sex-specific differences in capture likelihood (Gibbons 1990), in determining the observed population sex ratio. Nevertheless, there is no reason to think that species with GSD are more prone to sampling bias than species with TSD, but species-specific and sex-specific differences in behavior may increase the potential for sampling bias. The end result is that available data show there is no clear difference in mean population sex ratios between TSD and GSD species.

The lack of overall differences in population sex ratios between TSD and GSD groups of species appears surprising, although several key points should be kept in mind. First, as already noted, the primary sex ratio is less influential than differential survivorship between the sexes on the overall population sex ratio for long-lived species (Girondot and Pieau 1993). Second, primary sex ratios fluctuate from year to year around an average sex ratio (Janzen 1994a; Godfrey et al. 1996), and the standard deviation of population sex ratio is inversely proportional to the number of years of cohorts contributing to reproduction, and therefore to longevity. When longevity is very high, population sex ratio is dependent on the weighted average of primary sex ratio of many different cohorts. Thus longevity induces a buffer effect on population sex ratio that could be identically variable in TSD or GSD populations. The same argument can be used when short-lived span lizards lay several clutches during the year; the annual cohort concept should be changed to cohorts of eggs deposited the same month, for example. However, while the character of GSD or TSD may not definitely result in clear differences in population sex ratio and therefore population dynamics, it is likely that for specific situations, for example, during extended periods of rapid climate change, population sex ratios in TSD species would be vastly different from those of GSD species.

Recently, several models linking dispersal patterns and TSD have been proposed (Reinhold 1998; Julliard 2000; Freedberg and Wade 2001). These models and some tests are discussed in Valenzuela (Chapter 14).

TSD as a Factor of Decline

It is evident that if during sufficiently extended period of time only one sex were produced, the population would disappear. Temperature extremes in the environment can produce extreme biases in offspring sex ratios in species with TSD. For this reason, some authors have hypothesized that the occurrence of TSD in a species increases the probability of extinction in the face of a catastrophic environmental event. For instance, there have been suggestions that the large extinction event of many dinosaur species at the Cretaceous/Tertiary (K/T) boundary was likely due to large climatic changes coupled with the presence of TSD in these species (Bull 1983; Ferguson and Joanen 1982; Harvey and Partridge 1984; Janzen and Paukstis 1991a; Standora and Spotila 1985). However, several points relevant to this hypothesis must be kept in mind. First, the extinction events involved around 90% of all species existing during this period, and probably many species had GSD. Second, the sex-determining mechanism in dinosaurs remains entirely unknown; indeed the current phylogeny suggests that dinosaurs could have had TSD or GSD, because the only remaining within group (Aves) possesses GSD, and the remaining sister group (Crocodylia) possesses TSD (Janzen and Krenz, Chapter 13). Therefore, the hypothesis that Cretaceous dinosaurs had TSD is equally parsimonious with the hypothesis that they had GSD. Moreover, many groups for which TSD can be inferred by parsimony survived beyond the great extinction events of the K/T boundary (e.g., Carettochelyidae, Chelydridae, Sphenodontidae) (Rage 1998). Third, many groups with TSD have continued to exist since the late Eocene, when another severe climatic cooling event was observed (Pomerol and Premoli Silva 1986). Therefore, it seems implausible that TSD and a general cooling trend could have been the single major reason for the extinction of dinosaurs.

Another commonly cited threat to TSD species is rapid global warming due to the greenhouse effect (Deeming and Ferguson 1989b; Janzen 1994a; Mrosovsky et al. 1984b). Again, the idea is that increased incubation temperatures will result in extreme biases of sex ratios, leading to the inability of populations to reproduce. However, it should be noted that nest incubation temperatures usually vary across temporal or geographic ranges, so nesting females could potentially adjust their behavior (in space and/or time) to achieve certain offspring sex ratios. Indeed, a change in the nesting period or nest location can be a response of an individual (Morjan 2002). Therefore, even if mean temperatures rapidly increase, individuals still may be able to nest earlier or later in the season and therefore achieve incubation temperatures similar to those experienced before global warming. Behavioral changes as a means to achieve or maintain certain primary sex ratios may explain the difficulties to date in finding a significant relationship between latitude and pivotal temperature (the constant incubation temperature that produces a 50% offspring sex ratio) (Bull et al. 1982b; Ewert et al. 1994). Note that the fluctuation of primary sex ratio can be high for both small-sized populations (Girondot et al. 1994) and short-lived species (Janzen and Paukstis 1991b). At the same time, wide interannual fluctuation of the primary sex ratio is more likely to lead to extinction for a smaller population than for a larger one (Girondot et al. 1994).

TSD also has been linked with the decline of small populations by the way of the Allee effect (Berec et al. 2001). The Allee effect describes the general situation in which the growth rate of a population becomes negative when the population size falls below a certain threshold (the Allee limit; Courchamp et al. 1999). According to Berec et al. (2001), a population with TSD is more susceptible to the Allee effect due to its biased sex ratio, because the probability of finding a mate is reduced relative to a population with GSD and a balanced sex ratio. This hypothesis assumes that TSD results in a biased operational sex ratio as a direct result of biased primary sex ratios, but as discussed previously, the former does not necessarily follow from the latter. Additionally, the model of sex determination given by Berec et al. (2001), which uses directly the sex ratio versus constant incubation temperatures curve as a model of sex determination in natural conditions, is too simplistic to give realistic outputs (see also Valenzuela, Chapter 14). Nevertheless, it is an interesting avenue of approach to the study of the implications of TSD on population dynamics, and future studies along these lines should prove to be enlightening.

Another way by which a biased sex ratio can drive the decline of a population is by reduction of the effective population size, which in turn increases genetic drift. The effective size of the population (*Ne*) is the size of a hypothetical population that would have the same rate of change in homozygosity as the observed population (Crow and Kimura 1970). If *Nm* males and *Nf* females are present in the population, then $Ne = (4\ Nm\ Nf)/(Nm + Nf)$. If $Nm = Nf$, then $Ne = 2\ Nm$, but is lower if $Nm \neq Nf$. Therefore, populations with biased operational sex ratio are more susceptible to genetic drift and thus to accumulation of deleterious mutations (Charlesworth and Charlesworth 1998). However, the impact of this effect will be significant only for small population sizes over long periods of time. Rapid increases in population size due to female-biased sex ratio can overcome this problem (Wedekind 2002). However, in this particular case, there will be selection for masculinizing alleles (Girondot et al. 1994; Mrosovsky 1994) due to frequency-dependent selection for unbiased primary sex ratios (Shaw and Mohler 1953). Depending on the variability of sensitivity of sex ratios to temperature for a population, this effect can be neutral or deleterious due to an increase of males following limited resource availability (Girondot et al. 1994).

Interaction of TSD and Population Dynamics

Some interesting cases have been described in species with TSD in which the characteristics of population dynamics can modify the sex ratio at hatching, and therefore the consequences of TSD. One example is the sex ratio bias in offspring production due to nest destruction by the leatherback sea turtle *Dermochelys coriacea* nesting on Awala-Yalimapo beach in French Guiana (Girondot et al. 2002). There, the nesting season encompasses both the rainy and dry seasons. Therefore, nests laid earlier, during the rainy season, tend to produce a male-biased sex ratio, while nests laid later, during the dry season, tend to produce a female-biased sex ratio (Rimblot-Baly et al. 1986). However, nests laid earlier in the season are more likely to be accidentally destroyed during the nesting process by laying females, as nest destruction by subsequent nesting females is common on this high-density nesting beach (4 km long), where as many as 60,000 nests have been deposited during a single nesting season. Therefore, earlier male-producing nests produce fewer offspring than later female-producing nests, resulting in a feminizing bias in the overall sex ratio of hatchlings produced on the beach. Indeed, there is a positive relationship between the number of total nests laid and the female bias of the secondary sex ratio due to nest destruction. However, because of the higher nest destruction associated with higher nest density, there is no associated increase in female offspring with increased nesting effort.

Another example of direct interaction between TSD (and therefore sex ratio) and population density has been described for the alligator *Alligator mississippiensis* (Woodward and Murray 1993). Females of this species compete for and strongly defend their nests and nesting areas. Therefore, not all females are able to secure preferred nesting habitat, which is the wet marsh where cooler incubation temperatures produced nearly exclusively female hatchlings. Nesting females that are unable to find a location in the preferred areas will then attempt to nest in dry marsh or dry levees, where warmer incubation temperatures produce more males. A smaller reproductive population usually results in the possibility that all reproductive females are able to successfully secure locations in the preferred habitat, and thus the primary sex ratio will be female biased. Conversely, a larger reproductive population usually results in more nests being laid in dry warmer areas, thus leading to a more male-biased primary sex ratio. This mechanism facilitates the regulation of population size by altering the primary sex ratio: smaller populations produce relatively more females until the population grows sufficiently large and begins to produce relatively more males, which should slow population growth. However, given these hypotheses, the prevalence of observed female-biased primary sex ratios in populations is not an evolutionarily stable strategy because masculinizing alleles invade the population (Girondot and Pieau 1996; see also Valenzuela, Chapter 14).

Conclusions

Much emphasis has been given to the theoretical consequences of TSD for population dynamics (e.g., Head et al. 1987), but few experimental or observational data exist to test the strength of these hypotheses. Given that there is commonly much spatial and temporal variability in the thermal environment of nesting sites of TSD species and that many TSD species have relatively extended reproductive longevity, there is usually variation in mean primary sex ratios over a large time scale in TSD species. This observed variation is at odds with hypothetical models that often assume that TSD species have dramatic and unidirectional sex ratio biases (e.g., Freedberg and Wade 2001). Furthermore, the dynamics of a population are greatly affected by the operational sex ratio, which appears to be as variable in GSD species as it is in TSD species. Nevertheless, for small populations, the presence of TSD is likely associated with a higher risk of extinction than for small populations with GSD. Overall, there is little difference in the adaptive benefits of GSD and TSD for large populations with extended adult longevity and variable primary sex ratios across reproductive seasons (see Valenzuela, Chapter 14). In all other cases, TSD should be selected against, although its continued existence in smaller populations has necessitated the construction of alternative selection hypotheses (Charnov and Bull 1977).

Acknowledgements—We thank M. Tiwari and B. Godley for constructive comments.

NICOLE VALENZUELA

Conclusions

Missing Links and Future Directions

A great deal of information has been covered in the contributed chapters of this volume, allowing a synthetic view of temperature-dependent sex determination (TSD) in vertebrates. In this chapter, the content of each chapter is summarized, then several major conclusions that can be drawn from these reviews are presented. For additional conclusions and missing links, the reader is referred to individual chapters, where authors have provided their own perspective on these topics.

Book Overview

In Chapter 2, Conover brings up the need to use unambiguous terminology that allows distinction between TSD, GSD, and alternative phenomena in order to understand the diversity of sex-determining systems in fish. Definitions and methodology for TSD identification in fishes are explicitly presented, along with a summary of the incidence of TSD and other thermal sex ratio distortions in fish. It is pointed out that few studies have examined TSD and that therefore very little is known about the ecological relevance of laboratory results and the adaptive significance of TSD in fish (the genus *Menidia* being the major exception in this regard).

In Chapter 3, Ewert et al. review and expand the number of TSD cases in turtles by providing new data on previously unexamined species. TSD variation in terms of patterns, pivotal temperatures, and transitional ranges is also explored within and among species with the use of statistically rigorous criteria to characterize these traits. Data from well-studied taxa allowed the analysis of additional interesting parameters, such as the correlation between pivotal temperature and transitional ranges, and the functionality of different transitions and of geographic trends and their ecological correlates.

In Chapter 4, Deeming summarizes the results of studies that have characterized TSD in each of the crocodilian species examined thus far, highlighting the similarities and differences of the TSD systems among taxa. Thermally induced phenotypic effects other than sex in this group are also described, including morphological, behavioral, and physiological traits. Some postulated hypotheses to explain TSD in crocodilians are discussed, along with the relationship of TSD in crocodilians and sex determination in birds and dinosaurs.

In Chapter 5, Harlow reviews the incidence of TSD across lizard families and expands the number of TSD cases by providing new unpublished data on several species. Some previous reports are revisited, as they seem to be based on sample sizes too small to provide solid evidence for TSD, on sex misidentification, or on apparent mistaken citations that were perpetuated in the literature. A discussion of potential evolutionary explanations for the presence of TSD in this group is presented.

In Chapter 6, Nelson et al. describe TSD in the two extant tuatara species based not only on laboratory experiments but also on data from natural nests. Although much remains to be studied about the thermal ecology, molecular basis, and phenotypic effects of incubation temperature in this group, it seems from the existing data that tuataras posses TSD Ib, a mode also found among lizards (their most closely related group).

In Chapter 7, Chardard et al. present a detailed review of sex determination in amphibians. Current research suggests that the mode of amphibian sex determination is GSD (present in all cases studied thus far), with some species being susceptible to thermal sex reversal at atypically high or low temperatures. Sex differentiation and these thermal effects are described, including within and among species variation. It is pointed out that all existing data are restricted to the Anura and the Urodela: Gymnophiona remains unexamined. How widespread this thermal lability is among amphibians remains uncertain, as few taxa have been examined in this regard. A possible common mechanism for the action of temperature on gonadal sex differentiation is explored, and the need to elucidate the molecular basis of such mechanisms is stressed in order to reveal the evolutionary history of sex-determining mechanisms in vertebrates.

In Chapter 8, McCue provides a review on general thermal biology, including definitions of commonly used terminology, thermal limits, thermal sensitivity, and acute and chronic thermal effects on animals, with emphasis on ectotherms, at several levels of biological organization.

In Chapter 9, Georges et al. describe the existing models for TSD's thermal ecology, that is, the mode of action of temperature on development and sex ratio, under laboratory and field incubation conditions. Key challenges in this area are highlighted, including the need to be able to predict embryonic developmental trajectories, thermosensitive periods, and ultimately, sexual outcomes from nest temperature data. Prominence is given to the degree-hour approach, which attempts to reconcile results from controlled laboratory experiments and variable field temperature regimes by focusing on developmental rather than absolute time. An interesting deduction from this approach is the reevaluation of the concepts derived from laboratory studies and used to characterize TSD modes, such as the pivotal temperature. Explicit mathematical models are described, and freeware that should be useful to those interested in TSD and other phenotypic thermal effects is made available upon request.

In Chapter 10, Rhen and Lang describe the effects that temperature during development has on phenotypic traits other than sex in both TSD and GSD taxa, particularly in relation to fitness. Similarities and differences among taxa are considered. Given the inherent link between temperature and gonadal sex in TSD species, it is emphasized that these two effects must be decoupled to be able to attribute the correct cause to each phenotype. Thus, a difference is made between studies looking at thermal effects on natural versus hormonally reversed males and females, and those where the effect of temperature and gonadal sex are confounded. Clutch effects are also discussed. It is pointed out that the underlying basis (genetic, molecular, physiological, and developmental) of these phenotypic effects remains unknown, but some potential mechanisms are explored.

In Chapter 11, Place and Lance review current knowledge about the molecular networks associated with sex determination and differentiation in amniote vertebrates, starting with the best studied orders (mammals followed by birds), which correspond to GSD systems, and then contrasting this information with what is known about TSD reptiles. Each known molecular element in this network and its putative function are described. Emphasis is placed on the observation that many of these elements are common to these three vertebrate groups, yet differences are found in their sequence, sex specificity, and temporal expression during development. The expectation that many more common elements will be found in the future is highlighted, as is the prediction that diversity should reside at the regulation level.

In Chapter 12, Elf explores the link between yolk steroid hormones and sex determination in TSD reptiles. Unlike mammals and birds, reptiles represent a relatively new study group in terms of the effects of maternally derived hormones on sex differentiation. Hypotheses about the mode of action, and observations (published and unpublished) of the dynamics of yolk hormones during development in TSD taxa are described, providing an initial picture of the influence of estradiol and testosterone on sex determination/differentiation. Hormonal effects on development, growth, and behavior are also mentioned. Within and among species variation in yolk hormone levels and patterns of deposition as well as potential sources of such variation are discussed.

In Chapter 13, Janzen and Krenz report the results of an original phylogenetic analysis reexamining the question of the ancestry of sex-determining mechanisms among vertebrates. Conclusions are based on published and unpublished phylogenetic hypotheses about the relationship among and within the main extant vertebrate groups, sampling equal numbers of lepidosaurs and turtles among those with a known sex-determining system, and assuming that all transitions between sex-determining mechanisms are equally likely. Using parsimony, GSD appears to be ancestral under the two scenatios used: (1) using two main categories in the analysis (i.e., TSD and GSD), and (2) discriminating GSD modes (homomorphic sex chromosomes, and heteromorphic male and female heterogamety) rather than treating GSD as a single trait. On the other hand, TSD appears basal to sauropsids in the latter scenario, whereas the timing of its origin is equivocal in the former. The implications of these results for our understanding of TSD evolution are discussed.

In Chapter 14, Valenzuela reviews postulated hypotheses to explain the origin, diversification, maintenance, and loss of TSD among vertebrates by examining the existing theoretical and empirical support for or against each of these hypotheses. Adaptive and neutral alternatives are considered. Hypotheses are organized with respect to whether they relate to genetic or environmental, maternal or paternal effects on gonadal sex and other phenotypic traits, and whether they influence one or both fitness components (survival and fecundity) or sex ratios. Differences between short- and long-lived species are recognized in terms of the applicability and feasibility of gathering relevant data to test different hypotheses. The need to study TSD evolution through comparative approaches at different levels of biological organization is emphasized.

In Chapter 15, Girondot et al. present an improved age-structured model of population dynamics incorporating several parameters that account for some unique characteristics of TSD taxa. Studies reporting primary and secondary sex ratios in natural populations are reviewed, and the appropriateness of alternative methods for measuring these parameters is discussed. The influence of primary versus operational sex ratios on population dynamics is considered. Differences in population sex ratios between TSD and GSD taxa are explored, and the most probable sex-determining pattern is inferred for some unknown species from an unpublished phylogenetic analysis. The authors examine the reliability of contentions linking TSD as a factor of decline in several theoretical and empirical cases.

Overall Conclusions

Increasing numbers of studies have revealed many new cases of TSD in vertebrates, but the definite extent of prevalence of this sex-determining mechanism is still undetermined, as many taxa remain unstudied (reviewed in Chapters 2–7). Likewise, there is an urgent need for the study of TSD under natural conditions, even for many of the taxa already described from laboratory experiments. Therefore, some of the current hypotheses about TSD's ecological, physiological, and molecular mechanics and about TSD evolution may likely change as new examples and variants are discovered. To the extent that these issues receive continual attention a more clear understanding about the ecological and evolutionary significance of TSD should emerge.

Importantly, it seems evident from the contents of this volume that integrative approaches across hierarchical levels of biological organization will provide valuable insights into these questions and will complement the also indispensable unidisciplinary studies. Comparing TSD to GSD taxa directly should also help in obtaining a comprehensive view of sex-determining mechanisms in vertebrates. Understanding general thermal biology could reveal additional traits that may be important in TSD evolution. Besides empirical analysis, more theoretical work is also required. Some areas have been better targeted in this regard, such as the thermal mechanics of TSD, population dynamics, TSD/GSD evolution, and sex ratio evolution, but other aspects remain unexamined.

The discovery of TSD defies some traditional views about the functional mechanics, origins and evolutionary pathways of sex-determining systems in general, and the evolution of related traits (Valenzuela et al. 2003). Indeed, because TSD taxa show no genetic differences among the sexes (Solari 1994), TSD constitutes an unprecedented and yet largely unexplored evolutionary arena in some respects. For instance, the expression of sex-related traits, Fisherian rules of sex ratio evolution, and recombination rules apply differently to TSD than to GSD (Solari 1994; Valenzuela et al. 2003), and thus new theory and empirical testing on these topics is warranted.

An additional concern related to TSD relates to the challenges that TSD species face in terms of short-term and long-term persistence. This includes the potential effect of global warming, artificial incubation and translocations for management programs, habitat disturbance, and extinction of local populations. Some of these issues are treated by Girondot et al. (Chapter 15), particularly the interaction between TSD and population dynamics, and TSD as a factor of decline. TSD has prevailed over millions of years, enduring the drastic environmental changes in temperature that the earth has undergone, despite the fact that sex ratios under TSD are susceptible to drastic skews caused by the vagaries of the environment. However, global warming could still represent a threat for the short- and long-term survival of TSD taxa if the rate of climatic change is much faster than the potential rate of evolution of TSD, of nesting behavior, and/or of migration. Furthermore, habitat disturbance may impede the evolution of these traits to counterbalance climate change if no suitable habitats remain or if fragmentation renders them inaccessible. Further, small population sizes and short live-spans could also increase the risk of extinction of TSD taxa under these circumstances. Comparative modeling and simulation analyses of the consequences of TSD on population dynamics under both historical conditions and those predicted by global warming will be very helpful to understanding past and future processes.

Nonetheless, many TSD species are endangered, and although TSD itself could be at least partially responsible, as described above (see also Girondot et al., Chapter 15), their decline may be alleviated by concentrating conservation efforts on factors such as levels of exploitation of different age/sex classes and habitat disturbance that are more likely affecting population numbers in the short term. In fact, these factors are equally applicable to TSD and GSD endangered taxa. Management programs, however, must take into account the thermal dependence of the sex-determining system in TSD taxa, particularly if management involves translocation of nests or artificial incubation, in order to avoid the production of extreme sex ratio biases that will render conservation practices counterproductive (e.g., Dutton et al. 1985). Thus, knowledge about the thermal, physiological, and molecular mechanics of TSD will serve as an aid for conservation. An important consideration is that management practices that explicitly attempt to manipulate sex ratios (e.g., Vogt 1994, reviewed in Girondot et al. 1998) must be carefully evaluated because some designs can have undesirable consequences (Mrosovsky and Godfrey 1995; Lovich 1996; Girondot et al. 1998). In this regard, differential fitness effects of incubation temperature, which can be species specific (see Rhen and Lang, Chapter 10), should also be taken into account for each particular case.

Like other organisms, species exhibiting TSD represent integrated systems that are continuously challenged by the environment in which they live. Understanding TSD is understanding how TSD taxa cope in the short- and long-term with their particular biotic and abiotic circumstances at the genetic, molecular, physiological, behavioral, individual, population, and species levels and how their circumstances and strategies may or may not differ from those of their GSD counterparts. Such a comprehensive view of the ecological and evolutionary significance of TSD requires the effort of a multitude of researchers working on disparate but interrelated disciplines. This volume was intended as a compilation of the very extensive collective work that has been accomplished thus far, with the hope that it will inspire further and even more insightful research in this exciting field.

Literature Cited

Abell, A. J. 1997. Estimating paternity with spatial behavior and DNA fingerprinting in the striped plateau lizard, *Sceloporus virgatus* (Phrynosomatidae). *Behav. Ecol. Sociobiol.* 41:217–226.

Abinawanto, K. Shimada, K. Yoshida, and N. Saito. 1996. Effects of aromatase inhibitor on sex differentiation and levels of P450 (17 alpha) and P450 arom messenger ribonucleic acid of gonads in chicken embryos. *Gen. Comp. Endocrinol.* 102:241–246

Abucay, J. S., G. C. Mair, D. O. F. Skibinski, and J. A. Beardmore. 1999. Environmental sex determination: The effect of temperature and salinity on sex ratio in *Oreochromis niloticus* L. *Aquaculture* 173:219–234.

Adkins-Regan, E., M. A. Ottinger, and J. Park. 1995. Maternal transfer of estradiol to egg yolk alters sexual differentiation of avian offspring. *J. Exp. Zool.* 271:466–470.

Aida, T. 1936. Sex reversal in *Aplocheilus latipes* and a new explanation of sex differentiation. *Genetics* 21:136-153.

Akhmerov, R. N. 1986. Qualitative difference in mitochondria of endothermic and ectothermic animals. *FEBS* 198:251–255.

Alberts, A. C., A. M. Perry, J. M. Lemm, and J. A. Phillips. 1997. Effects of incubation temperature and water potential on growth and thermoregulatory behavior of hatchling Cuban rock iguanas *(Cyclura nubila)*. *Copeia* 1997:766–776.

Albrecht, K. H., and E. M. Eicher. 2001. Evidence that *Sry* is expressed in pre-Sertoli cells and Sertoli and granulosa cells have a common precursor. *Dev. Biol.* 240:92–107.

Allsteadt, J., and J. W. Lang. 1995a. Incubation temperature affects body size and energy reserves of hatchling American alligators *(Alligator mississippiensis)*. *Physiol. Zool.* 68:76–97.

———. 1995b. Sexual dimorphism in the genital morphology of young American alligators, *Alligator mississippiensis*. *Herpetologica* 51:314–325.

Anderson, A., and C. Oldham. 1988. Captive husbandry and propagation of the African fat-tail gecko, *Hemitheconyx caudicinctus*. *Intl. Herpetol. Symp. Captive Propagation and Husbandry* 10:75–85.

Andersson, M. 1994. *Sexual selection*. Princeton, NJ: Princeton University Press.

Andrews, J. E., C. A. Smith, and A. H. Sinclair. 1997. Sites of estrogen receptor and aromatase expression in the chicken embryo. *Gen. Comp. Endocrinol.* 108:182–190.

Andrews, R. M., and F. H. Pough. 1985. Metabolism of squamate reptiles: Allometric and ecological relationships. *Physiol. Zool.* 58:214–231.

Andrews, R. M., T. Mathies, and D. A. Warner. 2000. Effect of incubation temperature on morphology, growth and survival of juvenile *Sceloporus undulatus*. *Herptetol. Monogr.* 14:420–431.

Anttonen, M., I. Ketola, H. Parviainen, A. K. Pusa, and M. Heikinheimo. 2003. *Fog-2* and *Gata-4* are co-expressed in the mouse ovary and can modulate Müllerian-inhibiting substance expression. *Biol. Reprod.* 68:133–1340.

Arango N. A., R. Lovell-Badge, and R. R. Behringer. 1999. Targeted mutagenesis of the endogenous mouse Mis gene promoter: In vivo definition of genetic pathways of vertebrate sexual development. *Cell* 99:409–419.

Arnold, A. P., and R. A. Gorski. 1984. Gonadal steroid induction of structural sex differences in the central nervous system. *Ann. Rev. Neurosci.* 7:413–442.

Arnold, S. J., C. R. Peterson, and J. Gladstone. 1995. Behavioral variation in natural populations. VII. Maternal body temperature does not affect juvenile thermoregulation in a garter snake. *Anim. Behav.* 50:623–633.

Avery, R. A., J. D. Bedford, and C. P. Newcombe. 1982. The role of thermoregulation in lizard biology: Predatory efficiency in a temperate diurnal basker. *Behav. Ecol. Sociobiol.* 11:261–267.

Avery, H. W., J. R. Spotila, J. D. Congdon, R. U. Rischer, E. A. Standora, and S. B. Avery. 1993. Roles of diet protein and temperature in the growth and nutritional energetics of juvenile slider turtles, *Trachemys scripta. Physiol. Zool.* 66:902–925.

Avise, J. C. 2000. *Phylogeography*. Cambridge, MA: Harvard University Press.

Ayers, D. Y., and R. Shine. 1997. Thermal influences on foraging ability: Body size, posture and cooling rate of an ambush predator, the python *Morelia spilota. Funct. Ecol.* 11:342–347.

Badham, J. A. 1971. Albumen formation in eggs of the agamid *Amphibolurus barbatus. Copeia* 1971:543–545.

Baras, E., C. Prignon, G. Gohoungo, and C. Mélard. 2000. Phenotypic sex differentiation of blue tilapia under constant and fluctuating thermal regimes and its adaptive and evolutionary implications. *J. Fish Biol.* 57:210–223.

Baroiller, J. F., and H. D'Cotta. 2001. Environment and sex determination in farmed fish. *Comp. Biochem. Physiol. C* 130:399–409.

Baroiller, J. F., and Y. Guiguen. 2001. Endocrine and environmental aspects of sex differentiation in gonochoristic fish. In *Genes and mechanisms in vertebrate sex determination,* ed. G. Scherer and M. Schmid, EXS 91:177–201, Basel, Boston, Berlin: Birkhäuser.

Baroiller, J. F., D. Chourrout, A. Fostier, and B. Jalabert. 1995. Temperature and sex chromosomes govern sex ratios of the mouthbrooding cichlid fish *Oreochromis niloticus. J. Exp. Zool.* 273:216–223.

Baroiller, J. F., Y. Guigen, and A. Fostier. 1999. Endocrine and environmental aspects of sex differentiation in fish. *Cell. Mol. Life Sci.* 55:910–931.

Bartholomew, G. A., and R. C. Tucker. 1963. Control of changes in body temperature, metabolism and circulation by the agamid lizard, *Amphibolurus barbatus. Physiol. Zool.* 36:199–218.

Basolo, A. L. 1994. The dynamics of fisherian sex-ratio evolution: Theoretical and experimental investigations. *Am. Nat.* 144: 473–490.

Beamish, F. W. H. 1993. Environmental sex determination in southern brook lamprey, *Ichthyomyzon gagei. Can. J. Fish. Aquat. Sci.* 50:1299–1307.

Beardmore, J. A., G. C. Mair, and R. I. Lewis. 2001. Monosex male production in finfish as exemplified by tilapia: Applications, problems, and prospects. *Aquaculture* 197:283–301.

Beggs, K., J. E. Young, A. Georges, and P. West. 2000. Ageing the eggs and embryos of the pig-nosed turtle *Carettochelys insculpta. Can. J. Zool.* 78:373–392.

Bennett, A. F. 1980. The thermal dependence of lizard behavior. *Anim. Behav.* 28:752–762.

———. 1990. Thermal dependence of locomotor capacity. *Am. J. Physiol.* 259:R253–R258.

Bennett, A. F., and H. B. J. Alder. 1984. The effect of body temperature on the locomotory energetics of lizards. *J. Comp. Physiol. B* 155:21–27.

Bennett, A.F., and P. Licht. 1972. Anaerobic metabolism during activity in lizards. *J. Comp. Physiol.* 81:277–288.

———. 1974. Anaerobic metabolism during activity in amphibians. *Comp. Biochem. Physiol. A* 48:319–327.

Bently, P. J., and K. Schmidt-Nielsen. 1966. Cutaneous water loss in reptiles. *Science* 151:1547–1549.

Benton, M. J. 2000. *Vertebrate palaeontology*. 2nd ed. London: Blackwell Science.

Berec, L., D. S. Boukal, and M. Berec. 2001. Linking the Allee effect, sexual reproduction, and temperature-dependent sex determination via spatial analysis. *Am. Nat.* 157:217–230.

Berner, N. J. 1999. Oxygen consumption by mitochondria from an endotherm and an ectotherm. *Comp. Biochem. Physiol. B* 124:25–31.

Beuchat, C. A. 1986. Reproductive influences on the thermoregulatory behavior of a live-bearing lizard. *Copeia* 1986:971–979.

Beuchat, C. A., and S. Ellner. 1987. A qualitative test of life history theory: Thermoregulation by a viviparous lizard. *Ecol. Monogr.* 57:45–60.

Beuchat, C. A., F. H. Pough, and M. M. Stewart. 1984. Response to simultaneous dehydration and thermal stress in three species of Puerto Rican frogs. *J. Comp. Physiol. B* 1984:579–585.

Bickham, J. W. 1975. A cytosystematic study of the turtles of the genera *Clemmys, Mauremys* and *Sacalia. Herpetologica* 31:198–204.

Bickham, J. W., T. Lamb, P. Minx, and J. C. Patton. 1996. Molecular systematics of the genus *Clemmys* and the intergeneric relationships of emydid turtles. *Herpetologica* 52:89–97.

Bilinski, J. J., R. D. Reina, J. R. Spotila, and F. V. Paladino. 2001. The effects of nest environment on calcium mobilization by leatherback turtle embryos *(Dermochelys coriacea)* during development. *Comp. Biochem. Physiol. A* 130:151–162.

Binckley, C. A., J. R. Spotila, K. S. Wilson, and F. V. Paladino. 1998. Sex determination and sex ratios of Pacific leatherback turtles, *Dermochelys coriacea. Copeia* 1998:291–300.

Birchard, G. F., and G. C. Packard. 1997. Cardiac activity in supercooled hatchlings of the painted turtle *(Chrysemys picta). J. Herpetol.* 31:166–169.

Birk, O. S., D. E. Casiano, C. A. Wassif, T. Cogliati, L. Zhao, Y. Zhao, A. Grinberg, et al. 2000. The LIM homeobox gene *Lhx9* is essential for mouse gonad formation. *Nature* 403:909–913.

Blackburn, D. G., S. Kleis-San Francisco, and I. P. Callard. 1998. Histology of abortive egg sites in the uterus of a viviparous, placentotrophic lizard, the skink *Chalcides chalcides. J. Morphol.* 235:97–108.

Blackmore, M. S., and E. L. Charnov. 1989. Adaptive variation in environmental sex determination in a nematode. *Am. Nat.* 134:817–823.

Blázquez, M., P. T. Bosma, E. J. Fraser, K. J. W. Van Look, and V. L. Trudeau. 1998a. Fish as models for the neuroendocrine regulation of reproduction and growth. *Comp. Biochem. Physiol. C* 119:345–364.

Blázquez, M., S. Zanuy, M. Carillo, and F. Piferrer. 1998b. Effects of rearing temperature on sex differentiation in the European sea bass (*Dicentrarchus labrax* L.). *J. Exp. Zool.* 281:207–216.

Blouin-Demers, G., K. J. Kissner, and P. J. Weatherhead. 2000. Plasticity in preferred body temperature of young snakes in response to temperature during development. *Copeia* 2000:841–845.

Blumberg, M. S., S. J. Lewis, and G. Sokoloff. 2002. Incubation temperature modulates post-hatching thermoregulatory behavior in the Madagascar ground gecko, *Paroedura pictus*. *J. Exp. Biol.* 205:2777–2784.

Bobyn, M. L., and R. Brooks. 1994. Interclutch and interpopulation variation in the effects of incubation conditions on sex, survival and growth of hatchling turtles *(Chelydra serpentina)*. *J. Zool.* 233:233–257.

Bogart, M. H. 1987. Sex determination: A hypothesis based on steroid ratios. *J. Theor. Biol.* 128:349–357.

Bogert, C. M. 1949. Thermoregulation in reptiles, a factor in evolution. *Evolution* 3:195–211.

Böhme, W. 1995. Hemiclitoris discovered: A fully differentiated erectile structure in female monitor lizards (*Varanus* spp.) (Reptilia: Varanidae). *J. Zool. Syst. Evol. Res.* 33:129–132.

Bolten, A. B., K. A. Bjorndal, H. R. Martins, T. Dellinger, M.J. Biscoito, S. E. Encalada, and B. W. Bowen. 1998. Transatlantic developmental migrations of loggerhead sea turtles demonstrated by mtDNA sequence analysis. *Ecol. Appl.* 8: 1–7.

Booth, D. T. 1998. Effects of incubation temperature on the energetics of embryonic development and hatchling morphology in the Brisbane river turtle *Emydura signata*. *J. Comp. Physiol. B* 168:399–404.

———. 1999. Incubation temperature and growth of Brisbane river turtle *(Emydura signata)* hatchlings. *Proc. Linn. Soc. New South Wales* 121:45–52.

Booth, D. T., and K. Astill. 2001. Incubation temperature, energy expenditure and hatchling size in the green turtle *(Chelonia mydas)*, a species with temperature-sensitive sex determination. *Aus. J. Zool.* 49:389–396.

Booth, D. T., M. B. Thompson, and S. Herring. 2000. How incubation temperature influences the physiology and growth of embryonic lizards. *J. Comp. Physiol. B* 170:269–276.

Bowen, B. W., and S. A. Karl. 1997. Population genetics, phylogeography, and molecular evolution. In *The biology of sea turtles*, ed. P. L. Lutz and J.A. Musick, 29–50, Boca Raton, FL: CRC Press.

Bowden, R. M., M. A. Ewert, and C. E. Nelson. 2000. Environmental sex determination in a reptile varies seasonally and with yolk hormones. *Proc. Roy. Soc. Lond. B* 267:1745–1749.

Bowden, R. M., M. A. Ewert, J. L. Lipar, and C. E. Nelson. 2001. Concentrations of steroid hormones in layers and biopsies of Chelonian egg yolks. *Gen. Comp. Endocrinol.* 121:95–103.

Bowden, R. M., M. A. Ewert, S. Freedberg, and C. E. Nelson. 2002. Maternally derived yolk hormones vary in follicles of the painted turtle, *Chrysemys picta*. *J. Exp. Zool.* 293:67–72.

Boycott, R. C., and O. Bourquin. 2000. *The southern African tortoise book*. 2nd ed. Hilton, Kwazulu-Natal, South Africa: O. Bourquin.

Bradshaw, S. D. 1986. *Ecophysiology of desert reptiles*. Sydney, Australia: Academic Press.

Bragg, W. K., J. D. Fawcett, T. B. Bragg, and B. E. Viets. 2000. Nest-site selection in two eublepharid gecko species with temperature-dependent sex determination and one with genotypic sex determination. *Biol. J. Linn. Soc.* 69:319–332.

Braña, F., and X. Ji. 2000. Influence of incubation temperature on morphology, locomotor performance, and early growth of hatchling wall lizards *(Podarcis muralis)*. *J. Exp. Zool.* 286:422–433.

Brand, M. D., P. Couture, P. L. Else, K. W. Withers, and A. J. Hulbert. 1991. Evolution of energy metabolism: Proton permeability of the inner membrane of liver mitochondria is greater in a mammal than in a reptile. *Biochem. J.* 275:81–86.

Brattstrom, B. H. 1963. A preliminary review of the thermal requirements of amphibians. *Ecology* 44:238–255.

———. 1965. Body temperatures of reptiles. *Am. Mid. Nat.* 73: 376–422.

———. 1968. Thermal acclimation in anuran amphibians as a function of latitude and altitude. *Comp. Biochem. Physiol.* 24:93–111.

Brett, J. R. 1971. Energetic responses of salmon to temperature: A study of some thermal relations in the physiology and freshwater ecology of sockeye salmon *(Oncorhynchus nerka)*. *Am. Zool.* 11:99–113.

Broderick, A. C., B. J. Godley, S. Reece, and J. R. Downie. 2000. Incubation periods and sex ratios of green turtles: Highly female biased hatchling production in the eastern Mediterranean. *Mar. Ecol. Progr. Ser.* 202:273–281.

Broderick, A. C., B. J. Godley, and G. C. Hays. 2001. Trophic status drives interannual variability in nesting numbers of marine turtles. *Proc. Roy. Soc. Lond. B* 268:1481–1487.

Brooks, R. J., M. L. Bobyn, D. A. Galbraith, J. A. Layfield, and E. G. Nancekivell. 1991. Maternal and environmental influences on growth and survival of embryonic and hatchling snapping turtles *(Chelydra serpentina)*. *Can. J. Zool.* 69:2667–2676.

Brown, D. E., ed. 1982. Biotic communities of the American Southwest—United States and Mexico. *Desert Plants* 4:1–342.

Buehr M., S. Gu, and A. McLaren. 1993. Mesonephric contribution to testis differentiation in the fetal mouse. *Development* 117:273–281.

Bull, J. J. 1980. Sex determination in reptiles. *Q. Rev. Biol.* 55:3–21.

———. 1981. Evolution of environmental sex determination from genotypic sex determination. *Heredity* 47:173–184.

———. 1983. *Evolution of sex determining mechanisms*. Menlo Park, CA: Benjamin/Cummings.

———. 1985. Sex ratio and nest temperature in turtles: Comparing field and laboratory data. *Ecology* 66:1115–1122.

———. 1987. Temperature-sensitive periods of sex determination in a lizard: Similarities with turtles and crocodilians. *J. Exp. Zool.* 241:143–148.

———. 1989. Evolution and variety of sex-determining mechanisms in amniote vertebrates In *Evolutionary mechanisms in sex determination*, ed. S. S. Watchel, 57–65. Boca Raton, FL: CRC Press Inc.

Bull, J. J., and M. G. Bulmer. 1989. Longevity enhances selection of environmental sex determination. *Heredity* 63:315–320.

Bull, J. J., and E. L. Charnov. 1988. How fundamental are Fisherian sex ratios? *Oxford Surv. Evol. Biol.* 5:96–135.

———. 1989. Enigmatic reptilian sex ratios. *Evolution* 43:1561–1566.

Bull, J. J., and R. C. Vogt. 1979. Temperature-dependent sex determination in turtles. *Science* 206:1186–1188.

———. 1981. Temperature sensitive periods of sex determination in Emydid turtles. *J. Exp. Zool.* 218:435–440.

Bull, J. J., R. C. Vogt, and M. G. Bulmer. 1982a. Heritability of sex ratio in turtles with environmental sex determination. *Evolution* 36:333–341.

Bull, J. J., R. C. Vogt, and C. J. McCoy. 1982b. Sex determining temperatures in turtles: A geographic comparison. *Evolution* 36:326–332.

Bull, J. J., J. M. Legler, and R. C. Vogt. 1985. Non-temperature dependent sex determination in two orders of turtles. *Copeia* 1985:784–786.

Bull, J. J., W. H. N. Gutzke, and M. G. Bulmer. 1988. Nest choice in a captive lizard with temperature-dependent sex determination. *J. Evol. Biol.* 2:177–184.

Bull, J. J., T. Wibbels, and D. Crews. 1990. Sex-determining potencies vary among female incubation temperatures in a turtle. *J. Exp. Zool.* 256:339–341.

Bulmer, M. G., and J. J. Bull. 1982. Models of polygenic sex determination and sex ratio control. *Evolution* 36:13–26.

Burger, J. 1998a. Effects of incubation temperature on hatchling pine snakes: Implications for survival. *Behav. Ecol. Sociobiol.* 43:11–18.

———. 1998b. Antipredator behaviour of hatchling snakes: Effects of incubation temperature and simulated predators. *Anim. Behav.* 56:547–553.

———. 1990. Effects of incubation temperature on behavior of young Black Racers *(Coluber constrictor)* and Kingsnakes *(Lampropeltis getulus)*. *J. Herpetol.* 24:158–163.

———. 1991. Effects of incubation temperature on behavior of hatchling pine snakes: Implications for reptilian distribution. *Behav. Ecol. Sociobiol.* 28:297–303.

Burger, J., and R. T. Zappalorti. 1988. Effects of incubation temperature on sex-ratios in pine snakes: Differential vulnerability of males and females. *Am. Nat.* 132:492–505.

Burger, J., R. T. Zappalorti, and M. Gochfeld. 1987. Developmental effects of incubation temperature on hatchling pine snakes *Pituophis melanoleucus*. *Comp. Biochem. Physiol. A* 87(3): 727–732.

Burgoyne, P. S., M. Buehr, P. Koopman, J. Rossant, and A. McLaren. 1988. Cell-autonomous action of the testis-determining gene: Sertoli cells are exclusively XY in XX—XY chimaeric mouse testes. *Development* 102:443–450.

Burke, R. L. 1993. Adaptive value of sex determination mode and hatchling sex ratio bias in reptiles. *Copeia* 3:854–859.

Burke, R. L., M. A. Ewert, J. B. McLemore, and D. R. Jackson. 1996. Temperature-dependent sex determination and hatching success in the gopher tortoise *(Gopherus polyphemus)*. *Chelon. Cons. Biol.* 2:86–88.

Byskov, A. G. 1986. Differentiation of mammalian embryonic gonad. *Physiol. Rev.* 66:71–117.

Caccone, A., J. P. Gibbs, V. Ketmaier, E. Suatoni, and J. R. Powell. 1999. Origin and evolutionary relationships of giant Galápagos tortoises. *Proc. Natl. Acad. Sci. USA* 96:13223–13228.

Callan, H. G. 1986. *Lampbrush chromosomes.* Berlin: Springer.

Callard, I. P., V. A. Lance, A. R. Salhanick, and D. Barad. 1978. The annual ovarian cycle of *Chrysemys picta*: Correlated changes in plasma steroids and parameters of vitellogenesis. *Gen. Comp. Endocrinol.* 35:245–257.

Campos, Z. 1993. Effect of habitat on survival of eggs and sex ratio of hatchlings of *Caiman crocodilus yacare* in the Pantanal, Brazil. *J. Herpetol.* 27:127–132.

Cantor, A. B., and S. H. Orkin. 2001. Hematopoietic development: A balancing act. *Curr. Opin. Genet. Dev.* 11:513–519.

Capel, B. 1998. Sex in the 90s: *Sry* and the switch to the male pathway. *Ann. Rev. Physiol.* 60:497–523.

———. 2000. The battle of the sexes. *Mech. Dev.* 92:89–103.

Caricasole, A., A. Duarte, S. H. Larsson, N. D. Hastie, M. Little, G. Holmes, I. Todorov, et al. 1996. RNA binding by the Wilms tumor suppressor zinc finger proteins. *Proc. Natl. Acad. Sci. USA* 93:7562–7566.

Carr, A. 1964. Transoceanic migrations of the green turtle. *BioScience* 14: 49–52.

Carr, J. L., and J. W. Bickham. 1981. Sex chromosomes of the Asian black pond turtle, *Siebenrockiella crassicollis* (Testudines: Emydidae). *Cytogenet. Cell. Genet.* 31:178–183.

Cassuto, Y. 1971. Oxidative activities of liver mitochondria from mammals, birds, reptiles and amphibia as a function of temperature. *Comp. Biochem. Physiol. B* 39:919–923.

Castanet, J., D. G. Newman, and H. Saint-Girons. 1988. Skeletochronological data on the growth, age, and population structure of the tuatara, *Sphenodon punctatus*, on Stephens and Lady Alice Islands, New Zealand. *Herpetologica* 44:25–37.

Castaño-Mora, O. V., and M. Lugo-Rugeles. 1981. Estudio comparativo del comportamiento de dos especies de morrocoy: *Geochelone carbonaria* y *Geochelone denticulata* y aspectos comparables de su morfología externa. *Cespedesia* 10:55–122.

Caswell, H., and D. E. Weeks. 1986. Two-sex models: Chaos, extinction, and other dynamic consequences of sex. *Am. Nat.* 128:707–735.

Chardard, D., and C. Dournon. 1999. Sex reversal by aromatase inhibitor treatment in the newt *Pleurodeles waltl*. *J. Exp. Zool.* 283:43–50.

Chardard, D., A. Chesnel , C. Gozé, C. Dournon, and P. Berta. 1993. *PwSox1:* The first member of the *Sox* gene family in urodeles. *Nucl. Acids Res.* 21:3576.

Chardard D., G. Desvages, C. Pieau, and C. Dournon. 1995. Aromatase activity in larval gonads of *Pleurodeles waltl* (Urodele: Amphibia) during normal sex differentiation and during sex reversal by thermal treatment effect. *Gen. Comp. Endocrinol.* 99:100–107.

Charlesworth, B. 1978. Model for evolution of Y chromosomes and dosage compensation. *Proc. Natl. Acad. Sci. USA* 75:5618–5622.

———. 1980. Evolution in age-structured populations. Cambridge University Press.

———. 1991. The evolution of sex chromosomes. *Science* 251:1030-1033.

Charlesworth, B., and D. Charlesworth. 1998. Some evolutionary consequences of deleterious mutations. *Genetica* 102–103:3–19.

Charlesworth, B., J. A. Coyne, and N. H. Barton. 1987. The relative rates of evolution of sex chromosomes and autosomes. *Am. Nat.* 130:113–146.

Charlieu, J. P., S. Larsson, K. Miyagawa, V. van Heyningen, and N.D. Hastie. 1995. Does the Wilms' tumour suppressor gene, *Wt1*, play roles in both splicing and transcription? *J. Cell Sci. Suppl.* 19:95–99.

Charnier, M. 1966. Action de la température sur la sex-ratio chez l'embryon d'*Agama agama* (Agamidae: Lacertilien). *C. R. Séances Soc. Biol. l'Ouest Africain* 160:620–622.

Charnov, E. L., and J. J. Bull. 1977. When is sex environmentally determined? *Nature* 266:828–830.

Chen, B. 1990. The past and present situation of the Chinese alligator. *Asiatic Herpetol. Res.* 3:129–136.

Chevalier, J., M. H. Godfrey, M. Girondot. 1999. Significant difference of temperature-dependent sex determination between French Guiana (Atlantic) and Playa Grande (Costa-Rica, Pacific) leatherbacks *(Dermochelys coriacea)*. *Ann. Sci. Nat. Zool.* 20:147–152.

Choo, B. L., and L. M. Chou.1992. Does incubation-temperature influence the sex of embryos in *Trionyx sinensis*? *J. Herpetol.* 26:341–342.

Christian, K. A., and C. R. Tracy. 1983. Seasonal shifts in body temperature and use of microhabitats by Galapagos iguanas *(Conolophus pallidus)*. *Ecology* 64:463–468.

Churchill, T. A., and J. M. Storey. 1992a. Responses to freezing exposure of hatchling turtles *Trachemys scripta elegans:* Factors influencing the development of freeze tolerance by reptiles. *J. Exp. Biol.* 167:221–233.

Churchill, T. A., and K. B. Storey. 1992b. Freezing survival of the garter snake *Thamnophis sirtalis parietalis*. *Can. J. Zool.* 70:99–105.

Clarke, A. 1991. What is cold adaptation and how should we measure it? *Am. Zool.* 31:81–92.

Clarkson, M. J., and V. R. Harley. 2002. Sex with two *Sox* on: *Sry* and *Sox9* in testis development. *Trends Endocrinol. Metab.* 13:106–111.

Coen, E. 1983. Climate. In *Costa Rican natural history*, ed. D. H. Janzen, 35–46. University of Chicago Press.

Cogger, H. G. 1978. Reproductive cycles, fat body cycles and sociosexual behaviour in the mallee dragon, *Amphibolurus fordi* (Lacertilia: Agamidae). *Aus. J. Zool.* 26:653–672.

———. 2000. *Reptiles and amphibians of Australia*. Sydney: Reed New Holland.

Cogger, H. G., and H. Heatwole. 1981. The Australian reptiles: Origins, biogeography, distribution patterns and island evolution. In *Ecological biogeography of Australia*, ed. A. Keast, 1333–1373. The Hague, Netherlands: W. Junk.

Cohen, M. M., and C. Gans. 1970. The chromosomes of the order Crocodylia. *Cytogenetics* 9:81–105.

Cole, C. J., and C. Gans. 1987. Chromosomes of *Bipes*, *Mesobaena*, and other amphisbaenians (Reptilia), with comments on their evolution. *Am. Mus. Novit.* 2869:1–9.

Cole, C. J., H. C. Dessauer, C. R. Townsend, and M. G. Arnold. 1990. Unisexual lizards of the genus *Gymnophthalmus* (Reptilia: Teiidae) in the neotropics: Genetics, origin, and systematics. *Am. Mus. Novit.* 2994:1–29.

Cole, L. C. 1943. Experiments on toleration of high temperature in lizards with reference to adaptive coloration. *Ecology* 24:94–108.

Collins, T. M., P. H. Wimberger, and G. J. P. Naylor. 1994. Composition bias, character-state bias, and character-state reconstruction using parsimony. *Syst. Biol.* 43:482–496.

Collis-George, N., B. G. Davey, and D. E. Smiles. 1968. The soil and aerial environments. In *Fundamentals of modern agriculture*, ed. C. D. Blake, 497. Sydney University Press.

Congdon, J. D., R. U. Fischer, and R. E. Gatten, Jr. 1995. Effects of incubation temperatures on characteristics of hatchling American alligators. *Herpetologica* 51:497–504.

Conley, A. J., P. Elf, C. J. Corbin, S. Dubowsky, A. Fivizzani, and J. W. Lang. 1997. Yolk seroids decline during sexual differentiation in the alligator. *Gen. Comp. Endocrinol.* 107:191–200.

Conover, D. O. 1984. Adaptive significance of temperature-dependent sex determination in a fish. *Am. Nat.* 123:297–313.

Conover, D. O., and S. B. Demond. 1991. Absence of temperature-dependent sex determination in northern populations of two cyprinodontid fishes. *Can. J. Zool.* 69:530–533.

Conover, D. O., and M. Fleisher. 1986. The temperature-sensitive period of sex determination in *Menidia menidia*. *Can. J. Fish. Aquat. Sci.* 43:514–520.

Conover, D. O., and S. W. Heins. 1987a. The environmental and genetic components of sex ratio in *Menidia menidia*. *Copeia* 1987:732–743.

———. 1987b. Adaptive variation in environmental and genetic sex determination in a fish. *Nature* 326:496–498.

Conover, D. O., and B. E. Kynard. 1981. Environmental sex determination: Interaction of temperature and genotype in a fish. *Science* 213:577–579.

Conover, D. O., and D. A. Van Voorhees. 1990. Evolution of a balanced sex ratio by frequency-dependent selection in a fish. *Science* 250:1556–1558.

Conover, D. O., D. A. Van Voorhees, and A. Ehtisham. 1992. Sex ratio selection and changes in environmental sex determination in laboratory populations of *Menidia menidia*. *Evolution* 46:1722–1730.

Coomber, P., D. Crews, and F. Gonzalez-Lima. 1997. Independent effects of incubation temperature and gonadal sex on the volume and metabolic capacity of brain nuclei in the leopard gecko *(Eublepharis macularius)*, a lizard with temperature-dependent sex determination. *J. Comp. Neurol.* 380:409–421.

Coriat, A. M., E. Valleley, M. W. Ferguson, and P. T. Sharpe. 1994. Chromosomal and temperature-dependent sex determination: The search for a conserved mechanism. *J. Exp. Zool.* 270:112–116.

Courchamp, F., T. Clutton-Brock, and B. Grenfell. 1999. Inverse density dependence and the Allee effect. *Trends Ecol. Evol.* 14: 405–410.

Cowles, R. B. 1962. Semantics in biothermal studies. *Science* 135:670.

Craig, J. K., C. J. Foote, and C. C. Wood. 1996. Evidence for temperature-dependent sex determination in sockeye salmon *(Oncorhynchus nerka)*. *Can. J. Fish. Aquat. Sci.* 53:141–147.

Crawford, P. A., C. Dorn, Y. Sadovsky, and J. Milbrandt. 1998. Nuclear receptor DAX-1 recruits nuclear receptor corepressor N-CoR to steroidogenic factor 1. *Mol. Cell Biol.* 18:2949–2956.

Crawshaw, L. I. 1984. Low-temperature dormancy in fish. *Am. J. Physiol.* 246:R479–R486.

Cree, A. 1994. Low annual reproductive output in female reptiles from New Zealand. *N. Z. J. Zool.* 21:351–372.

Cree, A., C. H. Daugherty, S. F. Schafer, and D. Brown. 1991. Nesting and clutch size of tuatara *(Sphenodon guntheri)* on North Brother Island, Cook Strait. *Tuatara* 31:9–16.

Cree, A., M. B. Thompson, and C. H. Daugherty. 1995. Tuatara sex determination. *Nature* 375:543.

Crews, D. 1996. Temperature-dependent sex determination: The interplay of steroid hormones and temperature. *Zool. Sci.* 13:1–13.

Crews, D., and M. J. Bergeron. 1994. Role of reductase and aromatase in sex determination in the red-eared slider *(Trachemys scripta),* a turtle with temperature-dependent sex determination. *J. Endocrinol.* 143:279–289.

Crews, D., T. Wibbels, and W. H. N. Gutzke. 1989. Action of sex steroid hormones on temperature-induced sex determination in the snapping turtle *(Chelydra serpentina). Gen. Comp. Endocrinol.* 76:159–166.

Crews, D., J. J. Bull, and T. Wibbels. 1991. Estrogen and sex reversal in turtles: A dose-dependent phenomenon. *Gen. Comp. Endocrinol.* 81:357–364.

Crews, D., M. J. Bergeron, D. Flores, J. J. Bull, J. K. Skipper, A. Tousignant, and T. Wibbels. 1994. Temperature-dependent sex determination in reptiles: Proximate mechanisms, ultimate outcomes, and practical applications. *Dev. Genet.* 15:297–312.

Crews, D., A. R. Cantu, J. M. Bergeron, and T. Rhen. 1995. The relative effectiveness of androstenedione, testosterone, and estrone, precursors to estradiol, in sex reversal in the red-eared slider *(Trachemys scripta),* a turtle with temperature-dependent sex determination. *Gen. Comp. Endocrinol.* 100:119–127.

Crews, D., A. Cantu, T. Rhen, and R. Vohra. 1996a. The relative effectiveness of estrone, estra-diol-17beta, and estriol in sex reversal in the red-eared slider *(Trachemys scripta),* a turtle with temperature-dependent sex determination. *Gen. Comp. Endocrinol.* 102:317–326.

Crews, D., P. Coomber, R. Baldwin, N. Azad, and F. Gonzalez-Lima. 1996b. Effects of gonadectomy and hormone treatment on the morphology and metabolic capacity of brain nuclei in the leopard gecko *(Eublepharis macularius)* a lizard with temperature-dependent sex determination. *Horm. Behav.* 30:474–486.

Crews, D., J. Sakata, and T. Rhen. 1998. Developmental effects of intersexual and intrasexual variation in growth and reproduction in a lizard with temperature-dependent sex determination. *Comp. Biochem. Physiol. C* 119:229–241.

Crews, D., A. Fleming, E. Willingham, R. Baldwin, and J. K. Skipper. 2001. Role of steroidogenic factor 1 and aromatase in temperature-dependent sex determination in the red-eared slider turtle. *J. Exp. Zool.* 290:597–606.

Crow, J. F., and M. Kimura. 1970. *An introduction to population genetics theory.* New York: Harper and Row.

Crowley, S. R., and R. D. Pietruszka. 1983. Aggressiveness and vocalization in the leopard lizard *(Gambelia wislizennii):* The influence of temperature. *Anim. Behav.* 31:1055–1060.

Cury, P. 1994. Obstinate nature: An ecology of individuals. Thoughts on reproductive behavior and biodiversity. *Can. J. Fish. Aquat. Sci.* 51:1664–1673.

Dallwitz, M. J., and J. P. Higgins. 1992. User's guide to DEVAR. A computer program for estimating development rate as a function of temperature. *CSIRO Div. Entomol. Rep.* 2:1–23.

Daugherty, C. H. 1998. From dark days to a bright future: Survival of the tuataras. *ZooNooz* April: 8–13.

Davies, R. C., C. Calvio, E. Bratt, S. H. Larsson, A. I. Lamond, and N. D. Hastie. 1998. WT1 interacts with the splicing factor U2af65 in an isoform-dependent manner and can be incorporated into spliceosomes. *Genes Dev.* 12:3217–3225.

Davis, L. M., T. C. Glenn, D. C. Strickland, L. J. Guillette, R. M. Elsey, W. E. Rhodes, H. C. Dessauer, and R. H. Sawyer. 2002. Microsatellite DNA analyses support an east-west phylogeographic split of American alligator populations. *J. Exp. Zool.* 294:352–372.

Dawbin, W. H. 1982. The tuatara *Sphenodon punctatus*: Aspects of life history, growth and longevity. In *New Zealand herpetology*, ed. D. G. Newman, 237–250. New Zealand Wildlife Service Occasional Publication No. 2.

D'Cotta H., A. Fostier, Y. Guiguen, M. Govoroun, and J. F. Baroiller. 2001a. Aromatase plays a key role during normal and temperature-induced sex differentiation of Tilapia *Oreochromis niloticus. Mol. Reprod. Dev.* 59:265–276.

D'Cotta H., A. Fostier, Y. Guiguen, M. Govoroun, and J. F. Baroiller. 2001b. Search for genes involved in the temperature-induced gonadal sex differentiation in the tilapia, *Oreochromis niloticus. J. Exp. Zool.* 290:574–585.

de Candolle, A. P. 1855, *Géographique botanique raisonée.* Paris: Masson.

De Lisle, H. F. 1996. *The natural history of monitor lizards.* Malabar, FL: Krieger Publishing Co.

Deeming, D. C., and M. W. J. Ferguson. 1988. Environmental regulation of sex determination in reptiles. *Phil. Trans. Roy. Soc. Lond. B* 322:19–39.

———. 1989a. Effects of incubation temperature on the growth and development of embryos of *Alligator mississippiensis. J. Comp. Physiol. B* 159:183–193.

———. 1989b. The mechanism of temperature dependent sex determination in crocodilians: A hypothesis. *Am. Zool.* 29:973–985.

———. 1989c. In the heat of the nest. *New Sci.* 1657:33–38.

———. 1991a. Reduction in eggshell conductance to respiratory gases has no effect on sex determination in *Alligator mississippiensis. Copeia* 1991:240–243.

———. 1991b. Physiological effects of incubation temperature on embryonic development in reptiles and birds. In *Egg incubation: Its effects on embryonic development in birds and reptiles*, ed. D. C. Deeming and M. W. J. Ferguson, 147–171. Cambridge University Press.

Demuth, J. P. 2001. The effects of constant and fluctuating incubation temperatures on sex deterination, growth, and performance in the tortoise *Gopherus polyhemus. Can. J. Zool.* 79:1609–1620.

Desprez, D., and C. Mélard. 1998. Effect of ambient water temperature on sex determinism in the blue tilapia *Oreochromis aureus. Aquaculture* 162:79–84.

Desvages, G., and C. Pieau. 1992. Aromatase activity in gonads of turtle embryos as a function of the incubation temperature of eggs. *J. Steroid Biochem. Mol. Biol.* 41:851–853.

Desvages, G., M. Girondot, and C. Pieau. 1993. Sensitive stages for the effects of temperature on gonadal aromatase activity in embryos of the marine turtle Dermochelys coriacea. *Gen. Comp. Endocrinol.* 92:54–61.

Devlin, R. H., and Y. Nagahama. 2002. Sex determination and sex differentiation in fish: An overview of genetic, physiological, and environmental influences. *Aquaculture* 208:191–364.

Di Prisco, G., and B. Giardina. 1996. Temperature adaptation: Molecular aspects. In *Animals and temperature,* ed. I. A. Johnston and A. F. Bennett, 23–51. New York: Cambridge University Press.

Dickman, C. R., M. Letnic, and P. S. Mahon. 1999. Population dynamics of two species of dragon lizards in arid Australia: The effects of rainfall. *Oecologia* 119:357–366.

Doody, J. S. 1999. A test of the comparative influences of constant and fluctuating temperatures on phenotypes of hatchling turtles. Linnaeus Fund Research Report. *Chelon. Cons. Biol.* 3:529–531.

Dorazi, R., A. Chesnel, and C. Dournon. 1995. Opposite sex determination of gonads in two *Pleurodeles* species may be due to a temperature-dependent inactivation of sex chromosomes. *J. Heredity* 86:28–31.

Dournon, C., and C. Houillon. 1984. Démonstration génétique de l'inversion fonctionnelle du phénotype sexuel femelle sous l'action de la température d'élevage chez l'amphibien urodèle: *Pleurodeles waltlii* Michah. *Reprod. Nutr. Dév.* 24: 361–378.

———. 1985. Thermosensibilité de la différenciation sexuelle chez l'amphibien urodèle *Pleurodeles waltlii* Michah. Conditions pour obtenir l'inversion du phénotype sexuel de toutes les femelles génétiques sous l'action de la température d'élevage. *Reprod. Nutr. Dév.* 25:671–688.

Dournon, C., F. Guillet, D. Boucher, and J. C. Lacroix. 1984. Cytogenetic and genetic evidence of male sexual inversion by heat treatment in the newt *Pleurodeles poireti. Chromosoma* 90:261–264.

Dournon, C., A. Collenot, and M. Lauthier. 1988. Sex-linked peptidase-1 patterns in *Pleurodeles waltlii* Michah (Urodele Amphibian): Genetic evidence for a new codominant allele on the W sex chromosome and identification of ZZ, ZW and WW sexual genotypes. *Reprod. Nutr. Dév.* 28:979–987.

Dournon, C., C. Houillon, and C. Pieau. 1990. Temperature sex reversal in amphibians and reptiles. *Intl. J. Dev. Biol.* 34:81–92.

Downes, S. J., and R. Shine. 1999. Do incubation-induced changes in a lizard's phenotype influence its vulnerability to predators? *Oecologia* 120:9–18.

Du, W. G., and X. Ji. 2001. Influence of incubation temperature on embryonic use of material and energy in the Chinese soft-shelled turtle *(Pelodiscus sinensis). Acta Zool. Sinica* 47:512–517.

Du, X., P. Hublitz, T. Gunther, D. Wilhelm, C. Englert, and R. Schule. 2002. The LIM-only coactivator FHL2 modulates WT1 transcriptional activity during gonadal differentiation. *Biochim. Biophys. Acta* 1577:93–101.

Dubowsky, S., V. A. Lance, J. Lang, and A. Conley. 1995. Aromatase localization by immunocytochemistry in the ovary, but not testis, of the hatchling alligator a reptile with temperature dependent sex determination. *Proceedings of the Western Regional Conference on Comparative Endocrinology*, Seattle. [Abstract]

Dufaure, J. P., and J. Hubert. 1961. Table de développement du lézard vivipare: *Lacerta vivipara. Jacquin. Arch. Anat. Microsc. Morphol. Exp.* 50:309–328.

Dunlap, K. D., and J. W. Lang. 1990. Offspring sex ratio varies with maternal size in the common garter snake, *Thamnophis sirtalis. Copeia* 1990:568–570.

Dutton, P. H., C. P. Whitmore, and N. Mrosovsky. 1985. Masculinization of leatherback turtle dermochelys-coriacea hatchlings from eggs incubated in styrofoam boxes. *Biol. Cons.* 31:249–264.

Dutton, P. H., S. K Davis, T. Guerra, and D. Owens. 1996. Molecular phylogeny for marine turtles based on sequences of the ND4-leucine tRNA and control regions of mitochondrial DNA. *Mol. Phylogenet. Evol.* 5:511–521.

Eendeback, B. T. 1995. Incubation period and sex ratio of Herman's tortoise, *Testudo hermanni boettgerii. Chelon. Cons. Biol.* 1:227–231.

Eichenberger, P. 1981. Smaragdhagesissen kweken op bestelling. *Lacerta* 39:72–77.

El Mouden, E. H., M. Znair, and C. Pieau. 2001. Effects of incubation temperature on embryonic development and sex determination in the North African lizard, *Agama impalearis. Herpetol. J.* 11:101–108.

Elbrecht, A., and R. G. Smith. 1992. Aromatase enzyme activity and sex determination in chickens. *Science* 255:467–470.

Elf, P. K., and A. J. Fivizzani. 2002. Changes in sex steroid levels in yolks of the leghorn chicken, *Gallus domesticus*, during embryonic development. *J. Exp. Zool.* 293:594–600.

Elf, P. K., J. Allsteadt, J. W. Lang, and A. J. Fivizzani. 2001. The role of yolk steroid hormones in reptiles sex determination. In *Perspectives in comparative endocrinology*, ed. H. J. Th. Goos, R. K. Rastogi, H. Vaudry, and R. Pierantoni, 211–217. Bologna, Italy: Monduzzi Editore.

Elf, P. K., J. W. Lang, and A. J Fivizzani. 2002a. Yolk hormone levels in the eggs of snapping turtles and painted turtles. *Gen. Comp. Endocrinol.* 127:26–33.

———. 2002b. Dynamics of yolk steroid hormones during development in a reptile with temperature-dependent sex determination. *Gen. Comp. Endocrinol.* 127:34–39.

———. 2002c. Dynamics of exogenous ^{3}H-estradiol applied to snapping turtle eggs during embryonic development. *2002 Final Program and Abstracts*, 184. SICB Annual Meeting, Anaheim, CA. [Abstract]

———. 2003. Correlations between female age/body size and yolk steroid hormone levels in two species of TSD turtles. *2003 Final Program and Abstracts*. SICB Annual Meeting, Toronto, Canada. [Abstract]

Ellegren, H. 2000. Evolution of the avian sex chromosomes and their role in sex determination. *Trends Ecol. Evol.* 15:188–192.

———. 2002. Dosage compensation: Do birds do it as well? *Trends Genet.* 18:25–28.

Elliott, J. M. 1982. The effects of temperature and ration size on the growth and energetics of salmonids in captivity. *Comp. Biochem. Physiol. B* 73:81–91.

Elphick, M. J., and R. Shine. 1998. Long term effects of incubation temperatures on the morphology and locomotor performance of hatchling lizards (*Bassiana duperreyi*, Scincidae). *Biol. J. Linn. Soc.* 63:429–447.

———. 1999. Sex differences in optimal incubation temperatures in a scincid lizard species. *Oecologia* 118:431–437.

Else, P. L., and A. F. Bennett. 1987. The thermal dependence of locomotor performance and muscle contractile function in the salamander *Ambystoma tigrinum nebulosum*. *J. Exp. Biol.* 128:219–233.

Else, P. L., and A. J. Hulbert. 1981. Comparison of the "mammal machine" and the "reptile machine": Energy production. *Am. J. Physiol.* 240:R3–R9.

Emshwiller, M. G., and T. T. Gleeson. 1997. Temperature and effects on aerobic metabolism and terrestrial locomotion in American alligators. *J. Herpetol.* 31:142–147.

Etchberger, C. R. 1991. *Mechanistic and evolutionary considerations of temperature-dependent sex determination in turtles*. Unpubl. Ph.D. thesis, Indiana University, Bloomington, IN.

Etchberger, C. R., M. A. Ewert, J. B. Phillips, C. E. Nelson, and H. D. Prange. 1992a. Physiological responses to carbon dioxide in embryonic red-eared slider turtles, *Trachemys scripta*. *J. Exp. Zool.* 264:1–10.

Etchberger, C. R., M. A. Ewert, B. A. Raper, and C. E. Nelson. 1992b. Do low incubation temperatures yield females in painted turtles? *Can. J. Zool.* 70:391–394.

Etchberger, C. R., M. A. Ewert, J. B. Phillips, C. E. Nelson, and H. D. Prange. 1993. Environmental and maternal influences on embryonic pigmentation in a turtle *(Trachemys scripta elegans)*. *J. Zool.* 230:529–539.

Etchberger, C. R., M. A. Ewert, J. B. Phillips, and C. E. Nelson. 2002. Carbon dioxide influences environmental sex determination in two species of turtles. *Amphibia-Reptilia* 23:169–175.

Eubank, W. P., J. W. Atmar, and J. J. Ellington. 1973. The significance and thermodynamics of fluctuating versus static thermal environments on *Heliothis zea* egg development rates. *Entomol. Soc. Am.* 2:491–496.

Ewert, M. A. 1976. Nests, nesting and aerial basking of *Macroclemys* under natural conditions, and comparisons with *Chelydra* (Testudines: Chelydridae). *Herpetologica* 32:150–156.

———. 1979. The embryo and its egg: Development and natural history, In *Turtles: Perspectives and research*, ed. M. Harless and H. Morlock, 333–413. New York: John Wiley and Sons.

———. 1985. Embryology of turtles. In *Biology of the Reptilia*, vol. 14, ed. C. Gans, P. F. A. Maderson, and F. Billett, 75–267. New York: Wiley-Interscience.

———. 1991. Cold torpor, diapause, delayed hatching and aestivation in reptiles and birds. In *Egg incubation: Its effects on embryonic development in birds and reptiles*, ed. D. C. Deeming and M. J. W. Ferguson, 173–191. Cambridge University Press.

Ewert, M. A., and C. E. Nelson. 1991. Sex determination in turtles: Diverse patterns and some possible adaptive values. *Copeia* 1991:50–69.

———. 2003. Metabolic heating of embryos and sex determination in the American alligator *Alligator mississippiensis*. *J. Therm. Biol.* 28:159–165

Ewert, M. A., and D. S. Wilson. 1996. Seasonal variation of embryonic diapause in the striped mud turtle *(Kinosternon baurii)* and general considerations for conservation planning. *Chelon. Cons. Biol.* 2:43–54.

Ewert, M. A., C. R. Etchberger, and C. E. Nelson. 1990. An apparent co-occurrence of genetic and environmental sex determination in a turtle. *Am. Zool.* 30:56A.

Ewert, M. A., D. Jackson, and C. Nelson. 1994. Patterns of temperature-dependent sex determination in turtles. *J. Exp. Zool.* 270:3–15.

Eyring, H. 1935. The activated complex in chemical reactions. *J. Chem. Phys.* 3:107–115.

Falconer, D. S. 1989. *Introduction to quantitative genetics*. 3rd ed. New York: John Wiley and Sons.

Feder, M. E. 1978. Environmental variability and thermal acclimation in Neotropical and Temperate Zone salamanders. *Physiol. Zool.* 51:7–16.

———. 1983. Metabolic and biochemical correlated of thermal acclimation in the rough-skinned newt *Taricha granulosa*. *Physiol. Zool.* 56:513–552.

———. 1999. Organismal, ecological, and evolutionary aspects of heat-shock proteins and the stress response: Established conclusions and unresolved issues. *Am. Zool.* 39:857–864.

Feldman, C. R., and J. F. Parham. 2002. Molecular phylogenetics of emydine turtles: Taxonomic revision and the evolution of shell kinesis. *Mol. Phylogenet. Evol.* 22:388–398.

Feng, Z. M., A. Z. Wu, Z. Zhang, and C. L. Chen. 2000. GATA-1 and GATA-4 transactivate inhibin/activin beta-B-subunit gene transcription in testicular cells. *Mol. Endocrinol.* 14:1820–1835.

Ferguson, M. W .J. 1985. The reproductive biology and embryology of crocodilians. In *Biology of the Reptilia,* vol. 14, ed. C. Gans, F. S. Billett, and P. F. A. Maderson, 329–491. New York: Wiley and Sons.

Ferguson, M. W. J., and T. Joanen. 1982. Temperature of egg incubation determines sex in *Alligator mississippiensis*. *Nature* 296:850–853.

———. 1983. Temperature-dependent sex determination in *Alligator mississippiensis*. *J. Zool.* 200:143–177.

Festa-Bianchet, M. 1996. Offspring sex ratio studies of mammals: Does publication depend upon the quality of the research or the direction of the results? *Ecoscience* 3:42–44.

Fisher, R. A. 1930. *The genetical theory of natural selection*. Oxford Univiversity Press.

Flatt, T., R. Shine, P. A. Borges-Landaez, and S. J. Downes. 2001. Phenotypic variation in an oviparous montane lizard *(Bassiana duperreyi)*: The effects of thermal and hydric incubation environments. *Biol. J. Linn. Soc.* 74:339–350.

Fleming, A., and D. Crews. 2001. Estradiol and incubation temperature modulate regulation of steroidogenic factor 1 in the developing gonad of the red-eared slider turtle. *Endocrinology* 142:1403–1411.

Fleming, A., T. Wibbels, J. K. Skipper, and D. Crews. 1999. Developmental expression of steroidogenic factor 1 in a turtle with temperature-dependent sex determination. *Gen. Comp. Endocrinol.* 116:336–346.

Flores, D., A. Tousignant, and D. Crews. 1994. Incubation temperature affects the behavior of adult leopard geckos *(Eublepharis macularius). Physiol. Behav.* 55:1067–1072.

Foley, A. M. 1998. The nesting ecology of the loggerhead turtle *(Caretta caretta)* in the Ten Thousand Islands, Florida. Ph.D. dissertation, University of South Florida.

Frank, L. G., S. E. Glickman, and P. Licht. 1991. Fatal sibling aggression, precocial development, and androgens in neonatal spotted hyenas. *Science* 252:702–704.

Freedberg, S., and M. J. Wade. 2001. Cultural inheritance as a mechanism for population sex-ratio bias in reptiles. *Evolution* 55:1049–1055.

Freedberg, S., M. A. Ewert, and C. E. Nelson. 2001. Environmental effects on fitness and consequences for sex allocation in a reptile with environmental sex determination. *Evol. Ecol. Res.* 3:953–967.

Fu, J. 2000. Toward the phylogeny of the family Lacertidae: Why 4708 base pairs of mtDNA sequences cannot draw the picture. *Biol. J. Linn. Soc.* 71:203–217.

Fujimoto, T. 1979. Observations of primordial germ cells in the turtle embryo *(Caretta caretta):* Light and electron microscopic studies. *Dev. Growth Differ.* 21:3–10.

Fujioka, Y. 2001. Thermolabile sex determination in honmoroko. *J. Fish Biol.* 59:851–861.

———. 2002. Effects of hormone treatments and temperature on sex-reversal of Nigorobuna *Carassius carassius grandoculis. Fish. Sci.* 68:889–893.

Gabriel, W. N., B. Blumberg, S. Sutton, A. R. Place, and V. A. Lance 2001. Alligator aromatase cDNA sequence and its expression in embryos at male and female incubation temperatures. *J. Exp. Zool.* 290:439–448.

Gaffney, E. S. 1984. Historical analysis of theories of chelonian relationship. *Syst. Zool.* 33:283–301.

Gaffney, E. S., and P. A. Meylan. 1988. A phylogeny of turtles. In *The phylogeny and classification of the tetrapods*, ed. M. J. Benton, 157–219. Oxford: Clarendon.

Galbraith, D. A., B. N. White, R. J. Brooks, and P. T. Boag. 1993. Multiple paternity in clutches of snapping turtles *(Chelydra serpentina)* detected using DNA fingerprints. *Can. J. Zool.* 71:318–324.

Gallien, L. 1951. Sur la descendance unisexuée d'une femelle de *Pleurodeles waltlii* Michah ayant subi pendant sa phase larvaire l'action gynogène du benzoate d'oestradiol. *C. R. Acad. Sci. Paris D* 233:828–830.

Gallien, L., and M. Durocher. 1957. Table chronologique du développement chez *Pleurodeles waltl. Bull. Biol. Fr. Belg.* 91:97–114.

Ganesh, S., and R. Raman. 1995. Sex reversal by testosterone and not by estradiol or temperature in *Calotes versicolor*, the lizard lacking sex chromosomes. *J. Exp. Zool.* 271:139–144.

Gardom, T. 1993. *The natural history museum book of dinosaurs.* London: Carlton Books.

Gasca, S., J. Canizares, P. de Santa Barbara, C. Mejean, F. Poulat, P. Berta, and B. Boizet-Bonhoure. 2002. A nuclear export signal within the high mobility group domain regulates the nucleocytoplasmic translocation of SOX9 during sexual determination. *Proc. Natl. Acad. Sci. USA* 99:11199–11204.

Gatten, R. E. 1974. Effects of temperature and activity on aerobic and anaerobic metabolism and heart rate in the turtles *Pseudemys scripta* and *terrapene ornata. Comp. Biochem. Physiol. A* 48:619–648.

Gaze, P. 2001. *Tuatara recovery plan.* Threatened species recovery plan 47. New Zealand: Biodiversity Recovery Unit, Department of Conservation.

Geffeney, S., E. D. Brodie, Jr., P. C. Ruben, and E. D. Brodie III. 2002. Mechanisms of adaptation in a predator-prey arms race: TTX-resistant sodium channels. *Science* 297:1336–1339.

Geiser, F., B. T. Firth, and R. S. Seymour. 1992. Polyunsaturated dietary lipids lower the selected body temperature of a lizard. *J. Comp. Physiol. B* 162:1–4.

Georges, A. 1989. Female turtles from hot nests: Is it duration of incubation or proportion of development that matters? *Oecologia* 81:323–328.

———. 1992. Thermal-characteristics and sex determination in field nests of the pig-nosed turtle, *Carettochelys insculpta* (Chelonia: Carettochelydidae), from Northern Australia. *Aus. J. Zool.* 40:511–521.

Georges, A., and M. Adams. 1992. A phylogeny for the Australian chelid turtles based on allozyme electrophoresis. *Aus. J. Zool.* 40:453–476.

Georges, A., and S. McInnes. 1998. Temperature fails to influence hatchling sex in another genus and species of chelid turtle, *Elusor macrurus. J. Herpetol.* 32:596–598.

Georges, A., C. Limpus, and R. Stoutjesdijk. 1994. Hatchling sex in the marine turtle *Caretta caretta* is determined by proportion of development at a temperature, not daily duration of exposure. *J. Exp. Zool.* 270:432–444.

Georges, A., J. Birrell, K. M. Saint, W. McCord, and S. C. Donnellan. 1999. A phylogeny for side-necked turtles (Chelonia: Pleurodira) based on mitochondrial and nuclear gene sequence variation. *Biol. J. Linn. Soc.* 67:213–246.

Georges, A., K. Beggs, J. E. Young, and J. S. Doody. 2005. Modelling development of reptile embryos under fluctuating temperature regimes. *Physiol.Biochem. Zool.* 78(1). In press.

Gerum, S. V. 1999. Endogenous yolk steroids in five species of turtles: A comparison between sex determination mechanisms. M.Sc. thesis, University of North Dakota, Grand Forks.

Gibbons, J. W. 1990. Sex ratios and their significance among turtle populations In *Life history and ecology of the slider turtle,* 171–182. Washington, DC: Smithsonian Institution Press.

Gil, D., J. Graves, N. Hazon, and A.Wells. 1999. Male attractiveness and differential testosterone investment in zebra finch eggs. *Science* 286:126–128.

Girondot, M. 1999. Statistical description of temperature-dependent sex determination using maximum likelihood. *Evol. Ecol. Res.* 1:479–486.

Girondot, M., and C. Pieau. 1993. Effects of sexual differences of age at maturity and survival on population sex ratio. *Evol. Ecol.* 7:645–650.

———. 1996. On the limits of age-structured models for the

maintenance of environmental sex determination in crocodilians. *Ann. Sci. Nat.* 17:85–97.

———. 1999. A fifth hypothesis for the evolution of TSD in reptiles. *Trends Ecol. Evol.* 14:359–360.

Girondot, M., P. Zaborski, J. Servan, and C. Pieau. 1994. Genetic contribution to sex determination in turtles with environmental sex determination. *Genet. Res.* 63:117–127.

Girondot, M., H. Fouillet, and C. Pieau. 1998. Feminizing turtle embryos as a conservation tool. *Cons. Biol.* 12:353–362.

Girondot, M., A. D. Tucker, P. Rivalan, M. H. Godfrey, and J. Chevalier. 2002. Density-dependent nest destruction and population fluctuations of Guianan Letherback turtles. *Anim. Cons.* 5:75–84.

Glidewell, J. R., T. L. Beitinger, and L. C. Fitzpatrick. 1981. Heat exchange in submerged red-eared turtle, *Chrysemys scripta*. *Comp. Biochem. Physiol. A* 70:141–143.

Godfrey, M. H. 1997. Sex ratios of sea turtle hatchlings: Direct and indirect estimates. Unpubl. Ph.D. thesis, University of Toronto.

Godfrey, M. H., R. Baretto, and N. Mrosovsky. 1996. Estimating past and present sex ratios of sea turtles in Suriname. *Can. J. Zool.* 74:267–277.

Godfrey, M. H., A. F. D'Amato, M. Â. Marcovaldi, and N. Mrosovsky. 1999. Pivotal temperature and predicted sex ratios for hatchling hawksbill turtles from Brazil. *Can. J. Zool.* 77:1465–1473.

Godley, B. J., A. C. Broderick, and N. Mrosovsky. 2001. Estimating hatchling sex ratios of loggerhead turtles in Cyprus from incubation durations. *Mar. Ecol. Progr. Ser.* 210:195–201.

Godley, B. J., A. C. Broderick, F. Glen, and G. C. Hays. 2002. Temperature-dependent sex determination in Ascension Island green turtles. *Mar. Ecol. Progr. Ser.* 226:115–124.

Goode, J. M. 1991. Breeding semi-aquatic and aquatic turtles at the Columbus Zoo. In *First international symposium on turtles and tortoises: Conservation and captive husbandry, 1990*, ed. K. R. Beaman, S. Kaporaso, S. McKeown, and M. D. Graff, 66–76. Orange, CA: Chapman University.

Gorman, G. C. 1973. The chromosomes of the Reptilia, a cytotaxonomic interpretation. In *Cytotaxonomy and vertebrate evolution*, ed. A. B. Chiarelli and E. Capanna, 349–424. New York: Academic Press.

Goto, R., T. Mori, K. Kawamata, T. Matsubara, S. Mizuno, S. Adachi, and K. Yamauchi. 1999. Effects of temperature on gonadal sex determination in barfin flounder *Verasper moseri*. *Fish. Sci.* 65:884–887.

Goto, R., T. Kayaba, S. Adachi, and K. Yamauchi. 2000. Effects of temperature on sex determination in marbled sole *Limanda yokohamae*. *Fish. Sci.* 66:400–402.

Goy, R. W., and B. S. McEwen. 1980. *Sexual differentiation of the brain*. Cambridge, MA: MIT Press.

Graves, J. A. M. 2001. Sex from W to Z: Evolution of vertebrate sex chromosomes and sex determining genes. *J. Exp. Zool.* 290:449–462.

———. 2002. The rise and fall of *Sry*. *Trends Genet.* 18:259–264.

Graves, B. M., and D. J. Duvall. 1993. Reproduction, rookery use, and thermoregulation in free-ranging, pregnant *Crotalus v. viridis*. *J. Herpetol.* 27.

Greco, T. L., and A. H. Payne. 1994. Ontogeny of expression of the genes for steroidogenic enzymes P450 side-chain cleavage, 3 beta-hydroxysteroid dehydrogenase, P450 17 alpha-hydroxylase/C17-20 lyase, and P450 aromatase in fetal mouse gonads. *Endocrinology* 135:262–268.

Green, D. M. 1988. Cytogenetics of the endemic New Zealand frog, *Leiopelma hochstetteri*: Extraordinary supernumerary chromosome variation and a unique sex-chromosome system. *Chromosoma* 97:55–77.

Greenbaum, E., and J. L. Carr. 2001. Sexual differntiatioin in the spiny softshell turtle *(Apalone spinifera)*, a species with genetic sex determination. *J. Exp. Zool.* 290:190–200.

Greenwald, O. E. 1974. Thermal dependence of striking behavior and prey capture by gopher snakes. *Copeia* 1974:141–148.

Greenwald, O. E., and M. E. Kanter. 1979. The effects of temperature and behavioral thermoregulation on digestive efficiency rate in corn snakes *(Elaphe guttata guttata)*. *Physiol. Zool.* 52:398–408.

Greer, A. E. 1989. *The biology and evolution of Australian lizards*. Sydney: Surrey Beatty and Sons.

Grigg, G. C., and J. Alchin. 1976. The role of the cardiovascular system in thermoregulation of *Crocodylus johnstoni*. *Physiol. Zool.* 49:24–36.

Grismer, L. L. 1988. Phylogeny, taxonomy, classification, and biogeography of eublepharid geckos. In *Phylogenetic relationships of the lizard families*, ed. R. Estes and G. Pregill, 369–469. Stanford University Press.

Guillette, L. J., Jr., T. S. Gross, G. R. Masson, J. M. Matter, H. F. Percival, and A. R. Woodward. 1994. Developmental abnormalities of the gonad and abnormal sex hormone concentrations in juvenile alligators from contaminated and control lakes in Florida. *Env. Health Perspect.* 102:680–688.

Guillette, L. J., Jr., D. B. Pickford, D. A. Crain, A. A. Rooney, and H. F. Percival. 1996. Reduction in penis size and plasma testosterone concentrations in juvenile alligators living in a contaminated environment. *Gen. Comp. Endocrinal.* 101:32–42.

Guillette, L. J., Jr., A. R. Woodward, D. A. Crain, G. R. Masson, B. D. Palmer, M. C. Cox, Y-X. Qui, and E. F. Orlando. 1997. The reproductive cycle of female American alligator *(Alligator mississippiensis)*. *Gen. Comp. Endocrinol.* 108:87–101.

Gullberg, A., M. Olsson, and H. Tegelström. 1997. Male mating success, reproductive success and multiple paternity in a natural population of sand lizards: Behavioural and molecular genetics data. *Mol. Ecol.* 6:105–112.

Gurates, B., S. Sebastian, S. Yang, J. Zhou, M. Tamura, Z. Fang, T. Suzuki, et al. 2002. WT1 and DAX-1 inhibit aromatase P450 expression in human endometrial and endometriotic stromal cells. *J. Clin. Endocrinol. Metab.* 87:4369–4377.

Gutiérrez-Adán, A., M. Oter, B. Martínez-Madrid, B. Pintado, and J. de la Fuente. 2000. Differential expression of two genes located on the X chromosome between male and female *in vitro*-produced bovine embryos at the blastocyst stage. *Mol. Reprod. Dev.* 55:146–151.

Gutzke, W. H. N., and D. Crews. 1988. Embryonic temperature determines adult sexuality in a reptile. *Nature* 332:832–834.

Gutzke, W. H. N., and G. C. Packard. 1987a. The Influence of temperature on eggs and hatchlings of Blanding's turtles, *Emydoidea blandingii*. *J. Herpetol.* 21:161–163.

———. 1987b. Influence of the hydric and thermal environments on eggs and hatchlings of bull snakes *Pituophis melanoleucus*. *Physiol. Zool.* 60:9–17.

Gutzke, W. H. N., G. C. Packard, M. J. Packard, and T. J. Boardman. 1987. Influence of the hydric and thermal environments on eggs and hatchlings of painted turtles *(Chrysemys picta)*. *Herpetologica* 43:393–404.

Hagstrum, D. W., and W. R. Hagstrum. 1970. A simple device for producing fluctuating temperatures with an evaluation of the ecological significance of fluctuating temperatures. *Ann. Entomol. Soc. Am.* 63:1385–1389.

Hagstrum, D. W., and G. A. Milliken. 1991. Modeling differences in insect development times between constant and fluctuating temperatures. *Ann. Entomol. Soc. Am.* 84:369–379.

Hailey, A., and R. E. Willemsen. 2000. Population density and adult sex ratio of the tortoise *Testudo hermanni* in Greece: Evidence for intrinsic population regulation. *J. Zool.* 251:325–338.

Hairston, C. S., and P. M. Burchfield. 1992. The reproduction and husbandry of the water monitor at the Gladys Porter Zoo, Brownsville. *Intl. Zoo Yb.* 31:124–130.

Hall, W. P. 1970. Three probable cases of parthenogenesis in lizards (Agamidae, Chamaeleontidae, Gekkonidae). *Experientia* 26:1271–1273.

Hamilton, W. D. 1967. Extraordinary sex ratios. *Science* 156: 477–488.

Hanson, J., T. Wibbels, and R. E. Martin. 1998. Predicted female bias in sex ratios of hatchling loggerhead sea turtles from a Florida nesting beach. *Can. J. Zool.* 76:1850–1861.

Harding, J. H., and S. K. Davis. 1999. *Clemmys insculpta* (wood turtle) and *Emydoides blandingii* (Blanding's turtle) hybridization. *Herpetol. Rev.* 30:335–336.

Harlow, P. S. 1996. A harmless technique for sexing hatchling lizards. *Herpetol. Rev.* 27:71–72.

———. 2000. Incubation temperature determines hatchling sex in Australian rock dragons (Agamidae: genus *Ctenophorus*). *Copeia* 2000:958–964.

———. 2001. Ecology of sex-determining mechanisms in Australian agamid lizards. Unpubl. Ph.D. thesis, School of Biological Sciences, Macquarie University, Sydney, Australia.

Harlow, P. S., and M. F. Harlow. 1997. Captive reproduction and longevity in the eastern water dragon *(Physignathus lesueurii)*. *Herpetofauna* 27:14–19.

Harlow, P. S., and R. Shine. 1999. Temperature-dependent sex determination in the frillneck lizard, *Chlamydosaurus kingii* (Agamidae). *Herpetologica* 55:205–212.

Harlow, P. S., and J. E. Taylor. 2000. Reproductive ecology of the jacky dragon *(Amphibolurus muricatus)* an agamid lizard with temperature-dependent sex determination. *Aus. Ecol.* 25:640–652.

Harlow, P. S., D. J. Pearson, and M. Peterson. 2002. Reproduction and egg incubation in the western bearded dragon, *Pogona minor*. *West. Aus. Nat.* 23:181–185.

Harrington, R. W., Jr. 1967. Environmentally controlled induction of primary male gonochorists from eggs of self-fertilizing hermaphroditic fish *Rivulus marmoratus* Poey. *Biol. Bull.* 132: 174–199.

Harry, J. L., and D. A. Briscoe. 1988. Multiple paternity in loggerhead turtle *(Caretta caretta)*. *J. Heredity* 79:96–99.

Harvey, P. H., and L. Partridge. 1984. When deviants are favoured: Evolution of sex determination. *Nature* 307:689–691.

Harvey, P. H., A. J. Leigh Brown, J. Maynard Smith, and S. Nee. 1996. *New uses for new phylogenies*. New York: Oxford University Press.

Harwood, R. H. 1979. The effect of temperature on the digestive efficiency of three species of lizards, *Cnemidophorus tigris, Gerrhonotus multicarinatus,* and *Scleroporus occidentalis*. *Comp. Biochem. Physiol. A* 63:417–433.

Hayes, T. B. 1998. Sex determination and primary sex differentiation in amphibians: Genetic and developmental mechanisms. *J. Exp. Zool.* 281:373–399.

Head, G., R. M. May, and L. Pendleton. 1987. Environmental determination of sex in the reptiles. *Nature* 329:198–199.

Hedges, S. B., and L. L. Poling. 1999. A molecular phylogeny of reptiles. *Science* 283:998–1001.

Heppell, S.S. 1998. Application of life-history theory and population model analysis to turtle conservation. *Copeia* 1998:367–375.

Hertz, P. E., R. B. Huey, and E. Nevo. 1982. Fight versus flight: Body temperature influences defensive responses of lizards. *Anim. Behav.* 30:676–679.

Hewavisenthi, S., and C. J. Parmenter. 2000. Hydric environment and sex determination in the flatback turtle (*Natator depressus* Garman) (Chelonia: Cheloniidae). *Aus. J. Zool.* 48(6): 653–659.

———. 2001. Influence of incubation environment on the development of the Flatback turtle *(Natator depressus)*. *Copeia* 2001:668–682.

Hille, E. 1964. *Analysis*. New York: Blaisdell Publishing Company.

Hillis, D. M., and D. M. Green. 1990. Evolutionary changes of heterogametic sex in the phylogenetic history of amphibians. *J. Evol. Biol.* 3:49–64.

Hiraoka, Y., N. Komatsu, Y. Sakai, M. Ogawa, M. Shiozawa, and S. Aiso. 1997. *XLS13A* and *XLS13B: Sry*-related genes of *Xenopus laevis*. *Gene* 197:65–73.

Hochachka, P. W., and G. N. Somero. 2002. *Biochemical adaptation*. New York: Oxford University Press.

Hoggren, M., and H. Tegelstrom. 1995. DNA fingerprinting shows within-season multiple paternity in the Adder *(Vipera berus)*. *Copeia* 1995:271–277.

Holman, J. A., and U. Fritz. 2001. A new emydine species from the Middle Miocene (Barstovian) of Nebraska, USA, with a new generic arrangement for the species of *Clemmys* sensu McDowell (1964) (Reptilia: Testudines: Emydidae). *Zool. Abh. Mus. Tierkde. Dresden* 51:331–353.

Hori, T., S. Asakawa, Y. Itoh, N. Shimizu, and S. Mizuno. 2000. *Wpkci,* encoding an altered form of PKCI, is conserved widely on the avian W chromosome and expressed in early female embryos: Implication of its role in female sex determination. *Mol. Biol. Cell* 11:3645–3660.

Horrocks, J. A., and N. McA. Scott. 1991. Nest site location and nest success in the hawksbill turtle *Eretmochelys imbricata* in Barbados, West Indies. *Mar. Ecol. Progr. Ser.* 69:1–8.

Hostache, G., M. Pascal, and C. Tessier. 1995. Influence de la témperature d'incubation sur le rapport mâle: Femelle chez l'atipa, *Hoplosternum littorale* Hancock (1828). *Can. J. Zool.* 73:1239–1246.

Hotaling, E. C. 1990. Temperature-dependent sex determination: Factors affecting sex determination in nests of a New Jersey population of *Chelydra serpentina*. Unpubl. Ph.D. thesis, Rutgers University, Newark, NJ.

Hou, L. 1985. Sex determination by temperature for incubation in *Chinemys reevesii. Acta Herpetol. Sinica* 4:150.

Hsü, C. Y., N. W. Yu, and H. M. Liang. 1971. Induction of sex reversal in female tadpoles of *Rana catesbeiana* by temperature treatment. *Endocrinol. Jpn.* 18: 243–251.

Hubert, J. 1969. Origin and development of oocytes. In *Biology of the Reptilia*, vol. 14 ed. C. Gans, P. F. A. Maderson, and F. Billett. New York: Wiley-Interscience.

Huey, R. B. 1979. Integrating thermal physiology and ecology of ectotherms: A discussion of approaches. *Am. Zool.* 19:357–366.

Huey, R. B., and D. Berrigan. 1996. Testing evolutionary hypotheses of acclimation. In *Animals and temperature,* ed. I. A. Johnston and A. F. Bennett, 205–237. New York: Cambridge University Press.

Hultin, E. 1955. The influence of temperature on the rate of enzymatic processes. *Acta Chem. Scandanavia* 9:1700–1710.

Hutchison, V. H. 1961. Critical thermal maxima in salamanders. *Physiol. Zool.* 34:92–125.

Hutton, J. M. 1987. Incubation temperatures, sex ratios and sex determination in a population of Nile crocodiles *(Crocodylus niloticus). J. Zool. Lond.* 211:143–155.

Ikeda, Y., A. Swain, T. J. Weber, K. E. Hentges, E. Zanaria, E. Lalli, K. T. Tamai, et al. 1996. Steroidogenic factor 1 and *Dax-1* colocalize in multiple cell lineages: Potential links in endocrine development. *Mol. Endocrinol.* 10:1261–1272.

Ingraham, H. A., D. S. Lala, Y. Ikeda, X. Luo, M. W. Nachtigal, R. Abbud, J. H. Nilson, and K. L. Parker. 1994. The nuclear receptor steroidogenic factor 1 acts at multiple levels of the reproductive axis. *Genes Dev.* 8:2302–2312.

Iverson, J. B. 1989. The Arizona mud turtle, *Kinosternon flavescens arizonense* (Kinosternidae), in Arizona and Sonora. *Southwest. Nat.* 34:356–368.

———. 1992a. *A revised checklist with distribution maps of turtles of the world.* Richmond, IN: Privately printed.

———. 1992b. Phylogenetic hypotheses for the evolution of modern kinosternid turtles. *Herpetol. Monogr.* 4:1–27.

———. 1998. Molecules, morphology, and mud turtle phylogenetics (Kinosternidae). *Chelon. Cons. Biol.* 3:113–117.

———. 1999. Reproduction in the Mexican mud turtle *Kinosternon integrum. J. Herpetol.* 33:144–148.

———. 2002. Reproduction in female razorback musk turtles (*Sternotherus carinatus*: Kinosternidae). *Southwest. Nat.* 47:215–224.

Jaenicke, R. 1991. Protein stability and molecular adaptation to extreme conditions. *Eur. J. Biochem.* 202:715–728.

James, C., and R. Shine. 1985. The seasonal timing of reproduction: A tropical-temperate comparison in Australian lizards. *Oecologia* 67:464–474.

Janzen, F. J. 1992. Heritable variation for sex ratio under environmental sex determination in the common snapping turtle *(Chelydra serpentina). Genetics* 31:155–161.

———. 1993. The influence of incubation temperature and family on eggs, embryos, and hatchlings of the smooth softshell turtle *(Apalone mutica). Physiol. Zool.* 66:349–373.

———. 1994a. Climate change and temperature-dependent sex determination in reptiles. *Proc. Natl. Acad. Sci. USA* 91:7487–7490.

———. 1994b. Vegetational cover predicts the sex ratio of hatchling turtles in natural nests. *Ecology* 75:1593–1599.

———. 1995. Experimental evidence for the evolutionary significance of temperature-dependent sex determination. *Evolution* 49:864–873.

Janzen, F. J., and C. L. Morjan. 2001. Repeatability of microenvironment-specific nesting behavior in a turtle with environmental sex determination. *Anim. Behav.* 61:73–82.

———. 2002. Egg size, incubation temperature, and posthatching growth in painted turtles *(Chrysemys picta). J. Herpetol.* 36:308–311.

Janzen, F. J., and G. L. Paukstis. 1988. Environmental sex determination in reptiles. *Nature* 332:790.

———. 1991a. Environmental sex determination in reptiles: Ecology, evolution, and experimental design. *Q. Rev. Biol.* 66:149–179.

———. 1991b. A preliminary test of the adaptive significance of environmental sex determination in reptiles. *Evolution* 45: 435–440.

Janzen, F. J., M. E. Wilson, J. K. Tucker, and S. P. Ford. 1998. Endogenous yolk steroid hormones in turtles with different sex-determining mechanisms. *Gen. Comp. Endocrinol.* 111:306–317.

———. 2002. Experimental manipulation of steroid concentrations in circulation and in egg yolks of turtles. *J. Exp. Zool.* 293:58–66.

Jeyasuria, P., and A. R. Place. 1997. Temperature-dependent aromatase expression in developing diamondback terrapin *(Malaclemys terrapin)* embryos. *J. Steroid Biochem. Mol. Biol.* 61:415–425.

———. 1998. Embryonic brain-gonadal axis in temperature-dependent sex determination of reptiles: A role for P450 aromatase (CYP19). *J. Exp. Zool.* 281:428–449.

Jeyasuria, P., W. M. Roosenburg, and A. R. Place. 1994. Role of P-450 aromatase in sex determination of the diamondback terrapin, *Malaclemys terrapin. J. Exp. Zool.* 270:95–111.

Ji, X., and F. Braña. 1999. The influence of thermal and hydric environments on embryonic use of energy and nutrients, and hatchling traits in the wall lizards *(Podarcis muralis). Comp. Biochem. Physiol. A* 124:205–213.

Ji, X., and W. G. Du. 2001a. The effects of thermal and hydric environments on hatching success, embryonic use of energy and hatchling traits in a colubrid snake, *Elaphe carinata. Comp. Biochem. Physiol. A* 129:461–471.

———. 2001b. Effects of thermal and hydric environments on incubating eggs and hatchling traits in the Cobra, *Naja naja atra*. *J. Herpetol.* 35:186–194.

Ji, X., X. F. Xu, and Z. H. Lin. 1999. Influence of incubation temperature on characteristics of *Dinodon rufozonatum* (Reptilia: Colubridae) hatchlings, with comments on the function of residual yolk. *Zool. Res.* 20:342–346.

Ji, X., Q. B. Qui, and C. H. Diong. 2002. Influence of incubation temperature on hatching success, energy expenditure for embryonic development, and size and morphology hatchlings in the oriental garden lizard, *Calotes versicolor* (Agamidae). *J. Exp. Zool.* 292:649–659.

Jin, T., X. Zhang, H. Li, and P. E. Goss. 2000. Characterization of a novel silencer element in the human aromatase gene pII promoter. *Breast Cancer Res. Treat.* 62:151–159.

Joanen, T., L. McNease, and M. W. J. Ferguson. 1987. The effect of egg incubation temperature on post-hatching growth of American alligators. In *Wildlife management: Crocodiles and alligators*, ed. J. W. Webb, S. C. Manolis, and P. J. Whitehead, 533–537. Chipping Norton, NSW, Australia: Surrey Beatty and Sons.

John-Alder, H. B., M. C. Barnhart, and A. F. Bennett. 1989. Thermal sensitivity of swimming performance and muscle contraction in northern and southern populations of tree frogs *(Hyla crucifer)*. *J. Exp. Biol.* 142:357–372.

Johnson, F. H., and I. Lewin. 1946. The growth rate of *E. coli* in relation to temperature, quinine and coenzyme. *J. Cell. Comp. Physiol.* 28:47–75.

Johnston, I. A., V. L. A. Vieira, and J. Hill. 1996. Temperature and ontogeny in ectotherms: Muscle phenotype in fish. In *Animals and temperature,* ed. I. A. Johnston and A. F. Bennett, 153–204. New York: Cambridge University Press.

Johnston, I. A., V. L. A. Vieira, and G. K. Temple. 2001. Functional consequences and population differences in the developmental plasticity of muscle to temperature in Atlantic herring *Clupea harengus*. *Mar. Ecol. Progr. Ser.* 213:285–300.

Joss, J. M. P. 1989. Gonadal development and differentiation in *Alligator mississippiensis* at male and female producing incubation temperatures. *J. Zool. Lond.* 218:679–687.

Josseaume, B. 2002. Rôle des chéloniens dans la régénération des écosystèmes forestiers tropicaux. Exemple de la tortue denticulée *(Chelonoidis denticulata)* en Guyane française. Ph.D. thesis, Université Paris 6.

Jost, A. 1953. Studies on sex differentiation in mammals. *Rec. Progr. Horm. Res.* 8:379–418.

Jost, A., and S. Magre. 1988. Control mechanisms of testicular differentiation. *Phil. Trans. Roy. Soc. Lond. B* 322:55–61.

Jost, A., B. Vigier, J. Prepin, and J. P. Perchellet. 1973. Studies on sex differentiation in mammals. *Rec. Progr. Horm. Res.* 29:1–41.

Julliard, R. 2000. Sex-dispersal in spatially varying environments leads to habitat-dependent evolutionarily stable offspring sex ratios. *Behav. Ecol.* 11:421–428.

Jun-Yi, L., and L. Kau-Hung. 1982. Population ecology of the lizard *Japalura swinhonis formosensis* (Sauria: Agamidae) in Taiwan. *Copeia* 1982:425–434.

Kallman, K. D. 1984. A new look at sex determination in poeciliid fishes. In *Evolutionary genetics of fishes*, ed. B. J. Turner, 95–171. New York: Plenum Press.

Karl, S. A., B. W. Bowen, and J. C. Avise, 1992. Global population genetic structure and male-mediated gene flow in the green turtle *(Chelonia mydas):* RFLP analyses of anonymous nuclear loci. *Genetics* 131:163–173.

Karlin, S., and S. Lessard. 1986. *Theoretical studies on sex ratio evolution*, Princeton, NJ: Princeton University Press.

Kaska, Y., R. Downie, R. Tippett, and R. W. Furness. 1998. Temperatures of green turtle *(Chelonia mydas)* and loggerhead turtle *(Caretta caretta)*. *Can. J. Zool.* 76:723–729.

Kasparek, M., B. J. Godley, and A. C. Broderick. 2001. Nesting of the green turtle, *Chelonia mydas*, in the Mediterranean: A review of status and conservation needs. *Zool. Middle East* 24:45–74.

Katoh-Fukui, Y., R. Tsuchiya, T. Shiroishi, Y. Nakahara, N. Hashimoto, K. Noguchi, and T. Higashinakagawa. 1998. Male-to-female sex reversal in M33 mutant mice. *Nature* 393:688–692.

Kawabe, K., T. Shikayama, H. Tsuboi, S. Oka, K. Oba, T. Yanase, H. Nawata, and K. Morohashi. 1999. *Dax-1* as one of the target genes of Ad4BP/SF-1. *Mol. Endocrinol.* 13:1267–1284.

Kawano, K., S. Furusawa, H. Matsuda, M. Takase, and M. Nakamura. 2001. Expression of steroidogenic factor 1 in frog embryo and developing gonad. *Gen. Comp. Endocrinol.* 123:13–22.

Kent, J., S. C. Wheatley, J. E. Andrews, A. H. Sinclair, and P. Koopman. 1996. A male-specific role for *Sox9* in vertebrate sex determination. *Development* 122:2813–2822.

Kettlewell, J. R., C. S. Raymond, and D. Zarkower. 2000. Temperature-dependent expression of turtle *Dmrt1* prior to sexual differentiation. *Genesis* 26:174–178.

Kichler, K., M. T. Holder, S. K. Davis, R. Márquez-M, and D. W. Owens. 1999. Detection of multiple paternity in the Kemp's ridley sea turtle with limited sampling. *Mol. Ecol.* 8:819–830.

Kierszenbaum, A. L., and L. L. Tres. 2001. Primordial germ cell–somatic cell partnership: A balancing cell signaling act. *Mol. Reprod. Dev.* 60:277–280.

Kingston, R., and N. Bumstead. 1995. Monoclonal-antibodies identifying chick gonadal cells. *Brit. Poultry Sci.* 36:187–195.

Kitano, T., K. Takamune, T. Kobayashi, Y. Nagahama, and S.-I. Abe. 1999. Suppression of P450 aromatase gene expression in sex-reversed males produced by rearing genetically female larvae at a high water temperature during a period of sex differentiation in the Japanese flounder *(Paralichthys olivaceus)*. *J. Mol. Endocrinol.* 23:167–176.

Kitano, T., K. Takamune, Y. Nagahama, and S.-I. Abe. 2000. Aromatase inhibitor and 17 α-methyltestosterone cause sex-reversal from genetical females to phenotypic males and suppression of P450 aromatase gene expression in Japanese flounder *(Paralichthys olivaceus)*. *Mol. Reprod. Dev.* 56:1–5.

Kluge, A. G. 1987. Cladistic relationships in the Gekkonoidea (Squamata: Sauria). *Misc. Publ. Mus. Zool. Univ. Mich.* 173:1–54.

Knowles, T.W., and P. D. Weigl. 1990. Thermal dependence of anuran burst locomotor performance. *Copeia* 1990:796–802.

Kolbe, J., and F. J. Janzen. 2002. Impact of nest-site selection on nest success and nest temperature in natural and disturbed habitats. *Ecology* 83:269–281.

Komdeur, J., S. Daan, J. Tinbergen, and C. Mateman. 1997. Extreme adaptive modification of the Seychelles warbler's eggs. *Nature* 385:522–525.

Koopman, P. 2001a. The genetics and biology of vertebrate sex determination. *Cell* 105:843–847.

———. 2001b. *Sry, Sox9* and mammalian sex determination. In *Genes and mechanisms in vertebrate sex determination,* ed. G. Scherer and M. Schmid, EXS 91:25–56. Basel, Boston, Berlin: Birkhäuser.

Koopman, P., J. Gubbay, N. Vivian, P. Goodfellow, and R. Lovell-Badge. 1991. Male development of chromosomally female mice transgenic for *Sry. Nature* 351:117–121.

Koopman, P., M. Bullejos, and J. Bowles. 2001. Regulation of male sexual development by *Sry* and *Sox9. J. Exp. Zool.* 290:463–474.

Korpelainen, H. 1990. Sex ratios and conditions required for environmental sex determination in animals. *Biol. Rev.* 65:147–184.

———. 1998. Labile sex expression in plants. *Biol. Rev.* 73:157–180.

Koumoundouros, G., M. Pavlidis, L. Anezaki, C. Kokkari, A. Sterioti, P. Divanach, and M. Kentouri. 2002. Temperature sex determination in the European sea bass, *Dicentrarchus labrax* (L. 1758) (Teleostei, Perciformes, Moronidae): Critical sensitive ontogenetic phase. *J. Exp. Zool.* 2292:573–579.

Kraak, S. B. M., and I. Pen. 2002. Sex-determining mechanisms in vertebrates. In *Sex ratios: Concepts and research methods,* ed. I. C. W. Hardy, 158–177. Cambridge University Press.

Krackow, S. 1992. Sex ratio manipulation in wild house mice: The effect of fetal resorption in relation to the mode of reproduction. *Biol. Reprod.* 47:541–548.

Kreidberg, J. A., H. Sariola, J. M. Loring, M. Maeda, J. Pelletier, D. Housman, and R. Jaenisch. 1993. *Wt-1* is required for early kidney development. *Cell* 74:679–691.

Krenz, J. G., and F. J. Janzen. 2000. Turtle phylogeny: Insights from a nuclear gene. *Am. Zool.* 40:1092–1092. [Abstract]

Kumar, S., C. Tsai, and R. Nussinov. 2000. Factors enhancing protein thermostability. *Protein Eng.* 13:179–191.

Kuntz, S., A. Chesnel, M. Duterque-Coquillaud, I. Grillier-Vuissoz, M. Callier, C. Dournon, S. Flament, and D. Chardard. 2003. Differential expression of P450 aromatase during gonadal sex differentiation and sex reversal of the newt *Pleurodeles waltl. J. Steroid Biochem. Mol. Biol.* 84:89–100.

Kwon, J. Y., V. Haghpanah, L. M. Kogson-Hurtado, B. J. McAndrew, and D. J. Penman. 2000. Masculinization of genetic female Nile tilapia *(Oreochromis niloticus)* by dietary administration of an aromatase inhibitor during sexual differentiation. *J. Exp. Zool.* 287:46–53.

Kwon, J. Y., B. J. McAndrew, and D. J. Penman. 2001. Cloning of brain aromatase gene and expression of brain and ovarian aromatase genes during sexual differentiation in genetic male and female Nile tilapia *Oreochromis niloticus. Mol. Reprod. Dev.* 59:359–370.

Kwon, J. Y., B. J. McAndrew, and D. J. Penman. 2002. Treatment with an aromatase inhibitor suppresses high-temperature feminization of genetic male (YY) Nile tilapia. *J. Fish Biol.* 60:625–636.

Lacroix, J. C. 1970. Mise en évidence sur les chromosomes en écouvillon de *Pleurodeles poireti* Gervais, Amphibien Urodèle, d'une structure liée au sexe, identifiant le bivalent sexuel et marquant le chromosome W. *C. R. Acad. Sci. Paris D* 271: 102–104.

Lacroix, J. C., R. Azzouz, F. Simon, M. Bellini, J. Charlemagne, and C. Dournon. 1990. Characterization of the W and Z lampbrush heterochromosomes in the newt *Pleurodeles waltl.* W chromosome plays a role in sex female determination. *Chromosoma* 99:307–414.

Lacroix, J. C., R. Azzouz, D. Boucher, C. Abbadie, C. K. Pyne, and J. Charlemagne. 1985. Monoclonal antibodies to lampbrush chromosome antigens of *Pleurodeles waltlii. Chromosoma* 92:69–80.

Lagomarsino, I., and D. O. Conover. 1993. Variation in environmental and genetic sex determining mechanisms across a latitudinal gradient in the fish, *Menidia menidia. Evolution* 47:487–494.

Lance, V. A. 1989. Reproductive cycle of the American alligator. *Am. Zool.* 29:999–1018.

———. 1997. Sex determination in reptiles: An update. *Am. Zool.* 37:504–513.

Lance, V. A., and M. H. Bogart. 1991. Tamoxifen sex reverses alligator embryos at male producing temperature, but is an antiestrogen in female hatchlings. *Experientia* 47:263–266.

———. 1994. Studies on sex determination in the American alligator *Alligator mississippiensis. J. Exp. Zool.* 270:79–85.

Lance, V. A., N. Valenzuela, and P. von Hildebrand. 1992. A hormonal method to determine the sex of giant river turtles, *Podocnemis expansa*: Applications to endangered species research. *Am. Zool.* 32:16A. [Abstract]

Lance, V. A., R. M. Elsey, and J. W. Lang. 2000. Sex ratios of American alligators (Crocodylidae): Male or female biased? *J. Zool. Lond.* 252:71–78.

Lang, J. W. 1979. Thermophilic response of the American alligator and the American crocodile to feeding. *Copeia* 1979:48–59.

———. 1987. Crocodilian thermal selection. In *Wildlife management: Crocodiles and alligators*, ed. G .J. W. Webb, S. C. Manolis, and P. J. Whitehead, 301–317. Chipping Norton, NSW, Australia: Surrey Beatty and Sons.

Lang, J. W., and H. V. Andrews. 1994. Temperature-dependent sex determination in crocodilians. *J. Exp. Zool.* 270:28–44.

Lang, J. W., H. Andrews, and R. Whitaker. 1989. Sex determination and sex ratios in *Crocodylus palustris. Am. Zool.* 29:935–952.

Langerwerf, B. 1983. Uber die haltung und zucht von *Agama caucasia* (Sauria: Agamidae), nebst bemerkungen zur erfolgreichen zucht weiter palaearktischer echsen. *Salamandra* 19:11–12.

———. 1984. Techniques for large-scale breeding of lizards from temperate climates in greenhouse enclosures. *Acta Zool. Pathol. Antverp.* 78:163–176.

———. 1988. Management and breeding strategies at the Centro de Investigaciones Herpetólogicas in the Canary Islands. *Intl. Herpetol. Symp. Captive Propogation and Husbandry* 11:99–102.

Larsson, S. H., J. P. Charlieu, K. Miyagawa, D. Engelkamp, M. Rassoulzadegan, A. Ross, F. Cuzin, et al. 1995. Subnuclear localization of WT1 in splicing or transcription factor domains is regulated by alternative splicing. *Cell* 81:391–401.

Lee, C. H., O. S. Na, I. K. Yeo, H. J. Baek, and Y. D. Lee. 2000. Effects of sex steroid hormones and high temperatures on sex differentiation in black rockfish, *Sebastes schlegeli*. *J. Korean Fish. Soc.* 33:373–377.

Lemaire, R., J. Prasad, T. Kashima, J. Gustafson, J. L. Manley, and R. Lafyatis. 2002. Stability of a PKCI-1-related mRNA is controlled by the splicing factor ASF/SF2: A novel function for SR proteins. *Genes Dev.* 16:594–607.

Leshem, A., A. Ar, and R. A. Ackerman. 1991. Growth, water, and energy metabolism of the soft-shelled turtle *(Trionyx triunguis)* embryo: Effects of temperature. *Physiol. Zool.* 62:568–594.

Leslie, A. J., and J. R. Spotila. 2001. Alien plant threatens Nile crocodile *(Crocodylus niloticus)* breeding in Lake St. Lucia, South Africa. *Biol. Cons.* 98:347–355.

Letcher, B. H., and D. A. Bengtson. 1993. Effects of food density and temperature on feeding and growth of young inland silversides *(Menidia beryllina)*. *J. Fish Biol.* 43:671–686.

Lewis-Winokur, V., and R. M. Winokur. 1995. Incubation temperature affects sexual differentiation, incubation time, and posthatching survival in desert tortoises *(Gopherus agassizi)*. *Can. J. Zool.* 73:2091–2097.

Lillywhite, H. B., P. Licht, and P. Chelgren. 1973. The role of behavioral thermoregulation in the growth energetics of the toad, *Bufo boreas*. *Ecology* 54:375–383.

Limpus, C. J., and N. Nicholls. 1988. The Southern Oscillation regulates the annual numbers of green turtles *(Chelonia mydas)* breeding around Northern Australia. *Aus. J. Wildl. Res.* 15:157–161.

Limpus, C. J., P. C. Reed, and J. D. Miller. 1985. Temperature dependent sex determination in Queensland sea turtles: Intraspecific variation in *Caretta caretta*. In *Biology of Australian frogs and reptiles*, ed. G. Grigg, R. Shine, and H. Ehrmann, 343–351. Sydney: Surrey, Beatty and Sons.

Limpus, C. J., P. J. Couper, and K. L. D. Couper. 1993. Crab Island revisited: Reassessment of the world's largest flatback turtle rookery after twelve years. *Mem. Queensland Mus.* 33:277–289.

Lin, J. J., and G. N. Somero. 1995. Temperature-dependent changes in expression of thermostable and thermolabile isozymes of cytosolic malate dehydrogenase in the eurythermal goby fish *Gillichthys mirabilis*. *Physiol. Zool.* 68:114–128.

Lipar, J. L., E. D. Ketterson, V. Nolan, Jr., and J. M. Castro. 1999. Egg yolk layers vary in the concentration of steroid hormones in two avian species. *Gen. Comp. Endocrinol.* 115:220–227.

Liu, S. S., G. M. Zhang, and J. Zhu. 1995. Influence of temperature variations on rate of development in insects: Analysis of case studies from entomological literature. *Ann. Entomol. Soc. Am.* 88:107–119.

Losos, J. B. 1994. An approach to the analysis of comparative data when a phylogeny is unavailable or incomplete. *Syst. Biol.* 43:117–123.

Lovell-Badge, R., C. Canning, and R. Sekido. 2002. Sex-determining genes in mice: Building pathways. *Novartis Found. Symp.* 244:4–18; discussion 18–22, 35–42, 253–257.

Lovich, J. E. 1996. Possible demographic and ecologic consequences of sex ratio manipulation in turtles. *Chelon. Cons. Biol.* 2:114–117.

Lu, J. R., T. A. McKinsey, H. Xu, D. Z. Wang, J. A. Richardson, and E. N. Olson. 1999. FOG-2, a heart- and brain-enriched cofactor for GATA transcription factors. *Mol. Cell Biol.* 19:4495–4502.

Luckenbach, J. A., J. Godwin, H. V. Daniels, and R. J. Borski. 2003. Gonadal differentiation and effects of temperature on sex determination in southern flounder *(Paralichthys lethostigma)*. *Aquaculture* 216(1–4): 315–327.

Luo, X., Y. Ikeda, and K. L. Parker. 1994. A cell-specific nuclear receptor is essential for adrenal and gonadal development and sexual differentiation. *Cell* 77:481–490.

Lynch, K. W., and T. Maniatis. 1996. Assembly of specific SR protein complexes on distinct regulatory elements of the Drosophila doublesex splicing enhancer. *Genes Dev.* 10:2089–2101.

Maddison, D. R., and W. P. Maddison. 2000. *MacClade 4.0: Analysis of phylogeny and character evolution*. Version 4.0. Sunderland, MA: Sinauer Associates.

Maddock, M. B., and F. J. Schwartz. 1996. Elasmobranch cytogenetics: Methods and sex chromosomes. *Bull. Mar. Sci.* 58:147–155.

Magnusson, W. E., A. P. Lima, and R. M. Sampaio. 1985. Sources of heat for nests of *Paleosuchus trigonatus* and a review of crocodilian nest temperatures. *J. Herpetol.* 19:199–207.

Magnusson, W. E., A. P. Lima, J. M. Hero, T. M. Sanaiotti, and M. Yamakoshi. 1990. *Paleosuchus trigonatus* nests: Sources of heat for nests and embryo sex ratios. *J. Herpetol.* 24:397–400.

Magre, S., and A. Jost. 1984. Dissociation between testicular organogenesis and endocrine cytodifferentiation of Sertoli cells. *Proc. Natl. Acad. Sci. USA* 81:7831–7834.

———. 1991. Sertoli cells and testicular differentiation in the rat fetus. *J Electron Microsc. Tech.* 19:172–188.

Maheswaran, S., C. Englert, G. Zheng, S. B. Lee, J. Wong, D. P. Harkin, J. Bean, et al. 1998. Inhibition of cellular proliferation by the Wilm's tumor suppressor WT1 requires association with the inducible chaperone HSP70. *Genes Dev.* 12:1108–1120.

Manolis, S. C., G. J. W. Webb, and K. E. Dempsey. 1987. Crocodile egg chemistry. In *Wildlife management: Crocodiles and alligators*, ed. G. J. W. Webb, S. C. Manolis, and P. J. Whitehead, 445–472. Sydney: Surrey Beatty and Sons.

Manolis, S. C., G. J. W. Webb, and K. Richardson. 2000. Improving the quality of Australian crocodile skins. *Report to the Rural Industries Research and Development Corporation*. Publication No. 00/21.

Marcovaldi, M. A., M. Godfrey, and N. Mrosovsky. 1997. Estimating sex ratios of loggerhead turtles in Brazil from pivotal incubation durations. *Can. J. Zool.* 75:755–770.

Marshall, O. J., and V. R. Harley. 2001. Identification of an interaction between SOX9 and HSP70. *FEBS Lett.* 496:75–80.

Martineau, J., K. Nordqvist, C. Tilmann, R. Lovell-Badge, and B. Capel. 1997. Male-specific cell migration into the developing gonad. *Curr. Biol.* 7:958–968.

Martins, E. P. 1996. *Phylogenies and the comparative method in animal behavior*. New York: Oxford University Press.

Matsuda, M., Y. Nagahama, A. Shinomiya, T. Sato, C. Matsuda, T. Kobayashi, C. E. Morrey, et al. 2002. *DmY* is a Y-specific DM-domain gene required for male development in the medaka fish. *Nature* 417:559–563.

Matter, J. M., C. S. McMurray, A. B. Anthony, and R. L. Dickerson. 1998. Development and implementation of endocrine biomarkers of exposure and effects in American alligators *(Alligator mississippiensis)*. *Chemosphere* 37:1905–1914.

Mayer, L., S. Overstreet, C. Dyer, and C. Propper. 2002. Sexually dimorphic expression of steroidogenic factor 1 (SF-1) in developing gonads of the American bullfrog, *Rana catesbeiana*. *Gen. Comp. Endocrinol.* 127:40–47.

Mayhew, W. W. 1964. Taxonomic status of California populations of the lizard genus *Uma*. *Herpetologica* 20:170–183.

McBee, K., J. W. Bickham, and J. R. Dixon. 1987. Male heterogamety and chromosomal variation in Caribbean geckos. *J. Herpetol.* 21:68–71.

McCarrey, J. R., and U. K. Abbott. 1978. Chick gonad differentiation following excision of primordial germ cells. *Dev. Biol.* 66:256–265.

———. 1982. Functional differentiation of chick gonads following depletion of primordial germ cells. *J. Embryol. Exp. Morphol.* 68:161–174.

McDonald, J. H. 1999. Patterns of temperature adaptation in proteins from *Methanococcus* and *Bacillus*. *Mol. Biol. Evol.* 16:1785–1790.

McKnight, C. M., and W. H. N. Gutzke. 1993. Effects of the embryonic environment and of hatchling housing conditions on growth of the young snapping turtles *(Chelydra serpentina)*. *Copeia* 1993:475–482.

McLaren, A. 1995. Germ cells and germ cell sex. *Phil. Trans. Roy. Soc. Lond. B* 350:229–233.

McQueen, H. A., D. McBride, G. Miele, A. P. Bird, and M. Clinton. 2001. Dosage compensation in birds. *Curr. Biol.* 11:253–257.

Meeks, J. J., S. E. Crawford, T. A. Russell, K. I. Morohashi, J. Weiss, and J. L. Jameson. 2003. *Dax1* regulates testis cord organization during gonadal differentiation. *Development* 130:1029–1036.

Melville, J., J. A. Schulte, and A. Larson. 2001. A molecular phylogenetic study of ecological diversification in the Australian lizard genus *Ctenophorus*. *J. Exp. Zool.* 291:339–353.

Merchant-Larios, H. 2001. Temperature sex determination in reptiles: The third strategy. *J. Reprod. Dev.* 47:245–252.

Merchant-Larios, H., and N. Moreno-Mendoza. 1998. Mesonephric stromal cells differentiate into Leydig cells in the mouse fetal testis. *Exp. Cell. Res.* 244:230–238.

———. 2001. Onset of sex differentiation: Dialog between genes and cells. *Arch. Med. Res.* 32:553–558.

Métrailler, S., and G. Le Gratiet. 1996. Tortues continentales de Guyane française. Bramois, Switzerland: P.M.S. Editions.

Middaugh, D. P., and M. J. Hemmer. 1987. Influence of environmental temperature on sex-ratios in the tidewater silverside, *Menidia peninsulae* (Pisces: Atherinidae). *Copeia* 1987:958–964.

Milnes, M. R., R. N. Roberts, and L. J. Guillette. 2002. Effects of incubation temperature and estrogen exposure on aromatase activity in the brain and gonads of embryonic alligators. *Env. Health Perspect. Suppl. 3* 110:393–396.

Miranda, L. A., C. A. Strüssmann, and G. M. Somoza. 2001. Immunocytochemical identification of GtH1 and GtH2 cells during the temperature-sensitive period for sex determination in pejerrey, *Odontesthes bonariensis*. *Gen. Comp. Endocrinol.* 124: 45–52.

Miura, I., H. Ohtani, M. Nakamura, Y. Ichikawa, and K. Saitoh. 1998. The origin and differentiation of the heteromorphic sex chromosomes Z, W, X and Y in the frog *Rana rugosa*, inferred from the sequences of a sex-linked gene, ADP/ATP translocase. *Mol. Biol. Evol.* 15:1612–1619.

Miyamoto, N., M. Yoshida, S. Kuratani, I. Matsuo, and S. Aizawa. 1997. Defects of urogenital development in mice lacking *Emx2*. *Development* 124:1653–1664.

Miyata, S., K. Miyashita, and Y. Hosoyama. 1996. *Sry*-related genes in *Xenopus* oocytes. *Biochim. Biophys. Acta* 1308:23–28.

Mizusaki, H., K. Kawabe, T. Mukai, E. Ariyoshi, M. Kasahara, H. Yoshioka, A. Swain, et al. 2003. *Dax-1* (dosage-sensitive sex reversal-adrenal hypoplasia congenita critical region on the X Chromosome, gene 1) gene transcription is regulated by *Wnt4* in the female developing gonad. *Mol. Endocrinol.* 17:507–519.

Moffat, L. A. 1985. Embryonic development and aspects of reproductive biology in the tuatara, *Sphenodon punctatus*. In *Biology of the Reptilia*, vol. 14, ed. C. Gans, F. Billett, and P. F. A. Maderson, 494–521.

Mohanty-Hejmadi, P., S. K. Dutta, D. Dey, D. P. Rath, R. L. Rath, and S. Kar. 1999. Temperature-dependent sex determination in the salt-water crocodile, *Crocodylus porosus* Schneider. *Curr. Sci.* 76:695–696

Moniot, B., P. Berta, G. Scherer, P. Sudbeck, and F. Poulat. 2000. Male specific expression suggests role of *Dmrt1* in human sex determination. *Mech. Dev.* 91:323–325.

Montgomery, J. C., and J. A. MacDonald. 1990. Effects of temperature on nervous system: Implications for behavioral performance. *Am. J. Physiol.* 259:R191–R196.

Morales-Ramos, J. A., and J. R. Cate. 1993. Temperature-dependent development rates of *Catolaccus grandis* (Hymenoptera: Pteromalidae). *Environ. Entomol.* 22:226–233.

Moran, N. 1992. The evolutionary maintenance of alternative phenotypes. *Am. Nat.* 139:971–989.

Moreira, G. R. S. 1991. Observações sobre *Geochelone denticulata* (Linnaeus 1766) e *Geochelone carbonaria* (Spix 1824) na bacia do rio Uatumã, Amazônia central. *Bol. Mus. para Emilio Goeldi, Zool.* 7:183–188.

Morjan, C. L. 2002. Temperature-dependent sex determination and the evolutionary potential for sex ratio in the painted turtle, *Chrysemys picta*. Unpubl. Ph.D. thesis, Iowa State University, Ames, Iowa.

———. 2003. How rapidly can maternal behavior affecting primary sex ratio evolve in a reptile with enviornmental sex determination? *Am. Nat.* 162:205–219.

Morreale, S., G. Ruiz, J. Spotila, and E. Standora. 1982. Temperature-dependent sex determination: Current practices threaten conservation of sea turtles. *Science* 216:1245–1247.

Morrish, B. C., and A. H. Sinclair. 2002. Vertebrate sex determination: Many means to an end. *Reproduction* 124:447–457.

Morrison, S. F., J. S. Keogh, and I. A. W. Scott. 2002. Molecular determination of paternity in a natural population of the multiply mating polygynous lizard *Eulamprus heatwolei*. *Mol. Ecol.* 11:535–545.

Moskovits, D. K. 1985. The behavior and ecology of the two Amazonian tortoises, *Geochelone carbonaria* and *Geochelone denticulata*, in Northwestern Brasil, 328. University of Chicago.

Mrosovsky, N. 1980. Thermal biology of sea turtles. *Am. Zool.* 20:531–547.

———. 1988. Pivotal temperatures for loggerhead turtles *(Caretta caretta)* from northern and southern nesting beaches. *Can. J. Zool.* 66:661–669.

———. 1994. Sex ratio of sea turtles. *J. Exp. Zool.* 270:16–27.

Mrosovsky, N., and M. H. Godfrey. 1995. Manipulating sex ratios: Turtle speed ahead! *Chelon. Cons. Biol.* 1:238–240.

Mrosovsky, N., and C. Pieau. 1991. Transitional range of temperature, pivotal temperatures and thermosensitive stages for sex determination in reptiles. *Amphibia-Reptilia* 12:169–179.

Mrosovsky, N., and J. Provancha. 1992. Sex ratio of hatchling loggerhead sea turtles: Data and estimates from a 5-year study. *Can. J. Zool.* 70:530–538.

Mrosovsky, N., P. H. Dutton, and C. P. Whitmore. 1984a. Sex ratios of two species of sea turtle nesting in Suriname. *Can. J. Zool.* 62:2227–2239.

Mrosovsky, N., S. R. Hopkins-Murphy, and J. I. Richardson. 1984b. Sex ratio of sea turtles: Seasonal changes. *Science* 225: 739–741.

Mrosovsky, N., A. Bass, L. A. Corliss, J. I. Richardson, and T. H. Richardson. 1992. Pivotal temperatures for hawksbill turtles nesting in Antigua. *Can. J. Zool.* 70:1920–1925.

Mrosovsky, N., C. Baptistotte, and M. Godfrey. 1999. Validation of incubation duration as an index of the sex ratio of hatchling sea turtles. *Can. J. Zool.* 77:831–835.

Muller, H. J. 1914. A gene for the fourth chromosome of Drosophila. *J. Exp. Zool.* 17:325–336.

Murdock, C., and T. Wibbels. 2002. Steroidogenic factor-1 mRNA levels in the embryonic adrenal/kidney/gonadal complexes of *Trachemys scripta*. *2002 Final Program and Abstracts*, 314. SICB Annual Meeting, Anaheim, CA. [Abstract]

Murray, J. D., D. C. Deeming, and M. W. J. Ferguson. 1990. Size dependent pigmentation-pattern formation in embryos of *Alligator mississippiensis*: Time of initiation of pattern generation mechanism. *Proc. Roy. Soc. Lond. B* 239:279–293.

Murrish, D. E., and V. J. Vance. 1968. Physiological responses to temperature acclimation in the lizard *Uta mearnsi*. *Comp. Biochem. Physiol.* 27:329–337.

Muth, A. 1980. Physiological ecology of desert iguana *(Dipsosaurus dorsalis)* eggs: Temperature and water relations. *Ecology* 61: 1335–1343.

Muth, A., and J. J. Bull. 1981. Sex determination in desert iguanas: Does incubation temperature make a difference? *Copeia* 1981:869–870.

Muthukkaruppan, V., V. Kanakambika, V. Manickavel, and K. Veeraraghavan. 1970. Analysis of the development of the lizard, *Calotes versicolor*. *J. Morphol.* 130:479–490.

Muto, Y. 1961. The gonad of the toad, *Bufo vulgaris formosus*, cultured at high temperature. *Bull. Aichi Gakugei Univ.* 10:97–109.

Nager, R. G., P. Monaghan, R. Griffiths, D. C. Houston, and R. Dawson. 1999. Experimental demonstration that offspring sex ratio varies with maternal condition. *Proc. Natl. Acad. Sci. USA* 96:570–573.

Nakabayashi, O., H. Kikuchi, T. Kikuchi, and S. Mizuno. 1998. Differential expression of genes for aromatase and estrogen receptor during the gonadal development in chicken embryos. *J. Mol. Endocrinol.* 20:193–202.

Nakamura, M., T. Kobayashi, X. T. Chang, and Y. Nagahama. 1998. Gonadal sex differentiation in teleost fish. *J. Exp. Zool.* 281:362–372.

Nakayama, Y., T. Yamamoto, Y. Matsuda, and S. I. Abe. 1998. Cloning of cDNA for newt *Wt1* and the differential expression during spermatogenesis of the Japanese newt *Cynops pyrrhogaster*. *Dev. Growth Differ.* 40:599–608.

Nanda, I., E. Zend-Ajusch, Z. Shan, F. Grutzner, M. Schartl, D.W. Burt, M. Koehler, et al. 2000. Conserved synteny between the chicken Z sex chromosome and human chromosome 9 includes the male regulatory gene *Dmrt1:* A comparative (re)view on avian sex determination. *Cytogenet. Cell. Genet.* 89:67–78.

Nanda, I., M. Kondo, U. Hornung, S. Asakawa, C. Winkler, A. Shimizu, Z. Shan, et al. 2002. A duplicated copy of *Dmrt1* in the sex-determining region of the Y chromosome of the medaka, *Oryzias latipes*. *Proc. Natl. Acad. Sci. USA* 99:11778–11783.

Nanda, I., U. Hornung, M. Kondo, M. Schmid, and M. Schartl. 2003. Common spontaneous sex-reversed XX males of the Medaka *Oryzias latipes*. *Genetics* 163:245–251.

Navas, C. A. 1996. Metabolic physiology, locomotor performance, and thermal niche breadth in neotropical anurans. *Physiol. Zool.* 69:1481–1501.

———. 1997. Thermal extremes at high elevations in the Andes: Physiological ecology of frogs. *J. Therm. Biol.* 22:467–477.

Nelson, N. J. 1998. *Conservation of Brothers Island Tuatara, Sphenodon guntheri*. Unpubl. M.Con.Sc. thesis, Victoria University of Wellington, New Zealand.

———. 2001. *Temperature-dependent sex determination and artificial incubation of Tuatara, Sphenodon punctatus*. Unpubl. Ph.D. thesis, Victoria University of Wellington, New Zealand.

Newman, D. G. 1982. Tuatara, *Sphenodon punctatus*, and burrows, Stephens Island. In *New Zealand herpetology*, ed. D. G. Newman, 213–223. New Zealand Wildlife Service Occasional Publication No. 2.

Nomura, O., O. Nakabayashi, K. Nishimori, H. Yasue, and S. Mizuno. 1999. Expression of five steroidogenic genes including aromatase gene at early developmental stages of chicken male and female embryos. *J. Steroid Biochem. Mol. Biol.* 71:103–109.

Nomura, T., K. Arai, T. Hayashi, and R. Suzuki. 1998. Effect of temperature on sex ratios of normal and gynogenetic diploid leach. *Fish. Sci.* 64:753–758.

Noonan, B. P. 2000. Does the phylogeny of pelomedusoid turtles reflect vicariance due to continental drift? *J. Biogeogr.* 27:1245–1249.

O'Connell, D.J. 1998. The adaptive significance of environmental sex determination in the common snapping turtle, *Chelydra serpentina*. Ph.D dissertation, University of Texas–Arlington.

Ohe, K., E. Lalli, and P. Sassone-Corsi. 2002. A direct role of SRY and SOX proteins in pre-mRNA splicing. *Proc. Natl. Acad. Sci. USA* 99:1146–1151.

Ohno, S. 1967. *Sex chromosomes and sex-linked genes*. Berlin: Springer-Verlag.

———. 1979. *Major sex-determining genes*. Berlin: Springer-Verlag.

Olmo, E. 1986. Chordata 3, A. Reptilia. In *Animal cytogenetics*, vol. 4, ed. B. John, H. Bauer, H. Kayano, and A. Levan,1–100. Berlin, Germany: Gebrüder Borntraeger.

Olmo, E., G. Odierna, T. Capriglione, and A. Cardone. 1990. DNA and chromosome evolution in lacertid lizards. In *Cytogenetics of amphibians and reptiles*, ed. E. Olmo, 181–204. Boston: Birkhäuser Verlag.

Olsson, M. 1992. Contest success in relation to size and residency in male sand lizards *(Lacerta agilis)*. *Anim. Behav.* 44:386–388.

Olsson, M., and R. Shine. 2001. Facultative sex allocation in snow skink lizards *(Niveoscincus microlepidotus)*. *J. Evol. Biol.* 14:120–128.

Olsson, M., A. Gullberg, R. Shine, and H. Tegelstrom. 1994. Sperm competition in the sand lizard, *Lacerta agilis*. *Anim. Behav.* 48:93–200.

Olsson, M., A. Gullberg, R. Shine, T. Maden, and H. Tegelstrom. 1996. Paternal genotype influences incubation period, offspring size, and offspring shape in an oviparous reptile. *Evolution* 50:1328–1333.

O'Neill M., M. Binder, C. Smith, J. Andrews, K. Reed, M. Smith, C. Millar, et al. 2000. *Asw:* A gene with conserved avian W-linkage and female specific expression in chick embryonic gonad. *Dev. Genes Evol.* 210:243–249.

Oreal, E., C. Pieau, M. G. Mattei, N. Josso, J. Y. Picard, D. Carre-Eusebe, and S. Magre. 1998. Early expression of *Amh* in chicken embryonic gonads precedes testicular *Sox9* expression. *Dev. Dyn.* 212:522–532.

Osgood, D. W. 1970. Thermoregulation in water snakes studied by telemetry. *Copeia* 1970:568–571.

———. 1978. Effects of temperature on development of meristic characters in *Natrix fasciata*. *Copeia* 1978:33–37.

O'Steen, S. 1998. Embryonic temperature influences juvenile temperature choice and growth rate in snapping turtles, *Chelydra serpentina*. *J. Exp. Biol.* 201:439–449.

O'Steen, S., and F. J. Janzen. 1999. Embryonic temperature affects metabolic compensation and thyroid hormones in hatchling snapping turtles. *Physiol. Biochem. Zool.* 72:520–533.

Overall, K. L. 1994. Lizard egg environments. In *Lizard ecology: Historical and experimental perspectives*, ed. L. J. Vitt and E. R. Pianka, 51–72. Princeton, NJ: Princeton University Press.

Pace, H. C., and C. Brenner. 2003. Feminizing chicks: A model for avian sex determination based on titration of HINT enzyme activity and the predicted structure of an ASW-HINT heterodimer. *Genome Biol.* 4:R18:11–R18:16.

Packard, G. C., and M. J. Packard. 1990. Patterns of survival at subzero temperatures by hatchling painted turtles and snapping turtles. *J. Exp. Zool.* 254:23–236.

———. 1993. Hatchling painted turtles *(Chrysemys picta)* survive exposure to subzero temperatures during hibernation by avoiding freezing. *J. Comp. Physiol. B* 163:147–152.

———. 2001. Environmentally induced variation in size, energy reserves and hydration of hatchling painted turtles, *Chrysemys picta*. *Funct. Ecol.* 15:481–489.

Packard, M. J., G. C. Packard, J. D. Miller, M. E. Jones, and W. H. N. Gutzke. 1985. Calcium mobilization, water balance, and growth in embryos of the agamid lizard *Amphibolurus barbatus*. *J. Exp. Zool.* 235:349–357.

Packard, G. C., M. J. Packard, K. Miller, and T. J. Boardman. 1987. Influence of moisture, temperature, and substrate on snapping turtle eggs and embryos. *Ecology* 68:983–993.

———. 1988. Effects of temperature and moisture during incubation on carcass composition of hatchling snapping turtles *(Chelydra serpentina)*. *J. Comp. Physiol. B* 158:117–125.

Packard, G. C., M. J. Packard, and G. F. Birchard. 1989. Sexual differentiation and hatching success by painted turtles incubating in different thermal and hydric environments. *Herpetologica* 45:385–392.

Packard, G. C., M. J. Packard, and L. Benigan. 1991. Sexual differentiation, growth, and hatchling success by embryonic painted turtles incubated in wet and dry environments at fluctuating temperatures. *Herpetologica* 47:125–132.

Page, R. D. M., and E. C. Holmes. 1998. *Molecular evolution: A phylogenetic approach*, Oxford: Blackwell Science.

Pagel, M. 1999. The maximum likelihood approach to reconstructing ancestral character states of discrete characters on phylogenies. *Syst. Biol.* 48:612–622.

Paladino, F. V. 1985. Temperature effects on locomotion and activity bioenergetics of amphibians, reptiles, and birds. *Am. Zool.* 25:965–972.

Parker, K. L., and B. P. Schimmer. 2002. Genes essential for early events in gonadal development. *Ann. Med.* 34:171–178.

Patiño, R., K. B. Davis, J. E. Schoore, C. Uguz, C. A. Strüssmann, N. C. Parker, B. A. Simco, and C. A. Goudie. 1996. Sex differentiation of channel catfish gonads: Normal development and effects of temperature. *J. Exp. Zool.* 276:209–218.

Paukstis, G. L., and F. J. Janzen. 1990. Sex determination in reptiles: Summary of effects of constant temperatures of incubation on sex ratios of offspring. *Smithsonian Herpetol. Info. Serv.* 83:1–28.

Paukstis, G. L., W. H. N. Gutzke, and G. C. Packard. 1984. Effects of substrate water potential and fluctuating temperatures on sex ratios of hatchling painted turtles *Chrysemys picta*. *Can. J. Zool.* 62:1491–1494.

Pavlidis, M., G. Koumoundouros, A. Sterioti, S. Somarakis, P. Divanach, and M. Kentouri. 2000. Evidence of temperature-dependent sex determination in the European sea bass (*Dicentrarchus labrax* L.). *J. Exp. Zool.* 287:225–232.

Pearson, O. P. 1977. The effect of substrate and of skin color on thermoregulation of a lizard. *Comp. Biochem. Physiol. A* 58: 353–358.

Peccinini-Seale, D., and T. M. B. de Almeida. 1986. Chromosomal variation, nucleolar organizers and constitutive heterochromatin in the genus *Ameiva* and *Cnemidophorus* (Sauria: Teiidae). *Caryologia* 39:227–237.

Penrad-Mobayed, M., N. Moreau, and N. Angelier. 1998. Evidence for specific RNA/protein interactions in the differential segment of the W chromosome in the amphibian *Pleurodeles waltl. Dev. Growth Differ.* 40: 147–156.

Perret, M. 1996. Manipulation of sex ratio at birth by urinary cues in a prosimian primate. *Behav. Ecol. Sociobiol.* 38:259–266.

Petrie, M., H. Schwabl, N. Brande-Lavridsen, and T. Burke. 2001. Sex differences in avian yolk hormone levels. *Nature* 412:498.

Phelps, F. M. 1992. Optimal sex-ratio as a function of egg incubation-temperature in the crocodilians. *Bull. Math. Biol.* 54:123–148.

Phillips, J. A., and G. C. Packard. 1994. Influence of temperature and moisture on eggs and embryos of the white-throated savanna monitor *Varanus albigularis*: Implications for conservation. *Biol. Cons.* 69:131–136.

Pieau, C. 1971. Sur la proportion sexuelle chez les embryons de deux Chéloniens (*Testudo graeca* L. et *Emys orbicularis* L.): Issues d'oeufs incubés artificiellement. *C. R. Acad. Sci. Paris D* 272:3071–3074.

———. 1972. Effets de la température sur le développement des glandes génitales chez les embryons de deux Chéloniens, *Emys orbicularis* L. et *Testudo graeca* L. *C. R. Acad. Sci. Paris D.* 274: 719–722.

———. 1974. Sur la différenciation sexuelle chez des embryons d'*Emys orbicularis* L. (Chélonien) issus d'oeufs incubés dans le sol au cours de l'été 1973. *Bull. Soc. Zool. France* 99:363–376.

———. 1982. Modalities of the action of temperature on sexual differentiation in field-developing embryos of the European pond turtle Emys orbicularis (Emydidae). *J. Exp. Zool.* 220: 353–360.

———. 1996. Temperature variation and sex determination in reptiles. *Bioessays* 18:19–26.

Pieau, C., and M. Dorizzi. 1981. Determination of temperature sensitive stages for sexual differentiation of the gonads in embryos of the turtle, *Emys orbicularis*. *J. Morphol.* 170:373–382.

Pieau, C., M. Dorizzi, N. Richard-Mercier, and G. Desvasges. 1998. Sexual differentiation of gonads as a function of temperature in the turtle *Emys orbicularis*: Endocrine function, intersexuality and growth. *J. Exp. Zool.* 281:400–408.

Pieau, C., M. Dorizzi, and N. Richard-Mercier. 1999a. Temperature-dependent sex determination and gonadal differentiation in reptiles. *Cell. Mol. Life Sci.* 55:887–900.

Pieau, C., B. Belaïd, M. Dorizzi, and N. Richard-Mercier. 1999b. Intersexuality in turtles with temperature-dependent sex determination. In *Current studies in herpetology*, ed. C. Miaud and R. Guyétant, 17–21. Le Bourget du Lac, France: Societas Europaea Herpetologica (SEH).

Pieau, C., M. Dorizzi, and N. Richard-Mercier. 2001. Temperature-dependent sex determination and gonadal differentiation in reptiles. In *Genes and mechanisms in vertebrate sex determination*, ed. G. Scherer and M. Schmid, EXS 91:118–141. Boston: Birkhäuser.

Pierucci-Alves, F., A. M. Clark, and L. D. Russell. 2001. A developmental study of the desert hedgehog-null mouse testis. *Biol. Reprod.* 65:1392–1402.

Piferrer, F. 2001. Endocrine sex control strategies for the feminization of teleost fish. *Aquaculture* 197:229–281.

Pinheiro, M., G. Mourão, Z. Campos, and M. Coutinho.1997. Influence of the incubation temperature on the sex determination of the caiman *(Caiman crocodilus yacare)*. *Rev. Brasil Biol.* 57:383–391.

Piquet, J. 1930. Détermination du sexe chez les batraciens en fonction de la température. *Rev. Suisse Zool.* 37:173–281.

Place, A. R., J. Lang, S. Gavasso, and P. Jeyasuria. 2001. Expression of P450 (arom) in *Malaclemys terrapin* and *Chelydra serpentina:* A tale of two sites. *J. Exp. Zool.* 290:673–690.

Pomerol, C., and I. Premoli Silva. 1986. The Eocene-Oligocene transition: Events and boundary, In *Terminal Eocene events,* ed. C. Pomerol and I. Premoli Silva, 1–24. Amsterdam: Elsevier.

Porter, W. P., and C. R. Tracy. 1974. Modeling the effects of temperature changes on the ecology of the garter snake and leopard frog. *J. Heat Transfer* 96: 594–609.

Post, E., M. C. Forchhammer, N. C. Stenseth, and R. Langvatn. 1999. Extrinsic modification of vertebrate sex ratios by climatic variation. *Am. Nat.* 154:194–204.

Pough, F. H. 1980. The advantages of ectothermy for tetrapods. *Am. Nat.* 115:92–112.

Pough, F. H., R. M. Andrews, J. E. Cadle, M. L. Crump, A. H. Savitzky, and K. D. Wells. 1998. *Herpetology.* Upper Saddle River, NJ: Prentice Hall.

———. 2001. *Herpetology.* Upper Saddle River, NJ: Prentice Hall.

Preest, M. R., and F. H. Pough. 1989. Interaction of temperature and hydration on locomotion of toads. *Funct. Ecol.* 3:693–699.

Pritchard, P. C. H. 1976. Post-nesting movements of marine turtles (Cheloniidae and Dermochelyidae) tagged in the Guianas. *Copeia* 1976:749–754.

Qualls, C. P., and R. M. Andrews. 1999. Cold climates and the evolution of viviparity in reptiles: Cold incubation temperatures produce poor-quality offspring in the lizard, *Sceloporus virgatus*. *Biol. J. Linn. Soc.* 67:353–376.

Qualls, C. P., and R. Shine. 1996. Reconstructing ancestral reaction norms: An example using the evolution of reptilian viviparity. *Funct. Ecol.* 10:688–697.

Radder, R. S., B. A. Shanghag, and S. K. Saidapur. 2002. Influence of incubation temperature and substrate on eggs and embryos of the garden lizard, *Calotes versicolor* (Daud.). *Amphibia-Reptilia* 23:71–82.

Rage, J. C. 1998. Latest Cretaceous extinctions and environmental sex determination in reptiles. *Bull. Soc. Géol. France* 169:479–483.

Ragland, I. M., L. C. Wit, and J. C. Sellers. 1981. Temperature acclimation in the lizards *Cnemidophorus sexlineatus* and *Anolis carolinensis*. *Comp. Biochem. Physiol. A* 70:33–36.

Rajaguru, S. 2002. Thermal resistance time of estuarine fishes *Etroplus suratensis* and *Therapon jarbua*. *J. Therm. Biol.* 27:121–124.

Rao, K. P., and T. H. Bullock. 1954. Q_{10} as a function of size and habitat temperature in poikilotherms. *Am. Nat.* 88:33–44.

Raymond, C. S., C. E. Shamu, M. M. Shen, K. J. Seifert, B. Hirsch, J. Hodgkin, and D. Zarkower. 1998. Evidence for evolutionary conservation of sex-determining genes. *Nature* 391:691–695.

Raymond, C. S., J. R. Kettlewell, B. Hirsch, V. J. Bardwell, and D. Zarkower. 1999a. Expression of *Dmrt* in the genital ridge of mouse and chicken embryos suggests a role in vertebrate sexual development. *Dev. Biol.* 215:208–220.

Raymond, C. S., E. D. Parker, J. R. Kettlewell, L. G. Brown, D. C. Page, K. Kusz, J. Jaruzelska, et al. 1999b. A region of human chromosome 9p required for testis development contains two genes related to known sexual regulators. *Hum. Mol. Genet.* 8:989–996.

Raymond, C. S., M. W. Murphy, M. G. O'Sullivan, V. J. Bardwell, and D. Zarkower. 2000. *Dmrt1*, a gene related to worm and fly sexual regulators, is required for mammalian testis differentiation. *Genes Dev.* 14:2587–2595.

Raynaud, A., and C. Pieau. 1985. Embryonic development of the genital system. In *Biology of the Reptilia,* vol. 15, ed. C. Gans and F. Billett, 149–300, New York: Wiley-Interscience.

Reddy, J. C., and J. D. Licht. 1996. The *Wt1* Wilm's tumor suppressor gene: How much do we really know? *Biochim. Biophys. Acta* 1287:1–28.

Reece, S. E., A. C. Broderick, B. J. Godley, and S. A. West. 2002. The effects of incubation environment, sex and pedigree on the hatchling phenotype in a natural population of loggerhead turtles. *Evol. Ecol. Res.* 4:737–748.

Reibisch, J. 1902. Uber den Einfluss der Temperatur auf die Entwicklung von Fischeiern. *Wiss. Meeresuntersuch* 2:213–231.

Reichling, S. B., and W .H. N. Gutzke. 1996. Phenotypic consequences of incubation environment in the African elapid genus *Aspidelaps*. *Zoo Biol.* 15:301–308.

Reinhold, K. 1998. Nest-site philopatry and selection for environmental sex determination. *Evol. Ecol.* 12:245–250.

Rhen, T., and D. Crews. 1999. Embryonic temperature and gonadal sex organize male-typical sexual and aggressive behavior in a lizard with temperature-dependent sex determination. *Endocrinology* 140:4501–4508.

———. 2000. Organization and activation of sexual and agonistic behavior in the leopard gecko, *Eublepharis macularius*. *Neuroendocrinology* 71:252–261.

Rhen, T., and J. W. Lang. 1994. Temperature-dependent sex determination in the snapping turtle: Manipulation of the embryonic sex steroid environment. *Gen. Comp. Endocrinol.* 96: 234–254.

———. 1995. Phenotypic plasticity for growth in the common snapping turtle: Effects of incubation temperature, clutch, and their interaction. *Am. Nat.* 146:726–747.

———. 1998. Among-family variation for environmental sex determination in reptiles. *Evolution* 52:1514–1520.

———. 1999a. Embryonic and juvenile temperature independently influence growth in hatchling snapping turtles, *Chelydra serpentina*. *J. Therm. Biol.* 24: 33–41.

———. 1999b. Incubation temperature and sex affect mass and energy reserves of hatchling snapping turtles *(Chelydra serpentina)*. *Oikos* 86:311–319.

Rhen, T., P. K. Elf, A. J. Fivizzani, and J. W. Lang. 1996. Sex-reversed and normal turtles display similar sex steroid profiles. *J. Exp. Zool.* 274:221–226.

Rhen, T., E. Willingham, J. T. Sakata, and D. Crews. 1999. Incubation temperature influences sex steroid levels in juvenile red-eared slider turtles, *Trachemys scripta*, a species with temperature-dependent sex determination. *Biol. Reprod.* 61: 1275–1280.

Rhodes, W. E., and J. W. Lang. 1995. Sex ratios of naturally incubated alligator hatchlings: Field techniques and initial results. *Proc. Ann. Conf. Southeast. Assoc. Fish Wildl. Agencies,* 49:640–646.

———. 1996. Alligator nest temperatures and hatchling sex ratios in coastal South Carolina. *Proc. Ann. Conf. Southeast. Assoc. Fish Wildl. Agencies,* 50:520–531.

Rimblot-Baly, F., J. Lescure, J. Fretey, and C. Pieau. 1986. Sensibilité à la température de la différenciation sexuelle chez la tortue Luth, *Dermochelys coriacea* (Vandelli 1761): Application des données de l'incubation artificielle à l'étude de la sex-ratio dans la nature. *Ann. Sci. Nat. Zool.* 13: 277-290.

Robert, K. A., and M. B. Thompson. 2001. Sex determination: Viviparous lizard selects sex of embryos. *Nature* 412:698–699.

Robert, K. A., M. B. Thompson, and F. Seebacher. 2003. Facultative sex allocation in the viviparous lizard *Eulamprus typanum,* a species with temperature-dependent sex determination. *Aust. J. Zool.* 51:367–370.

Roff, D.A. 1996. The evolution of threshold traits in animals. *Q. Rev. Biol.* 71:3–35.

Rogulska, T., W. Ozdzenski, and A. Komar. 1971. Behaviour of mouse primordial germ cells in the chick embryo. *J. Embryol. Exp. Morphol.* 25:155–164.

Rol'nik, V. V. 1970. *Bird embryology*. Jerusalem: Weiner Bindery.

Rome, L. C. 1983. The effect of long-term exposure to different temperatures on the mechanical performance of frog muscle. *Physiol. Zool.* 56:33–40.

———. 1990. Influence of temperature on muscle recruitment and muscle function in vivo. *Am. J. Physiol.* 259:R210–R222.

Römer, U., and W. Beisenherz. 1996. Environmental determination of sex in *Apistogramma* (Cichlidae) and two other freshwater fishes (Teleostei). *J. Fish Biol.* 48:714–725.

Roosenburg, W. M. 1996. Maternal condition and nest site choice: An alternative for the maintenance of environmental sex determination? *Am. Zool.* 36:157–168.

Roosenburg, W. M., and K. C. Kelley. 1996. The effect of egg size and incubation temperature on growth in the turtle, *Malaclemys terrapin*. *J. Herpetol.* 30:198–204.

Roosenburg, W. M., and P. Niewiarowski. 1998. Maternal effects and the maintenance of environmental sex determination. In *Maternal effects as adaptations*, ed. T. A. Musseau and C.W. Fox, 307–322. New York: Oxford University Press.

Rubin, D. A. 1985. Effect of pH on sex ratios in cichlids and a poecilliid (Teleostei). *Copeia* 1985:233–235.

Rudloff, W. 1981. *World-climates*. Stuttgart: Wissenschaftlische Verlagsgesellschaft.

Ryan, K. M. 1990. Effects of egg incubation condition on the post-hatching growth and performance of the snapping turtle, *Chelydra serpentina*. M.A. thesis, State University of New York College–Buffalo.

Saillant, E., A. Fostier, P. Haffray, B. Menu, J. Thimonier, and B. Chatain. 2002. Temperature effects and genotype-temperature interactions on sex determination in the European sea bass (*Dicentrarchus labrax* L.). *J. Exp. Zool.* 292:494–505.

Sakata, J. T., T. Rhen, and D. Crews. 1998. Ontogeny of secondary sex structures and gonadal steroids in the leopard gecko. *Am. Zool.* 38:86A. [Abstract 297]

Sakata, J. T., P. Coomber, F. Gonzalez-Lima, and D. Crews. 2000. Functional connectivity among limbic brain areas: Differential effects of incubation temperature and gonadal sex in the leopard gecko, *Eublepharis macularius*. *Brain Behav. Evol.* 55:139–151.

Salame-Mendez A., J. Herrera-Munoz, N. Moreno-Mendoza, and H. Merchant-Larios. 1998. Response of diencephalon but not the gonad to female-promoting temperature with elevated estradiol levels in the sea turtle *Lepidochelys olivacea*. *J. Exp. Zool.* 280:304–313.

SAS Institute. 1988. *SAS/STAT User's guide, Release 6.03 Edition*. Cary, NC: SAS Institute, Inc.

Scheib, D. 1983. Effects and role of estrogens in avian gonadal differentiation. Differentiation 23(Suppl.):S87–S92.

Scherer, G. 2002. The molecular genetic jigsaw puzzle of vertebrate sex determination and its missing pieces. *Novartis Found. Symp.* 244:225–236; discussion 236–229, 253–227.

Scherrer, S. P., D. A. Rice, and L. L. Heckert. 2002. Expression of steroidogenic factor 1 in the testis requires an interactive array of elements within its proximal promoter. *Biol. Reprod.* 67: 1509–1521.

Schmahl, J., E. M. Eicher, L. L. Washburn, and B. Capel. 2000. *Sry* induces cell proliferation in the mouse gonad. *Development* 127:65–73.

Schmid, M., and C. Steinlein. 2001. Sex chromosomes, sex-linked genes, and sex determination in the vertebrate class Amphibia. In *Genes and mechanisms in vertebrate sex determination*, ed. G. Scherer and M. Schmid, EXS 91:143–176. Berlin: Birkhäuser Verlag.

Schmidt, W., K. Tamm, and E. Wallikewitz. 1994a. *Chameleons*. Vol. 2, *Care and breeding*. Neptune City, NJ: TFH Publications Inc.

———. 1994b. *Chameleons*. Vol. 1, *Species*. Neptune City, NJ: TFH Publications Inc.

Schoolfield, R. M., P. J. H. Sharpe, and C. E. Magnuson. 1981. Non-linear regression of biological temperature-dependent rate models based on absolute reaction-rate theory. *J. Theor. Biol.* 88:719–731.

Schultz, R. J. 1993. Genetic regulation of temperature-mediated sex ratios in the livebearing fish *Poeciliopsis lucida*. *Copeia* 1993: 1148–1151.

Schulz, J. P. 1975. Sea turtles nesting in Surinam. *Zool. Verh.* 143:1–143.

Schwabl, H. 1993. Yolk is a source of maternal testosterone for developing birds. *Proc. Natl. Acad. Sci. USA* 90:11446–11450.

———. 1996a. Maternal testosterone in the avian egg enhances postnatal growth. *Comp. Biochem. Physiol. A* 114:271–276.

———. 1996b. Environment modifies the testosterone levels of a female bird and its eggs. *J. Exp. Zool.* 276:157–163.

Schwabl, H., D. W. Mock, and J. A. Gieg. 1997. A hormonal mechanism for parental favouritism. *Nature* 386:231.

Schwartzkopf, L., and R. J. Brooks. 1985. Sex determination in northern painted turtles: Effect of incubation at constant and fluctuating temperatures. *Can. J. Zool.* 63:2543–2547.

Schwerdtfeger, W. 1976. *World survey of climatology*, vol. 12. Amsterdam: Elsevier Scientific.

Scott, J. R., C. R. Tracy, and D. Pettus. 1982. A biophysical analysis of daily and seasonal utilization of climate space by a montane snake. *Ecology* 63:482–493.

Seipp, R., and F. W. Henkel. 2000. *Rhacodactylus: Biology, natural history and husbandry*. Frankfurt am Main, Germany: Edition Chimaira.

Serb, J. M., C. A. Phillips, and J. B. Iverson. 2001. Molecular phylogeny and biogeography of *Kinosternon flavescens* based on complete mitochondrial control region sequences. *Mol. Phylogenet. Evol.* 18:149–162.

Shaffer, H. B., P. Meylan, and M. L. McKnight. 1997. Tests of turtle phylogeny: Molecular, morphological and paleontological approaches. *Syst. Biol.* 46:235–268.

Sharma, P. M., M. Bowman, S. L. Madden, F. J. Rauscher III, and S. Sukumar. 1994. RNA editing in the Wilm's tumor susceptibility gene, *Wt1*. *Genes Dev.* 8:720–731.

Sharpe, P. J. H., and D. W. DeMichele. 1977. Reaction kinetics of poikilotherm development. *J. Theor. Biol.* 64:649–670.

Shaver, D. J., D. W. Owens, A. H. Chaney, C. W. Caillouet, R. Burchfield, Jr., and R. Marquez. 1988. Styrofoam box and beach temperatures in relation to incubation and sex ratios of Kemp's ridley sea turtles. In *Proceedings of the 8th Annual Workshop on Sea Turtle Conservation and Biology*, ed. B. Schroeder, 103–108. NOAA Technical Memorandum NMFS-SEFC-214.

Shaw, R. F., and J. D. Mohler. 1953. The selective significance of the sex ratio. *Am. Nat.* 87:337–342.

Shawlot, W., and R. R. Behringer. 1995. Requirement for *Lim1* in head-organizer function. *Nature* 374:425–430.

Shibata, K., M. Takase, and M. Nakamura. 2002. The *Dmrt1* expression in sex-reversed gonads of amphibians. *Gen. Comp. Endocrinol.* 127:232–241.

Shine, R. 1983. Reptilian reproductive modes: The oviparity-viviparity continuum. *Herpetologica* 39:1–8.

———. 1995. A new hypothesis for the evolution of viviparity in reptiles. *Am. Nat.* 145:809–823.

———. 1999. Why is sex determined by nest temperature in many reptiles? *Trends Ecol. Evol.* 14:186–189.

Shine, R., and M. J. Elphick. 2000. The effect of short-term weather fluctuations on temperatures inside lizard nests, and on the phenotypic traits of hatchling lizards. *Biol. J. Linn. Soc.* 72:555–565.

Shine, R., and P. S. Harlow. 1996. Maternal manipulation of offspring phenotypes via nest-site selection in an oviparous lizard. *Ecology* 77:1808–1817.

Shine, R., M. J. Elphick, and P. S. Harlow. 1995. Sisters like it hot. *Nature* 378:451–452.

———. 1997a. The influence of natural incubation environments on the phenotypic traits of hatchling lizards. *Ecology* 78:2559–2568.

Shine, R., T. R. L. Madsen, M. J. Elphick, and P. S. Harlow. 1997b. The influence of nest temperature and maternal brooding on hatchling phenotypes in water phytons. *Ecology* 78:1713–1721.

Shine, R., M. J. Elphick, and S. Donnellan. 2002. Co-occurrence of multiple, supposedly incompatible modes of sex determination in a lizard population. *Ecol. Lett.* 5:486–489.

Shoemaker, V. H., M. A. Baker, and J. P. Loveridge. 1989. Effect of water balance on thermoregulation in waterproof frogs *(Chiromantis* and *Phyllomedusa). Physiol. Zool.* 62:133–146.

Shoop, C. R., C. A. Ruckdeschel, and R. D. Kenney. 1998. Female biased sex ratio of juvenile loggerhead sea turtles in Georgia. *Chelon. Cons. Biol.* 3:93–96.

Sievert, L. M., and P. Andreadis. 1999. Specific dynamic action and postprandial thermophily in juvenile northern water snakes, *Nerodia sipedon. J. Therm. Biol.* 24:51–55.

Siggers, P., L. Smith, and A. Greenfield. 2002. Sexually dimorphic expression of *Gata-2* during mouse gonad development. *Mech. Dev.* 111:159–162.

Silverman, E., S. Eimerl, and J. Orly. 1999. CCAAT enhancer-binding protein beta and GATA-4 binding regions within the promoter of the steroidogenic acute regulatory protein (StAR) gene are required for transcription in rat ovarian cells. *J. Biol. Chem.* 274:17987–17996.

Sinclair, A. H., P. Berta, M. S. Palmer, J. R. Hawkins, B. L. Griffiths, M. J. Smith, J. W. Foster, et al. 1990. A gene from the human sex-determining region encodes a protein with homology to a conserved DNA-binding motif. *Nature* 346:240–244.

Skoczylas, R. 1970. Influence of temperature on gastric digestion in the grass snake, *Natrix natrix* L. *Comp. Biochem. Physiol.* 33: 793–804.

Smith, A.M.A. 1987. The sex and survivorship of embryos and hatchlings of the Australian freshwater crocodile, *Crocodylus johnstoni*. Unpubl. Ph.D. thesis, School of Botany and Zoology, Australian National University, Canberra, Australia.

Smith, C. A., and J. M. P. Joss. 1994. Sertoli-cell differentiation and gonadogenesis in *Alligator mississippiensis. J. Exp. Zool.* 270:57–70

Smith, C. A., and A. H. Sinclair. 2001. Sex determination in the chicken embryo. *J. Exp. Zool.* 290:691–699.

Smith, C. A., P. K. Elf, J. W. Lang, and J. M. P. Joss. 1995. Aromatase enzyme activity during gonadal sex differentiation in alligator embryos. *Differentiation* 58:281–290.

Smith, C. A., J. E. Andrews, and A. H. Sinclair. 1997. Gonadal sex differentiation in chicken embryos: Expression of estrogen receptor and aromatase genes. *J. Steroid Biochem. Mol. Biol.* 60:295–302.

Smith, C. A., P. J. McClive, P. S. Western, K. J. Reed, and A. H. Sinclair. 1999a. Conservation of a sex-determining gene. *Nature* 402:601–602.

Smith, C. A., M. J. Smith, and A. H. Sinclair. 1999b. Expression of chicken steroidogenic factor-1 during gonadal sex differentiation. *Gen. Comp. Endocrinol.* 113:187–196.

———. 1999c. Gene expression during gonadogenesis in the chicken embryo. *Gene* 234:395–402.

Smith, C. A., V. Clifford, P. S. Western, S. A. Wilcox, K. S. Bell, and A. H. Sinclair. 2000. Cloning and expression of a *Dax1* homologue in the chicken embryo. *J. Mol. Endocrinol.* 24:23–32.

Smith, C. A., M. Katz, and A. H. Sinclair. 2003. *Dmrt1* is upregulated in the gonads during female-to-male sex reversal in ZW chicken embryos. *Biol. Reprod.* 68:560–570.

Smith, E. N. 1976. Heating and cooling rates of the American alligator, *Alligator mississippiensis. Physiol. Zool.* 49:37–48.

Snyder, G. K. 1971. Influence of temperature and hematocrit on blood viscosity. *Am. J. Physiol.* 220:1667–1672.

Sockman, K., H. Scwabl, and P. J. Sharp. 2001. Regulation of yolk-androgen concentrations by plasma prolactin in the American kestrel. *Horm. Behav.* 40:462–471.

Solari, A. J. 1994. *Sex chromosomes and sex determination in vertebrates*. Boca Raton, FL: CRC Press Inc.

Somero, G.N. 1995. Proteins and temperature. *Ann. Rev. Physiol.* 57:43–68.

Somero, G. N., E. Dahlhoff, and J. J. Lin. 1996. Stenotherms and eurytherms: Mechanisms establishing thermal optima and tolerance ranges. In *Animals and temperature*, ed. I. A. Johnston and A. F. Bennett, 53–78. New York: Cambridge University Press.

Souza, R. R., and R. C. Vogt. 1994. Incubation temperature influences sex and hatchling size in the neotropical turtle *Podocnemis unifilis. J. Herpetol.* 28:453–464.

Spencer, R., M. B. Thompson, and P. B. Banks. 2001. Hatch or wait? A dilemma in reptilian incubation. *Oikos* 93:401–406.

Spotila, J. R., and E. A. Standora. 1986. Sex determination in the desert tortoise: A conservative management strategy is needed. *Herpetologica* 42:67–72.

Spotila, J. R., L. C. Zimmerman, C. A. Binckley, J. S. Grumbles, D. C. Rostral, A. List, Jr., E. C. Beyer, et al. 1994. Effects of incubation conditions on sex determination, hatching success and growth of hatchling desert tortoises, *Gopherus agassizii. Herpetolog. Monogr.* 8:103–116.

Spotila, L. D., and S. E. Hall. 1998. Expression of a new RNA-splice isoform of *Wt1* in developing kidney-gonadal complexes of the turtle, *Trachemys scripta. Comp. Biochem. Physiol. B* 119:761–767.

Spotila, L. D., J. R. Spotila, and S. E. Hall. 1998. Sequence and expression analysis of *Wt1* and *Sox9* in the red-eared slider turtle, *Trachemys scripta. J. Exp. Zool.* 281:417–427.

St. Clair, R. C. 1995. How developmental environment affects life history in box turtles. Ph.D. dissertation, University of Oklahoma–Norman.

———. 1998. Patterns of growth and sexual size dimorphism in two species of box turtles with environmental sex determination. *Oecologia* 115:501–507.

Stamps, J. A. 1983. Sexual selection, sexual dimorphism, and territoriality. In *Lizard ecology: Studies of a model organism*, ed.

R. B. Huey, E. R. Pianka, and T. W. Schoener, 169–204. Cambridge, MA: Harvard University Press.

Standora, E. A., and J. R. Spotila. 1985. Temperature dependent sex determination in sea turtles. *Copeia* 1985:711–722.

Steyermark, A. C., and J. R. Spotila. 2000. Effects of maternal identity and incubation temperature on snapping turtle *(Chelydra serpentina)* metabolism. *Physiol. Biochem. Zool.* 73: 298–306.

———. 2001. Effects of maternal identity and incubation temperature on hatching and hatchling morphology in snapping turtles, *Chelydra serpentina. Copeia* 2001:129–135.

Stockley, P. 1999. Sperm selection and genetic incompatibility: Does relatedness of mates affect male success in sperm competition? *Proc. Roy. Soc. Lond. B* 266:1663–1669.

Storey, K. B., and J. M. Storey. 1984. Biochemical adaptation for freezing tolerance in the wood frog, *Rana sylvatica. J. Comp. Physiol. B* 155:29–36.

Strüssmann, C. A., S. Moriyama, E. F. Hanke, J. C. Calsina Cota, and F. Takashima. 1996a. Evidence of thermolabile sex determination in pejerrey. *J. Fish Biol.* 48:643–651.

Strüssmann, C. A., J. C.Calsina Cota, G. Phonlor, H. Higuchi, and F. Takashima. 1996b. Temperature effects on sex differentiation of two South American atherinids, *Odontesthes argentinensis* and *Patagonina hatcheri. Env. Biol. Fishes* 47:143–154.

Strüssmann, C. A., T. Saito, M. Usui, H. Yamada, and F. Takashima. 1997. Thermal thresholds and critical period of thermolabile sex determination in two altherinid fishes, *Odontesthes bonariensis* and *Patagonina hatcheri. J. Exp. Zool.* 278:167–177.

Sugita J., M. Takase, and M. Nakamura. 2001. Expression of *Dax-1* during gonadal development of the frog. *Gene* 280:67–74.

Sullivan, J. A., and R. J. Schultz. 1986. Genetic and environmental basis of variable sex ratios in laboratory strains of *Poeciliopsis lucida. Evolution* 40:152–158.

Sumida, M., and M. Nishioka. 2000. Sex-linked genes and linkage maps in amphibians. *Comp. Biochem. Physiol. B* 126:257–270.

Svensson, E. C., R. L. Tufts, C. E. Polk, and J. M. Leiden. 1999. Molecular cloning of *Fog-2*: A modulator of transcription factor GATA-4 in cardiomyocytes. *Proc. Natl. Acad. Sci. USA* 96:956–961.

Svensson, E. C., G. S. Huggins, F. B. Dardik, C. E. Polk, and J. M. Leiden. 2000. A functionally conserved N-terminal domain of the friend of GATA-2 (FOG-2) protein represses GATA4-dependent transcription. *J. Biol. Chem.* 275:20762–20769.

Swain, A., and R. Lovell-Badge. 1999. Mammalian sex determination: A molecular drama. *Genes Dev.* 13:755–767.

Swingland, I. R., J. Adams, and P. J. Greenwood. 1990. Sex determination and sex ratios in patchy environments. In *Living in a patchy environment,* ed. B. Shorrocks and I. R. Swingland, 219–240. New York: Oxford University Press.

Takase, M., S. Noguchi, and M. Nakamura. 2000. Two *Sox9* messenger RNA isoforms: Isolation of cDNAs and their expression during gonadal development in the frog *Rana rugosa. FEBS Lett.* 466:249–254.

Teranishi, M., Y. Shimada, T. Hori, O. Nakabayashi, T. Kikuchi, T. Macleod, R. Pym, et al. 2001. Transcripts of the MHM region on the chicken Z chromosome accumulate as non-coding RNA in the nucleus of female cells adjacent to the *Dmrt1* locus. *Chromosome Res.* 9:147–165.

Thapliyal, J. P., K. S. Singh, and A. Chandola. 1973. Pre-laying stages in the development of Indian garden lizards, *Calotes versicolor. Ann. Embryol. Morphol.* 6:253–259.

Thompson, M. 1993. Estimate of the population structure of the eastern water dragon, *Physignathus lesueurii* (Reptilia: Agamidae), along riverside habitat. *Wildl. Res.* 20:613–619.

Thompson, M. B. 1990. Incubation of eggs of tuatara, *Sphenodon punctatus. J. Zool. Lond.* 222:303–318.

Thompson, M. B., D. G. Newman, and P. R. Watson. 1991. Use of oxytocin in obtaining eggs from tuatara *(Sphenodon punctatus). J. Herpetol.* 25:101–104.

Thompson, M. B., G. C. Packard, M. J. Packard, and B. Rose. 1996. Analysis of the nest environment of tuatara *Sphenodon punctatus. J. Zool. Lond.* 238:239–251.

Thorbjarnarson, J. 1997. Are crocodilian sex ratios female biased? The data are equivocal. *Copeia* 1997:451–455.

Tokunaga, S. 1985. Temperature-dependent sex determination in *Gekko japonicus* (Gekkonidae: Reptilia). *Dev. Growth Differ.* 27: 117–120.

Torres Maldonado, L. C., A. Landa Piedra, N. Moreno Mendoza, A. Marmolejo Valencia, A. Meza Martinez, and H. Merchant Larios. 2002. Expression profiles of *Dax1, Dmrt1,* and *Sox9* during temperature sex determination in gonads of the sea turtle *Lepidochelys olivacea. Gen. Comp. Endocrinol.* 129:20–26.

Tousignant, A., and D. Crews. 1995. Incubation-temperature and gonadal sex affect growth and physiology in the leopard gecko *(Eublepharis macularius),* a lizard with temperature-dependent sex determination. *J. Morphol.* 224:159–170.

Tremblay, J. J., and R. S. Viger. 1999. Transcription factor GATA-4 enhances Müllerian inhibiting substance gene transcription through a direct interaction with the nuclear receptor SF-1. *Mol. Endocrinol.* 13:1388–1401.

———. 2001a. GATA factors differentially activate multiple gonadal promoters through conserved GATA regulatory elements. *Endocrinology* 142:977–986.

———. 2001b. Nuclear receptor *Dax-* represses the transcriptional cooperation between GATA-4 and SF-1 in Sertoli cells. *Biol. Reprod.* 64:1191–1199.

Tremblay, J. J., N. M. Robert, and R. S. Viger. 2001. Modulation of endogenous GATA-4 activity reveals its dual contribution to Mullerian inhibiting substance gene transcription in Sertoli cells. *Mol. Endocrinol.* 15:1636–1650.

Tribe, M., and F. Brambell. 1932. The origin and migration of the primordial germ cells of *Sphenodon punctatus. Q. J. Micr. Sci.* 75:251–282.

Trivers, R. L., and D. E. Willard. 1973. Natural selection of parental ability to vary the sex ratio of offspring. *Science* 179: 90–92.

Tryon, B. W. 1979. Notes on the reproduction of the lizard *Cophotis ceylonica. Brit. J. Herpetol.* 5:845–847.

Tsai, C. L., L. H. Wang, and L. S. Fang. 2001. Estradiol and para-chlorophenylalanine downregulate the expression of brain aromatase and estrogen receptor-alpha mRNA during the critical period of feminization in tilapia *(Oreochromis mossambicus)*. *Neuroendocrinology* 74:325–334.

Tsugawa, K. 1980. Thermal dependence in kinetic properties of lactate dehydrogenase from the African clawed toad, *Xenopus laevis*. *Comp. Biochem. Physiol. B* 66:459–466.

Tsunekawa, N., M. Naito, Y. Sakai, T. Nishida, and T. Noce. 2000. Isolation of chicken vasa homolog gene and tracing the origin of primordial germ cells. *Development* 127:2741–2750.

Uchida, T. 1937a. Studies on the sexuality of amphibians. II. Sexual induction in a sexually semi-differentiated salamander. *J. Fac. Sci. Hokkaido Imp. Univ. (Zool.)* 6:35–58.

———. 1937b. Studies on the sexuality of amphibia. III. Sex transformation in *Hynobius retardatus* by the function of high temperature. *J. Fac. Sci. Hokkaido Imp. Univ. (Zool.)* 6:59–70.

U.S. Department of Commerce. 1966. *World weather records, 1951–1960,* vol. 3. Washington, DC: U.S. Government Printing Office.

Vaillant, S., M. Dorizzi, C. Pieau, and N. Richard-Mercier. 2001. Sex reversal and aromatase in chicken. *J. Exp. Zool.* 290: 727–740.

Vainio, S., M. Heikkila, A. Kispert, N. Chin, and A. P. McMahon. 1999. Female development in mammals is regulated by WNT-4 signalling. *Nature* 397:405–409.

Valenzuela, N. 2000. Multiple paternity in side-neck turtles *Podocnemis expansa*: Evidence from microsatellite DNA data. *Mol. Ecol.* 9:99–105.

———. 2001a. Maternal effects on life history traits in the Amazonian giant river turtle *Podocnemis expansa*. *J. Herpetol.* 35: 368–378.

———. 2001b. Constant, shift, and natural temperature effects on sex determination in *Podocnemis expansa* turtles. *Ecology* 82: 3010–3024.

Valenzuela, N., and F. J. Janzen. 2001. Nest-site philopatry and the evolution of temperature-dependent sex determination. *Evol. Ecol. Res.* 3:779–794.

Valenzuela, N., R. Botero, and E. Martínez. 1997. Field study of sex determination in *Podocnemis expansa* from Colombian Amazonia. *Herpetologica* 53:390–398.

Valenzuela, N., D. C. Adams, and F. J. Janzen. 2003. Pattern does not equal process: Exactly when is sex environmentally determined? *Am. Nat.* 161:676–683.

Valleley, E. M. A., U. Muller, M. W. J. Ferguson, and P. T. Sharpe. 1992. Cloning and expression analysis of two *Zfy*-related zinc finger genes from *Alligator mississippiensis*, a species with temperature-dependent sex determination. *Gene* 119:221–228.

Valleley, E. M., E. J. Cartwright, N. J. Croft, A. F. Markham, and P. L. Coletta. 2001. Characterization and expression of *Sox9* in the Leopard gecko, *Eublepharis macularius*. *J. Exp. Zool.* 291:85–91.

Van Damme, R., D. Bauwens, and R. F. Verheyen. 1990. Evolutionary rigidity of thermal physiology: The case of the cool temperate lizard *Lacerta vivipara*. *Oikos* 57:61–67.

———. 1991. The thermal dependence of feeding behavior, food consumption and gut passage time in the lizard *Lacerta vivipara* Jacquin. *Funct. Ecol.* 5:507–517.

Van Damme, R., D. Bauwens, F. Braña, and R. F. Verheyen. 1992. Incubation temperature differentially affects hatching time, egg survival, and hatchling performance in the lizard *Podarcis muralis*. *Herpetologica* 48:220–228.

Vanhooydonck, B., R.Van Damme, T. J. M. Van Cooren, and D. Bauwens. 2001. Proximate causes of intraspecific variation in locomotor performance in the lizard *Gallotia galloti*. *Physiol. Biochem. Zool.* 74:937–945.

Van Mierop, L. H. S., and S. M. Barnard. 1978. Further observations on thermoregulation in the brooding female *Python molurus bivittatus* (Serpentes: Biodae). *Copeia* 1978:615–621.

Vidal, V. P., M. C. Chaboissier, D. G. de Rooij, and A. Schedl. 2001. *Sox9* induces testis development in XX transgenic mice. *Nat. Genet.* 28:216–217.

Viets, B. E., A. Tousignant, M. A. Ewert, C. E. Nelson, and D. Crews. 1993. Temperature-dependent sex determination in the leopard gecko, *Eublepharis macularius*. *J. Exp. Zool.* 265: 679–683.

Viets, B. E., M. A. Ewert, L. G. Talent, and C. E. Nelson. 1994. Sex-determining mechanisms in squamate reptiles. *J. Exp. Zool.* 270:45–56.

Viger, R. S., C. Mertineit, J. M. Trasler, and M. Nemer. 1998. Transcription factor *Gata-4* is expressed in a sexually dimorphic pattern during mouse gonadal development and is a potent activator of the Mullerian inhibiting substance promoter. *Development* 125:2665–2675.

Vogt, R. C. 1980. Natural history of the map turtles *Graptemys pseudogeographica* and *G. ouachitensis* in Wisconsin. *Tulane Stud. Zool. Bot.* 22:17–48.

———. 1990. Reproductive parameters of *Trachemys scripta venusta* in southern Mexico. In *Life history and ecology of the slider turtle*, ed. J. W. Gibbons, 162–168. Washington, DC: Smithsonian Institution Press.

———. 1994. Temperature controlled sex determination as a tool for turtle conservation. *Chelon. Cons. Biol.* 1:159–162.

Vogt, R. C., and J. J. Bull. 1984. Ecology of hatchling sex ratio in map turtles. *Ecology* 65:582–587.

Vogt, R. C., and O. Flores Villela. 1992. Effects of incubation-temperature on sex determination in a community of Neotropical freshwater turtles in southern Mexico. *Herpetologica* 48: 265–270.

Vogt, R. C., V. H. Cantarelli, and A. G. de Carvalho. 1994. Reproduction of the cubecado, *Peltocephalus dumerilianus*, in the biological reserve of Rio Trombetas, Para, Brazil. *Chelon. Cons. Biol.* 1:145–148.

Volpe, E. P. 1957. Embryonic temperature tolerance and rate of development in *Bufo valliceps*. *Physiol. Zool.* 30:164–176.

Wagner, E. 1980. Temperature-dependent sex determination in a gekko lizard. *Q. Rev. Biol.* 55:21(Appendix).

Wagner, T. L., H. Wu, P. J. H. Sharpe, R. M. Schoolfield, and R.N. Coulson. 1984. Modeling insect development rates: A literature review and application of a biophysical model. *Ann. Entomol. Soc. Am.* 77:208–225.

Waldschmidt, S. R., S. M. Jones, and W. P. Porter. 1986. The effect of body temperature and feeding regime on activity, passage time, and digestive coefficient in the lizard *Uta stansburiana*. *Physiol. Zool.* 59:376–383.

Walker, D., P. E. Moler, K. A. Buhlmann, and J. C. Avise. 1998. Phylogeographic patterns in *Kinosternon subrubrum* and *K. bau-*

rii based on mitochondrial DNA restriction analyses. *Herpetologica* 54:174–184.

Wallace, H., and B. M. N. Wallace. 2000. Sex reversal of the newt *Triturus cristatus* reared at extreme temperatures. *Intl. J. Dev. Biol.* 44:807–810.

Wallace, H., G. M. I. Badawy, and B. M. N. Wallace. 1999. Amphibian sex determination and sex reversal. *Cell. Mol. Life Sci.* 55:901–909.

Wang, L. H., and C. L. Tsai. 2000. Effects of temperature on the deformity and sex differentiation of tilapia, *Oreochromis mossambicus*. *J. Exp. Zool.* 286:534–537.

Wapstra, E., M. Olssson, R. Shine, A. Edwards, R. Swain, and J. M. P. Joss. 2004. Maternal basking determines offspring sex in a viviparous reptile. Biology Letters, *Proc. Roy. Soc. London B (Suppl.)* 271:S230–S232.

Watanabe, K., T. R. Clarke, A. H. Lane, X. Wang, and P. K. Donahoe. 2000. Endogenous expression of Mullerian inhibiting substance in early postnatal rat sertoli cells requires multiple steroidogenic factor-1 and GATA-4-binding sites. *Proc. Natl. Acad. Sci. USA* 97:1624–1629.

Weathers, W. W., and F. N. White. 1971. Physiological thermoregulation in turtles. *Am. J. Physiol.* 221:704–710.

Webb, G. J. W., and H. Cooper-Preston. 1989. Effects of incubation temperature on crocodiles and the evolution of reptilian oviparity. *Am. Zool.* 29:953–971.

Webb, G. J. W., and A. M. A. Smith. 1984. Sex ratio and survivorship in the Australian freshwater crocodile *Crocodylus johnstoni*. In *The structure, development and evolution of reptiles*, ed. M. W. J. Ferguson, 319–355. London: Academic Press.

Webb, G. J. W., R. Buckworth, and S. C. Manolis. 1983. *Crocodylus johnstoni* in the McKinlay River area, NT. VI. Nesting biology. *Aus. Wildl. Res.* 10:607–637.

Webb, G. J. W., D. Choquenot, and P. Whitehead. 1986. Nests, eggs and embryonic development of *Carettochelys insculpta* (Chelonia: Carettochelidae) from northern Australia. *J. Zool. Lond.* 1B:521–550.

Webb, G. J. W., A. M. Beal, S. C. Manolis, and K. E. Dempsey. 1987. The effects of incubation temperature on sex determination and embryonic development rate in *Crocodylus johnstoni* and *C. porosus*. In *Wildlife management: Crocodiles and alligators*, ed. G. J. W. Webb, S. C. Manolis, and P. J. Whitehead, 507–531. Chipping Norton, NSW, Australia: Surrey Beatty and Sons.

Webb, G. J. W., S. C. Manolis, and H. Cooper-Preston. 1990. Crocodile management and research in the Northern Territory: 1988–1990. In *Proceedings of the 10th Working Meeting of the Crocodile Specialist Group*, 253–273. IUCN–World Conservation Union, Gainesville, Florida.

Webb, G. J. W., S. C. Manolis, B. Ottley, and A. Heyward. 1992. Crocodile management and research in the Northern Territory: 1990–1992. In *Proceedings of the 11th Working Meeting of the Crocodile Specialist Group*, 203–246. IUCN–World Conservation Union, Gainesville, Florida.

Webb, J. K., G. P. Brown, and R. Shine. 2001. Body size, locomotor speed and antipredator behaviour in a tropical snake (*Tropidonophis mairii*, Colubridae): The influence of incubation environments and genetic factors. *Funct. Ecol.* 15:561–568.

Wedekind, C. 2002. Manipulating sex ratios for conservation: Short-term risks and long-term benefits. *Anim. Cons.* 5:13–20.

Wernstedt, F. L. 1959. *World climatic data*, vol. 2. Ann Arbor, MI: Edwards Brothers, Inc.

Werren, J. H., and E. L. Charnov. 1978. Facultative sex ratios and population dynamics. *Nature* 272:349–350.

Werren, J. H., M. J. Hatcher, and H. C. J. Godfray. 2002. Maternal-offspring conflict leads to the evolution of dominant zygotic sex allocation. *Heredity* 88:102–111.

West, S. A, C. M. Lively, and A. F. Read. 1999. A pluralist approach to sex and recombination. *J. Evol. Biol.* 12:1003–1012.

Western, P. S., and A. H. Sinclair. 2001. Sex, genes and heat: Triggers of diversity. *J. Exp. Zool.* 290:624–631.

Western, P. S., J. L. Harry, J. A. M. Graves, and A.H. Sinclair. 1999. Temperature-dependent sex determination: Upregulation of *Sox9* expression after commitment to male development. *Dev. Dyn.* 214:171–177.

Western, P. S., J. L. Harry, J. A. Marshall Graves, and A. H. Sinclair. 2000. Temperature-dependent sex determination in the American alligator: Expression of *Sf1, Wt1* and *Dax1* during gonadogenesis. *Gene.* 241:223–232.

White, M. J. D. 1973. *Animal cytology and evolution*. Cambridge University Press.

White, R. B., and P. Thomas. 1992a. Adrenal-kidney and gonadal steroidogenesis during sexual differentiation of a reptile with temperature-dependent sex determination. *Gen. Comp. Endocrinol.* 88:10–19.

———. 1992b. Whole-body and plasma concentrations of steroids in the turtle, *Trachemys scripta*, before, during, and after the temperature-sensitive period for sex determination. *J. Exp. Zool.* 264:159–166.

Whitehead, P. J., and R. S. Seymour. 1990. Patterns of metabolic rate in embryonic crocodilians *Crocodylus johnstoni* and *Crocodylus porosus*. *Physiol. Zool.* 63:334–352.

Whitehead, P. J., J. T. Puckridge, C. M. Leigh, and R. S. Seymour. 1989. Effect of temperature on jump performance of the frog *Limnodynastes tasmaniensis*. *Physiol. Zool.* 62:937–949.

Whitmore, C., P. Dutton, and N. Mrosovsky. 1985. Sexing sea turtles: Gross appearance versus histology. *J. Herpetol.* 19:430–431.

Wibbels, T., and D. Crews. 1995. Steroid-induced sex determination at incubation temperatures producing mixed sex ratios in a turtle with TSD. *Gen. Comp. Endocrinol.* 100:53–60.

Wibbels, T., J. J. Bull, and D. Crews. 1991a. Synergism between temperature and estradiol: A common pathway in turtle sex determination? *J. Exp. Zool.* 260:130–134.

———. 1991b. Chronology and morphology of temperature-dependent sex determination. *J. Exp. Zool.* 260:371–381.

Wibbels, T., F. C. Killebrew, and D. Crews. 1991c. Sex determination in Cagle's map turtle: Implications for evolution, development, and conservation. *Can. J. Zool.* 69:2693–2696.

Wibbels, T., R. E. Martin, D. W. Owens, and M. S. Amoss. 1991d. Female-biased sex ratio of immature loggerhead sea turtles inhabiting the Atlantic coastal water of Florida. *Can. J. Zool.* 69: 2973–2977.

Wibbels, T., J. J. Bull, and D. Crews. 1994. Temperature-dependent sex determination: A mechanistic approach. *J. Exp. Zool.* 270:71–78.

Wibbels, T., D. Rostral, and R. Byles. 1998. High pivotal temperature in the sex determination of the olive ridley sea turtle, *Lepidochelys olivacea,* from Plays Nancite, Costa Rica. *Copeia* 1998: 1086–1088.

Wilhelm, D., and C. Englert. 2002. The Wilm's tumor suppressor WT1 regulates early gonad development by activation of *Sf1*. *Genes Dev.* 16:1839–1851.

Wilhoft, D. C, E. Hotaling, and P. Franks. 1983. Effects of temperature on sex determination in embryos of the snapping turtle, *Chelydra serpentina*. *J. Herpetol.* 17:38–42.

Willingham, E. and D. Crews. 1999. Sex reversal effects of environmentally relevant xenobiotic concentrations on the red-eared slider turtle, a species with temperature-dependent sex determination. *Gen. Comp. Endocrinol.* 113:429–435.

Willingham, E., R. Baldwin, J. K. Skipper, and D. Crews. 2000. Aromatase activity during embryogenesis in the brain and adrenal-kidney-gonad of the red-eared slider turtle, a species with temperature-dependent sex determination. *Gen. Comp. Endocrinol.* 119:202–207.

Wilson, D. S. 1998. Nest-site selection: Microhabitat variation and its effects on the survival of turtle embryos. *Ecology* 79: 1884–1892.

Wilson, M. A., and A. C. Echternacht. 1987. Geographic variation in the critical thermal minimum of the green anole, *Anolis carolinesis* (Sauria: Iguanidae), along a latitudinal gradient. *Comp. Biochem. Physiol. A* 87:757–760.

Winge, O. 1934. The experimental alteration of sex chromosomes into autosomes and vice versa, as illustrated by Lebistes. *C. R. Trav. Lab. Carlsburg. Ser. Physiol.* 21:1–49.

Winge, O., and E. Ditlevsen. 1948. Colour inheritance and sex determination in *Lebistes*. *C. R. Trav. Lab. Carlsberg Ser. Physiol.* 24:227–248.

Witschi, E. 1914. Experimentelle Untersuchungen über die Entwicklungsgeschichte der Keimdrusen von *Rana temporaria*. *Arch. Mikr. Anat. Entw.* 85:9–113.

———. 1929a. Studies on sex differentiation and sex determination in amphibians. II. Sex reversal in female tadpoles of *Rana sylvatica* following the application of high temperature. *J. Exp. Zool.* 52:267–291.

———. 1929b. Studies on sex differentiation and sex determination in amphibians. III. Rudimentary hermaphroditism and Y chromosome in *Rana temporaria*. *J. Exp. Zool.* 54:157–223.

———. 1930. Studies on sex differentiation and sex determination in amphibians. IV. The geographical distribution of the sex races of the European grass frog (*Rana temporaria* L.). A contribution to the problem of the evolution of sex. *J. Exp. Zool.* 56:149–165.

———. 1959. Age of sex-determining mechanisms in vertebrates. *Science* 130:372–375.

Witten, G. J., and A. J. Coventry. 1984. A new lizard of the genus *Amphibolurus* (Agamidae) from southern Australia. *Proc. Roy. Soc. Vict.* 96:155–159.

Wolfner, M. F. 2003. Sex determination: Sex on the brain? *Curr. Biol.* 13:R101–103.

Wood, J. R., and F. E. Wood. 1980. Reproductive biology of captive green sea turtles *(Chelonia mydas)*. *Am. Zool.* 20:499–505.

Woodward, D. E., and J. D. Murray. 1993. On the effect of temperature-dependent sex determination on sex ratio and survivorship in corocodilians. *Proc. Roy. Soc. Lond. B* 252:149–155.

Yadava, M. R. 1980. Hatching time for the eggs of soft-shell turtle, *Kachuga dhongoka* (Gray) at various temperatures. *Indian Forrester* 106:721–725.

Yamahira, K., and D. O. Conover. 2003. Interpopulation variability in temperature-dependent sex determination of the tidewater silverside *Menidia peninsulae* (Pisces: Atherinidae). *Copeia* 2003:156–159.

Yamamoto, E. 1999. Studies on sex-manipulation and production of cloned populations of hirame, *Paralichthys olivaceus* (Temminck et Schlegel). *Aquaculture* 173:235–246.

Yamamoto, T. 1969. Sex differentiation. In *Fish physiology*, vol. 3, ed. W. S. Hoar and D. J. Randall 171–175. New York: Academic Press Inc.

Yntema, C. L. 1968. A series of stages in the embryonic development of *Chelydra serpentina*. *J. Morphol.* 125:219–252.

———. 1979. Temperature levels and periods of sex determination during incubation of eggs of *Chelydra serpentina*. *J. Morphol.* 159:17–28.

Yntema, C. L., and N. Mrosovsky. 1982. Critical periods and pivotal temperatures for sexual differentiation in loggerhead sea turtles. *Can. J. Zool.* 60:1012–1016.

Yoshikura, M. 1959. The action of pituitary in sex differentiation and sex reversal in amphibians. II. Effects of high temperature on the gonads of hypophysectomized frog larvae. *Kumamoto J. Sci. B* 4:69–101.

———. 1963. Influence of high temperature on the development of gonads of thiourea-treated frog tadpoles. *Kumamoto J. Sci. B* 6:79–101.

Young, J. E., A. Georges, J. S. Doody, P. B. West, and R. L. Alderman. 2005. Pivoltal range and thermosensitive period of the pig-nosed turtle, *Carrettochelys insculpta* (Testudines: Carettochelydidae) from northern Australia. *Can. J. Zool.* In Press.

Yu, R. N., M. Ito, T. L. Saunders, S. A. Camper, and J. L. Jameson. 1998. Role of Ahch in gonadal development and gametogenesis. *Nat. Genet.* 20:353–357.

Zaborski, P. 1986. Temperature and estrogen dependent changes of sex phenotype and HY antigen expression in gonads of a newt. In *Progress in developmental biology*, Part A, 163–169. New York: Alan R. Liss, Inc.

Ziegler, T., and W. Böhme. 1996. Zur Hemiclitoris der squamaten Reptilien: Auswirkungen auf einige Methoden der Geschlechtsunterscheidung. *Herpetofauna* 18:11–19.

Zug, G. R., L. J. Vitt, and J. P. Caldwell. 2001. *Herpetology. An introductory biology of amphibians and reptiles,* 2nd edition. San Diego: Academic Press.

Contributors

Kerry Beggs
Applied Ecology Research Group
University of Canberra
ACT 2601, Australia
beggs@aerg.canberra.edu.au

J. J. Bull
Section of Integrative Biology
University of Texas
1 University Station C0930
Austin, TX 78712-0253, USA
bull@bull.biosci.utexas.edu

Dominique Chardard
Laboratoire de Biologie Expérimentale—Immunologie
EA 3442: Génétique, Signalisation, Différenciation
Université Henri Poincaré-Nancy 1, B.P. 239
54506 Vandoeuvre-lès-Nancy cedex, France
dominique.chardard@scbiol.uhp-nancy.fr

Amand Chesnel
Laboratoire de Biologie Expérimentale—Immunologie
EA 3442: Génétique, Signalisation, Différenciation
Université Henri Poincaré-Nancy 1, B.P. 239
54506 Vandoeuvre-lès-Nancy cedex, France
amand.chesnel@scbiol.uhp-nancy.fr

David Conover
Marine Sciences Research Center
State University of New York
Stony Brook, NY 11794-5000, USA
dconover@notes.cc.sunysb.edu

Franck Courchamp
UMR 8079, Ecologie, Systématique et Evolution
Université Paris-Sud, Orsay et CNRS, Bât. 362
91405 Orsay cedex, France
franck.courchamp@ese.u-psud.fr

Alison Cree
Department of Zoology
University of Otago
P. O. Box 56
Dunedin, New Zealand
alison.cree@stonebow.otago.ac.nz

Charles H. Daugherty
School of Biological Sciences
Victoria University of Wellington
P. O. Box 600
Wellington, New Zealand
charles.daugherty@vuw.ac.nz

D. C. Deeming
Hatchery Consulting and Research
9 Eagle Drive
Welton, Lincoln, LN2 3LP, United Kingdom
charlie@deemingdc.freeserve.co.uk

Virginie Delmas
UMR 8079, Ecologie, Systématique et Evolution
Université Paris-Sud, Orsay et CNRS, Bât. 362
91405 Orsay cedex, France
virginie.delmas@ese.u-psud.fr

Sean Doody
Applied Ecology Research Group
University of Canberra
ACT 2601, Australia
doody@aerg.canberra.edu.au

Christian Dournon
Laboratoire de Biologie Expérimentale—Immunologie
EA 3442: Génétique, Signalisation, Différenciation
Université Henri Poincaré-Nancy 1, B.P. 239
54506 Vandoeuvre-lès-Nancy cedex, France
christian.dournon@scbiol.uhp-nancy.fr

Pamela K. Elf
University of Minnesota Crookston
Crookston, MN 56716-5100
University of North Dakota
Grand Forks, ND 58202, USA
pelf@mail.crk.umn.edu

Cory R. Etchberger
Department of Biology
Johnson County Community College
12345 College Blvd.
Overland Park, KS 66210, USA
turtleman@kc.rr.com

Michael A. Ewert.
Dept. of Biology
1001 East Third Street
Indiana University
Bloomington, IN 47405-3700, USA
mewert@bio.indiana.edu

Arthur Georges
Applied Ecology Research Group
University of Canberra
ACT 2601, Australia
georges@aerg.canberra.edu.au

Marc Girondot
UMR 8079, Ecologie, Systématique et Evolution
Université Paris-Sud, Orsay et CNRS, Bât. 362
91405 Orsay cedex, France
marc.girondot@ese.u-psud.fr

Matthew H. Godfrey
UMR 8079, Ecologie, Systématique et Evolution
Université Paris-Sud, Orsay et CNRS, Bât. 362
91405 Orsay cedex, France
Current address:
North Carolina Wildlife Resources Commission
307 Live Oak Street
Beaufort, NC 28516, USA

Peter S. Harlow
School of Biological Sciences
Macquarie University
NSW 2109, Australia
pharlow@zoo.nsw.gov.au

Fredric J. Janzen
Department of Zoology and Genetics
Iowa State University
Ames, IA 50011-3223, USA
fjanzen@iastate.edu

Susan N. Keall
School of Biological Sciences
Victoria University of Wellington
P. O. Box 600
Wellington, New Zealand
susan.keall@vuw.ac.nz

James G. Krenz
Department of Ecology and Evolutionary Biology
University of Arizona
Tucson, AZ 85721, USA
jkrenz@email.arizona.edu

Valentine A. Lance
Center for the Reproduction of Endangered Species (CRES)
P.O. Box 120551
San Diego CA 92112, USA
lvalenti@sunstroke.sdsu.edu

Jeffrey W. Lang
2369 Bourne Avenue
St. Paul, MN 55108-1618, USA
jeff_lang@und.nodak.edu

Marshall D. McCue
Department of Ecology and Evolutionary Biology
University of California, Irvine
321 Steinhaus Hall,
Irvine, CA 92697-2525, USA
mmccue@uci.edu

May Penrad-Mobayed
Laboratoire de Biochimie du Développement
UMR 7592: Institut Jacques Monod
CNRS-Universités Paris 6 et 7
2 Place Jussieu
75251 Paris cedex 05, France
penrad@ijm.jussieu.fr

Craig E. Nelson
Department of Biology
1001 East Third Street
Indiana University
Bloomington, IN 47405-3700, USA
nelson1@indiana.edu

Nicola J. Nelson
School of Biological Sciences
Victoria University of Wellington
P. O. Box 600
Wellington, New Zealand
nicola.nelson@vuw.ac.nz

Claude Pieau
Laboratoire de Biochimie du Développement
UMR 7592: Institut Jacques Monod
CNRS-Universités Paris 6 et 7
2 Place Jussieu
75251 Paris cedex 05, France
pieau@ijm.jussieu.fr

Allen R. Place
Center of Marine Biotechnology
Columbus Center, Suite 236
701 East Pratt Street
Baltimore, MD 21202, USA
place@umbi.umd.edu

Anne-Caroline Prévot-Julliard
UMR 8079, Ecologie, Systématique et Evolution
Université Paris-Sud, Orsay et CNRS, Bât. 362
91405 Orsay cedex, France
anne-caroline.Julliard@ese.u-psud.fr

Turk Rhen
Department of Biology
Box 9019
University of North Dakota
Grand Forks, ND 58202, USA
turk.rhen@und.nodak.edu

Philippe Rivalan
UMR 8079, Ecologie, Systématique et Evolution
Université Paris-Sud, Orsay et CNRS, Bât. 362
91405 Orsay cedex, France
philippe.rivalan@ese.u-psud.fr

Michael B. Thompson
School of Biological Sciences and Wildlife Research Institute
Heydon-Laurence Building (A08)
University of Sydney
NSW 2006, Australia
thommo@bio.usyd.edu.au

Nicole Valenzuela
Department of Ecology, Evolution, and Organismal Biology
Iowa State University
Ames, IA 50011, USA
nvalenzu@iastate.edu

Jeanne Young
Applied Ecology Research Group
University of Canberra
ACT 2601, Australia
youngj@aerg.canberra.edu.au

Index

Page numbers in *italics* indicate illustrations or tables